**Natural Born Monads**

# Natural Born Monads

On the Metaphysics of Organisms
and Human Individuals

Edited by
Andrea Altobrando and
Pierfrancesco Biasetti

DE GRUYTER

ISBN 978-3-11-099670-8
e-ISBN (PDF) 978-3-11-060466-5
e-ISBN (EPUB) 978-3-11-060366-8

**Library of Congress Control Number: 2020938882**

**Bibliographic information published by the Deutsche Nationalbibliothek**
The Deutsche Nationalbibliothek lists this publication in the Deutsche Nationalbibliografie;
detailed bibliographic data are available on the Internet at http://dnb.dnb.de.

© 2022 Walter de Gruyter GmbH, Berlin/Boston
This volume is text- and page-identical with the hardback published in 2020.
Printing and binding: CPI books GmbH, Leck

www.degruyter.com

# Table of Contents

Andrea Altobrando, Pierfrancesco Biasetti
**Natural Born Monads —— 1**

Antonio M. Nunziante
**Between Laws and Norms. Genesis of the Concept of Organism in Leibniz and in the Early Modern Western Philosophy —— 11**

Hugh Desmond, Philippe Huneman
**The Ontology of Organismic Agency: A Kantian Approach —— 33**

Michela Bordignon
**Teleology, Backward Causation and Contradiction. Hegel's Dialectical Account of Organic Nature —— 65**

Luca Illetterati
**Being Rational: Hegel on the Human Way of Being —— 89**

Federico Sanguinetti
**Hegel and the Question "What Characterizes Human Beings qua Animal Organisms of a Specific Sort?" —— 111**

Yusuke Akimoto
**Marx's Philosophy on Natural History —— 137**

Caroline Angleraux
**From Monads to Monera —— 153**

Robert Kocis
**Idealism and Darwin – Rejection, Accommodation, Appropriation: James Hutchison Stirling and David George Ritchie —— 177**

Yūjin Itabashi
**Biology and the Philosophy of History: Nishida Kitarō and the Philosophy of "Necessity that Includes Freedom" —— 199**

Takeshi Morisato
**Tanabe Hajime and the Concept of Species: Approaching Nature as a Missing Shade in the *Logic of Species*** —— 213

Andrea Gambarotto
**Teleology, Life, and Cognition: Reconsidering Jonas' Legacy for a Theory of the Organism** —— 243

Pierfrancesco Biasetti
**Dialectical Thinking and Science: The Case of Richard Lewontin, Dialectical Biologist** —— 265

Lenny Moss
**Can Normativity be the Force of Nature that Solves the Problem of *Partes Extra Partes?* Episode IV – A New Hope – Natural Detachment and the Case of the Hybrid Hominin** —— 293

Tom Rockmore
**Towards a Constructivist Approach to Human Nature** —— 315

Index of names —— 333

# Andrea Altobrando, Pierfrancesco Biasetti
## Natural Born Monads

> The concrete highway was edged with a mat of tangled, broken, dry grass, and the grass heads were heavy with oat beards to catch on a dog's coat, and foxtails to tangle in a horse's fetlocks, and clover burrs to fasten in sheep's wool; sleeping life waiting to be spread and dispersed, every seed armed with an appliance of dispersal, twisting darts and parachutes for the wind, little spears and balls of tiny thorns, and all waiting for animals and for the wind, for a man's trouser cuff or the hem of a woman's skirt, all passive but armed with appliances of activity, still, but each possessed of the anlage of movement.
>
> John Steinbeck, The Grapes of Wrath

Numerical identity is the essential feature of individuality: to identify something as an individual, we need to be able to track its survival as it changes in space and time. Organisms, in this sense, can be considered as a peculiar kind of individuals. Whenever we analyze a living being, we inevitably end up considering two key elements: its internal structure, and its relationship with the surrounding environment. On the one hand, this happens because the spatio-temporal survival of an organism depends on its *inner articulation* – on its functional parts, and on the way they work together. In this way, it is possible to give a simple definition of an organism as an individual whose numerical identity is preserved thanks to the work of its parts. Yet, on the other hand, this is not enough. This "work of the parts" invariably involves some sort of exchange with what is placed outside the border of the organism. Organisms entertain a fundamental relationship with their environment, one on which their individuality depends: it is this *ecological relationship* that further defines the subset of organisms inside the larger set of individuals.

Throughout the history of ideas, this basic idea of what an organism is has been applied to different scales and levels of organization. In fact, the embedding of an organism in an environment does not solely concern its relationship with the medium of its movement, with the climate, resources, and other non-living parts of a habitat. It also concerns its interactions with other organisms, be they conspecific or allospecific. In this respect, it has always been tempting to consider the various communities formed at different biological levels, from pairs of symbionts to networks of species sharing an ecosystem, as organisms in themselves. There is a similar temptation with regard to our stance on human communities: in fact, until recently, holistic sociological and political

---

**Andrea Altobrando,** University of Padova / China University of Political Science and Law
**Pierfrancesco Biasetti,** Leibniz Institute for Zoo and Wildlife Research

theories, giving priority to wholes rather than individuals to explain social and even psychological phenomena, have not been rare.

According to such views, the individual organism at a "lower" level becomes just a part of a larger organism in the "upper" lever. Its life – and, more radically, its *significance* – can be explained, and can actually subsist, only in relation to the function carried out for the complex whole of which it is part. Concerning human communities, this social holism easily results in a negation of any intrinsic value of simple individuals as individuals – i.e., besides, or beyond, their social roles and functions. For most of our modern societies, this seems unacceptable. On the other hand, symmetrical "atomistic" alternatives seem likewise untenable, as they do not seem capable of accounting for the "natural" human propensity to live inside different kinds of groups.

This volume is motivated by the wish to contribute to a better understanding of this peculiar situation of human individuals. We believe that only a thorough and careful assessment of the concept of the organism, and of its relations to the environment and to itself, can allow us better to understand human groups, and the relations between human individuals and their environment.

We are convinced that the conceptual history of the terms at stake is highly important. For this reason, we have gathered several scholars specialized in different fields of philosophy in order to sketch a specific line of conceptual development concerning the organism that runs into the current debates in metaphysics, philosophical anthropology, social philosophy, and philosophy of biology. It is a story that starts with Leibniz and his understanding of *organism* and *human nature*, and continues its arc across the philosophy of German Idealism, before branching into several post-Darwinian redefinitions of this legacy. Its main thread is, as the title of the volume suggests, the concept of the *monad*. But another important protagonist is *dialectics*. What monadology and dialectics share, in our view, is that both concepts entail a fundamental relationship between an individual and its otherness – be this its natural or social environment. This may sound odd given the current meaning of "monad" – a term usually associated with solipsistic views of subjects and substances in general. We believe this latter reading to be extremely narrow and unrepresentative of the original concept. In this regard, one should not forget that Leibniz' monadology was devised to formulate a metaphysically and ontologically pluralistic view of the universe, and that a monad, in Leibniz' understanding, is "made" of its relationships with all other monads.

The volume is divided in two parts. In the first part the analysis is centered on the core figures that form the "trunk" of the tradition we would like to discuss: Leibniz, Kant, and Hegel. In the second part, the volume focuses instead on some of the post-Darwinian ramifications of the monadological and dialectical legacy, showing how the concepts first formulated at the end of seventeenth century, and constantly reassessed since then, still represent the theoretical hard core one has to confront when reflecting on what an organism is.

The first article of the volume, written by Antonio M. Nunziante, sets up the discussion, by focusing on the genesis of the concept of the organism in Leibniz and in the Early Modern Western philosophy. Leibniz is one of the first authors to try a tentative systematization of the discoveries made by the emerging life sciences. He does this by re-inserting teleology into the mechanistic paradigm of Early Modern philosophy, and by providing, with his concept of the monad, a metaphysical account of life and organism. Nunziante shows how the terminological and conceptual invention of Leibniz is indebted both to the Cartesian tradition of the "machine", and to the Aristotelian tradition of the "form". In this sense, Leibniz's attempt is exemplar of the Early Modern tension between, on the one hand, the need to explain the universe in purely physical terms, and, on the other hand, the need to understand the mind of the "designer" teleologically.

Teleology, and its place in our understanding of nature, is famously a central issue of Kant's Third Critique, where Leibniz's original attempt to merge teleological discourse with the modern mechanistic view of nature is given a new formulation. In the second paper of the volume, Hugh Desmond and Andreas Huneman analyze the role and the legitimacy of agential explanations in current biology, navigating between the Scylla and Charybdis of "ontic" and "reductionistic" interpretations of agency. In this way, they show how Kant's reflection on the apparently unavoidable use of teleological descriptions of organisms is still of the greatest importance for our understanding of the living world.

The teleological issue raised by Leibniz forms the basis not only of Kant's, but also of Hegel's view on life sciences. In her article, Michela Bordignon shows how Hegel, via Kant, develops his dialectical account of the concept of life in order to overcome the logico-ontological issues posed by modern science. Bordignon illustrates Hegel's dialectic as a process of self-determination implying a fundamental relation with – or more precisely, an assimilation of – alterity. She also shows how, by means of "dialectical thinking", Hegel aims at overcoming Kant's understanding of natural purposiveness as having a merely regulative value. Bordignon's contribution thus shows how Hegel's account of dialectic is, at least partially, born from the need to explain the dynamic nature of an organism. He achieves this by reassessing, and, in part, by overcoming, the monadological ideas of Leibniz.

The introduction of dialectical thinking operated by Hegel constitutes a fundamental step in our conceptual history, a passage that reveals how every philosophical reflection on the organism and on the proper understanding of human individuals has to take into account the issues regarding self-determination and both the natural and the social environment. In their contributions, Luca Illetterati and Federico Sanguinetti specifically try to work out Hegel's ideas concerning human nature. Both Illetterati and Sanguinetti point out how, for Hegel, thought is what differentiates human individuals from other non-human animals. Illetterati focuses on the classical characterization of humans as rational animals, and shows how, when considered according to Hegel's philosophical system, rationality marks human beings as essentially open to others and to alterity in general. In this sense, the essence of human subjectivity consists in the "natural" capacity to realize oneself as a rationally self-determining subject – i.e., as a *free* subject. In turn, the implementation of the subject's freedom essentially involves a relationship with otherness.

Sanguinetti too insists on the rational essence of humans, as well as on the fundamental link that Hegel draws between rational self-determination and freedom. He does so by carrying out an analysis of the very meaning of freedom in Hegel's *Philosophy of Subjective Spirit*. This analysis shows a fundamental connection between the capacity to be free and self-conscious responsiveness to norms. This latter is reached through a dialectical process, which coincides with the liberation from nature achieved by the spirit in different stages. However, drawing from John McDowell's recent works, Sanguinetti claims that Hegel's account of the process should not be read in evolutionary terms. In this way, he anticipates some issues from the second part of the volume, namely the possibility of reconciling the Leibniz-Kant-Hegel legacy with the post-Darwinian naturalistic account of the organism in general, and of the human individual in particular. According to Sanguinetti, from a Hegelian/McDowellian perspective, the possibility of a conciliation relies on a clear distinction of fields and methodologies, insofar as there is a fundamental difference between the philosophical and the scientific understandings of "what characterizes human beings as individual organisms of a specific sort". More specifically, all evolutionary, scientific, anthropogenetic understandings of human beings are not properly able to grasp the fundamental, constitutive, and thus causally inexplicable, self-referential nature of human beings.

Hegel is the figure in modern philosophy who most clearly stretches the understanding of the inner structure of living beings and of their relationship to the environment beyond the realm of purely biological sciences. Signally, Hegel points out that the dialectical relationship of human individuals with both their natural and their social environments is different from that of other animals to their purely biological complexion, as well as to their groups and species. One could say that Hegel tries to work out a holistic understanding of the totality of being, which can account for the dependency of human individuals both on their societies and on their species, but that is also able to safeguard their individual independence and freedom. In this way, Hegel tries to avoid a reduction of human individuals to mere pawns within a society understood as a big organism. The next step in our conceptual history is thus, unsurprisingly, represented by Marx. By confronting recent debates in Marxian scholarship, Yusuke Akimoto shows how Marx's thought, contrary to what has long been assumed, is neither anthropocentric nor unable to understand the importance of the natural environment. Akimoto stresses that Marx's account of labor as a basic factor for the development of human nature is based on a triadic structure, which links together human individuals, society, and natural environment. This shows, according to Akimoto, that neither an organicist understanding of society, nor the assimilation of human individuals to purely natural organisms, would properly grasp what human nature, as it were, is.

Other historical successors of Hegel took seriously the need to avoid a reductionist understanding of human life. Among them, the so-called British Idealists tried to cope with the new view of nature sparked by Charles Darwin's theory of evolution. To them, Darwinism came as a renewed assault on the citadel of teleology, and, as such, as something deeply problematic. In his paper, Robert Kocis shows how two authors from this tradition, James Hutchinson Stirling and David George Ritchie, devoted considerable reflections to the issue. Stirling mainly criticized Darwinism, considering it a biased theory, incompatible with any sound philosophical system. At the opposite end of the spectrum, Ritchie tried to appropriate Darwinism, and integrate it into a fully-fledged philosophical system – one deeply indebted not only to Hegel, but even more to Kant's reflection on the nature of teleology. In this way, Kocis presents the philosophy of Ritchie as a significant philosophical achievement, capable of reconciling, even in a post-Darwinian era, teleology and life sciences.

An interesting alternative to the "Hegelian" development of Leibniz's ideas on organism and life, which, nevertheless, crosses in many aspects the trajectory of German Classic Philosophy, was developed by Haeckel, one of the greatest propagators of Darwin's idea in Germany. As explained by Caroline Angleraux, Haeckel, in his attempt to give a systematic account of the new discoveries in biology, and to achieve an almost metaphysical view of reality, incorporated Schelling's interpretation of Leibniz into his views. In particular, Haeckel transformed the metaphysical concept of the monad into the biological-physical, i.e., naturalistic, concept of monera. In fact, in his doctrine of monera as the smallest living parts of the universe, Haeckel tried to combine, in a Leibnizian fashion, philosophical reflection and scientific investigations in a vitalist view of the universe. Haeckel, in this sense, can be considered as a thinker at the crossroads of a new era, when natural sciences became more and more independent from philosophical reflections. On the one hand, he still attempts to offer a metaphysical perspective on the universe. On the other hand, perhaps due to influence from Schelling, he shuns dialectics, and firmly anchors his vitalism to a naturalistic ground, interpreting the basic constituents of reality as purely natural entities.

This same need, i.e., to reach a holistic, yet purely "natural" understanding of life and reality, lies at the core of the first works of Japanese modern philosophy. At its beginning, Japanese modern philosophers were deeply influenced by the dialectical reading of life and reality derived from Hegel, often through the filters of Green and Bosanquet, i.e. two important figures of British Idealism. Although British Idealism has long been as good as forgotten (at least until very recently), one should not forget that, at its time, it was hugely influential. In fact, while British Idealism's star was quickly waning from the anglophone world, giving rise to the anti-idealistic stance of the first generation of analytic philosophers, dialectical explanations of life and human nature were brought forward in an original manner – and, for the first time, beyond "Western borders" – by the Japanese philosophers of the first half of the twentieth century. This is especially evident in the so-called Kyoto School, whose beginnings are conventionally placed in 1911 with the publication of Kitaro Nishida's seminal work *An Inquiry into the Good*. At the time of their massive encounter with Western philosophical and scientific traditions, Japanese philosophers attempted to carry out an original form of synthesis, with the ambitious aim of overcoming all impasses of previous speculative systems, while, at the same time, integrating the most recent scientific theories. As Yujin Itabashi shows, with regard to the issues of the organism and the human individual, Nishida tries somehow to conciliate, or rather to sublate, Hegel's dialectical thinking with the then recent biological theories of John Scott Haldane (the father of J.B.S. Haldane). Itabashi poignantly shows that the Japanese thinkers, and notably the main representatives of the Kyoto School, were especially interested in a philosophical understanding of life, in all its forms.

This is shown also by the fact that, more or less in the same period, Hajime Tanabe, who was one of the first students of Nishida, and one of the most prominent and original ones, started to develop a "logic of species". Tanabe devoted thirteen volumes to such an enterprise. Strangely enough, he does as good as never really tackle Darwin's *Origin of Species*. This perplexing fact notwithstanding, Takeshi Morisato attempts to show how Tanabe's highly metaphysical and speculative thoughts on the relationship between species and individual can fruitfully be employed for a philosophical and systematic understanding of organism which, at least in Tanabe's view, should overcome the one-sidedness of both Kant and Hegel.

Naturalism has been the dominant metaphysics of most of contemporary Western philosophy. Perhaps predictably, however, there is much debate on its definition and scope, especially when it comes to discussing the status of "normative" phenomena in the world. In this regard, contemporary naturalism can be divided in two broad categories. On the one hand, the so-called "scientific", "strong", and "reductionist" naturalism, in which normative phenomena are reduced to non-normative ones. On the other hand, "liberalist" and "non-reductionist" naturalism. This latter view has the advantage of being more inclusive, but, at the same time, it needs a plausible explanation of how it could be possible to account for normative phenomena without resorting to the non-natural. In his paper, Andrea Gambarotto analyzes Hans Jonas' attempt to overcome the "antinomical" stance on teleology that goes back at least to Kant, replacing it with a perspective grounded on the reciprocal co-implication of life and cognition. Gambarotto's intention is to assess this approach, and see if it can be of use in the development of a non-reductive yet naturalistic philosophy of biology. In the end, Gambarotto shows that, by putting an excessive emphasis on the first-person perspective, Jonas's account of the organism and teleology steps into "super-naturalistic" territory. Gambarotto believes, however, that freed from this excess, Jonas' ideas could be of high interest in the development of a naturalized account of normative phenomena.

As a matter of fact, reductionism seems, to a higher or lesser degree, to be a significant element of many contemporary approaches to the organism. In fact, according to biologists Richard Lewontin and Richard Levins, it could be even counted as one of the constitutive traits of contemporary biology. Building on the Marxist tradition of dialectics (and, in particular, on Engel's dialectics of nature), Lewontin and Levins have crafted an alternative approach, devoid of the "ideological assumptions" beneath the machine metaphor that shapes contemporary views of the organism and of its relation to its development and environment. In his paper, Pierfrancesco Biasetti tries to assess this approach, principally by adopting Lewontin's constructivist views as a test case for dialectical biology. His conclusions are that while it is unquestionable that Lewontin's characterization of the organism stands as an inspiring conceptualization for many system-centered approaches to development and evolution, it remains debatable whether these approaches can be considered "dialectical", at least in the articulate sense of the term framed by Lewontin and Levins.

Besides the tradition of dialectical materialism, many works and researches have recently been carried out to show how Hegel's *Naturphilosophie* can still be aptly deployed in order to advance our understanding of nature and human life. Sanguinetti, as already mentioned, insists that Hegel would be against any form of reduction of genuinely philosophical questions to historical-causal questions concerning human specificity. In the penultimate chapter of this volume Lenny Moss advances a quite opposite view, pushing forward with certain Hegelian ideas in order to tackle fundamental issues in philosophical anthropology – and, we would say, also in metaphysics, understood as the doctrine concerning the basic structure and elements of reality. Although he avoids a causal account of human nature, Moss illustrates how some recent discoveries in paleontology and anthropology seem to be in agreement with a dialectical understanding of life and nature. Moreover, he points out that, while both Leibniz and Kant have considered the responsivity to norms that characterizes human beings as somehow static, without development, and, thus, as something without history, Hegel, thanks to his dialectical thinking, allows us not only to recognize stages within the development of human mind, but also to fill the gap between nature and spirit. Moss recognizes that Hegel's *Phenomenology*, by presupposing mindedness as somehow there "from the beginning", cannot properly satisfy contemporary needs for a naturalistic account of mindedness. Moss, however, believes that some of Hegel's ideas could be used to such a purpose.

In this way, Moss' article clearly shows how the conceptual history we sketch in this volume is far from being dead, or accomplished. It is instead a living history, and a source of ideas and thoughts that, like living organisms, can keep living only when (self-)moving and developing. In other words, the life and liveliness of such tradition depends on our capacity to further develop it. According to such tradition, indeed, certain natural organisms understand themselves as freely self-moving and self-determining unities – this is what Leibniz would call *spiritual monads*. Spiritual monads are "naturally" disposed to build themselves respective to norms, and not only to natural laws. As a consequence, the topicality and the outliving of the conceptual history portrayed here depends on the very existence of subjects that are able to understand themselves as spiritual monads – and this means as subjects for which such a history is not simply assumed, or passively inherited, but also actively worked out, analyzed, and even transformed.

That the conceptual history inaugurated by Leibniz' monads and further enriched by Kant's reflections on teleology and by Hegel's dialectical thinking is still alive and kicking is lastly shown by another contemporary philosopher, with whose contribution this volume ends: Tom Rockmore. The validity of Kant's voice within the contemporary debates in philosophy of biology, especially as regards teleology, is pointed out in Chapter Two by Desmond and Huneman. In this closing chapter of the volume, we can see how a certain development of Kant's views can help us also to understand the specific issue of human nature. Rockmore claims that the Copernican revolution operated by Kant should be implemented further into the form of a constructivist approach to human sciences in general. These sciences do not just ascertain what we, as human beings, are. Rather, they participate in the making of ourselves. Indeed there is one fundamental "fact" the human sciences especially need to recognize in order to be both meaningful and truthful, namely that our nature consists in a kind of self-construction, or, to say it better, that our natural determination consists in the need, and somehow also the duty, to determine who, and perhaps even what, we are. No concept of human nature can be considered compelling. It goes without saying, though, that such self-construction is neither totally free, nor arbitrary – and it is definitely not to be understood in individualistic terms, being it always a social endeavor.

In conclusion, we take the liberty of saying that, if the conceptual history we have sketched here is correct, then it is also correct to say that we, as humans, are *natural born monads*. We are individual unities, that naturally strive to preserve ourselves as unities, but in a fundamental – and we would even say "essential" – relationship with an environment, both natural and social, which goes far beyond ourselves – and without which we would have no life. To Leibniz, the "external world" was somehow necessary in order to have something "going on" in the monad, since the monad, in the end, is a representation of its environment – and, ultimately, of the whole world. More "naturalistically", all kinds of living organisms can be preserved only through constant exchange and communication with their environment. In the case of human organisms, communication and exchange concern not only one's natural environment, including one's own bodily constitution, but also the social environment. Indeed, the "moral", or existential, determination of the human individual as a self-determining agent would make no sense, if the normativity to which one is supposed to be responsive were not embedded in an environment that, from time to time, urges one to exert one's self-determination. Something "alien" is required to keep self-determination really, i.e., actively, alive. We could, in this sense, say, that the understanding of the fundamental "monadicity" of our existence was enriched first by Kant's reflections on different types of teleology, and then by Hegel's insistence on the necessity of alterity for the very possibility of one's own genuine self-determination. If *this* is what we are, namely if we are self-consciously self-determining individuals in an environment that exceeds ourselves, and if such an environment, indeed, also partakes in making what we are, we could then paraphrase Sartre, and say that we are *naturally* condemned to "freely" – and this means also consciously – determine ourselves as self-determining living beings. Such a condemnation is both practical, in as much as it regards our actions and our drive of self-conservation, and theoretical, because we cannot help but try to find a clarified and comprehensible explanation of ourselves that is in agreement with the aforementioned natural determination to self-determination. And all this means that we are naturally born monads.

We cannot choose to be otherwise. We are born as monads, and our birth does not depend on us. We are only able partially to decide what kind of monads we want to be – and this implies, first of all, not so much what kind of identity we want, but rather what kind of environment we build for our self-preservation. We are mirrors of the universe – mirrors whose identity depends on our power to shape, at least partially, the universe itself, and on the shape we manage to give to it.

## Acknowledgments

The idea of this project raised on the occasion of two workshops organized at the Hokkaido University in 2014 2015. We would like to thank the Japan Society for the Promotion of Science (Kakanhi Grant Nr. JP17K02153) for supporting those workshops, and the National Social Science Fund of China (Grant No. 17BZX085) for supporting part of the later work for this volume.

It was a real pleasure to work with de Gruyter publishers, Berlin. Our sincere thanks go to Christoph Schirmer and Tim Vogel. A special thanks, finally, goes to Philippa Adrych for her extraordinary editorial work on all texts of this volume (aside of this acknowledgment!).

We dedicate this volume to the memory of Lorenzo Calabi.

Antonio M. Nunziante
# Between Laws and Norms. Genesis of the Concept of Organism in Leibniz and in the Early Modern Western Philosophy

**Abstract:** The word "organism" represents an original keyword of the early-modern philosophical world. As it was first developed by Leibniz, it seems to blend together two different conceptual paradigms: the Cartesian model of the "machines" and the Aristotelian legacy of the "individual natures". According to the first, nature represents itself the prototype of any good mechanical functioning, but at the same time its inner development is explained by the occurrence of a normative dimension that rules the world of primitive forces in the dynamics. For such reasons, the "organism" lexicon is affected by an internal stress that is extremely interesting to analyze for it seems to posit a normative turn acting from the within of a mechanically conceived notion of life.

## 1 Introduction

As it was first construed, between late Scholasticism and early Modernity, the concept of the organism carried within itself a characteristic dichotomy. This is because two distinct epistemological models, each based on different ontological intuitions, coexisted within it. Using a remark pointed out by Wilfrid Sellars, we can say that the first model was typical of the Aristotelian and late-Scholastic tradition (the so called "thing-nature framework");[1] whereas the second belonged to the nascent tradition of natural sciences and modern mechanics ("event-law lexicon").[2]

---

[1] In this model, the core elements of the framework consist of individuals endowed with their own "nature" (e. g., a fig plant, a labrador dog, an individual named Anthony). According to Sellars, the Aristotelian world is characterized by a realistic intuition regarding individuals: there are flowers, animals, humans, planets; in other words, there are natural individualities endowed with specific forms. Cf. Sellars 1949, p. 565–566.

[2] In this case, the dominant aspect is physical legality. For modern scientists there are no "things", but processes governed by mechanical laws. Sellars 1949, p. 566.

---

**Antonio M. Nunziante,** University of Padova

https://doi.org/10.1515/9783110604665-002

The purpose of this paper is to analyze the tension subsisting between these two models, namely between the idea of a processuality governed by mechanical laws and the idea of a self-normativity rooted in the nature of individuals. In other words, the idea is to show how the dichotomy between the "thing-nature" and the "event-law" frameworks needs to be better explained, but also made less abstract by considering it from within, analyzing the very way in which the concept of organism had originally been construed.

Taking this internal perspective, we will see how both teleological elements and mechanical categories coexist in this newborn early-Modern concept; it will be also shown, however, that it is somewhat misleading to regiment the question in terms of an abstract dualism.[3]

The topic discussed here thus deals in some ways with a technical issue that would seem to be peripheral to the great themes of modern philosophy (the word "organism" appears only rarely in Leibniz's writings, for instance, and with a very different meaning from subsequent uses of the term.) In actual fact, however, it is by no means a lesser issue, because the semantics of the term have long-lasting metaphysical and epistemological effects that only become fully apparent in the post-Darwinian age.

"Perhaps it will take a long time," wrote Heidegger strikingly, "to realize that the idea of organism and of organic is a purely modern, mechanical-technical concept, so that what grows naturally by itself is interpreted as an artifact that produces itself" (Heidegger 1976, p. 255). The philosopher who is better able than others to help us sketch the details of this theoretical story is undoubtedly Leibniz.

## 2 The First Occurrences of the Term: The Shift Between Its Countable and Uncountable Meanings

Despite its simplicity, Sellars's remark that in early Modernity the application of the concept of "individuality" to natural phenomena had been based on two different epistemic models has the merit of highlighting an ontological ques-

---

**3** This is something that Sellars does not do, whereas McDowell hypostatizes (probably to an excessive degree) the dichotomy between "space of reasons" and "realm of law" (cf. McDowell 1996, p. 71).

tion that has often engendered many controversies.⁴ In some ways, it brings to mind the ancient, never-ending dispute over universals, except that, in this case, the focus of the discussion is not on the ontological status of genera and species. This is because, as Sellars puts it, modern scientists did not know what kinds of things existed until they succeeded in formulating a law that could subsequently be translated into the language of things (Sellars 1949, p. 566).

While the thing-nature framework, in fact, took for granted the existence of natural individualities endowed with properties (the appropriate ontological status of which would have subsequently been discussed), the event-law lexicon questioned the very existence of what common sense presents to us as individual, while concentrating on the procedural dimension of the event and of the laws-of-nature dimension. For the scientist, that there are "things" that have individual shape is part of the immediacy that forms our ordinary experience, but what lies at the deep, primary level of the natural world is the legality of the principles that cause the appearance of such events. Such principles concern the motion of bodies, which could be understood with the aid of phoronomy, kinematics, dynamics, and which pertained, more generally, to the domain of disciplines that comprised the framework of classical mechanics.

The other relevant aspect of this dichotomy regards its characteristic epistemic asymmetry. Although these two levels were often confused in early modern philosophy (at least in Sellars's opinion), the dominant mode of understanding was soon to become the event-law lexicon. The ancient Aristotelian world of forms, substances and properties would in fact regress until it was eventually replaced by Darwinian evolutionary theories.⁵ Taking for granted, therefore, the efficacy of Sellars's descriptive tool, which seems to work especially well if applied to long-term historical trends (his analyses are aligned in some ways with Hus-

---

4 Of course, the diversity of epistemic models described by Sellars cannot be confined in the straitjacket of a merely dichotomous contraposition. This is particularly evident from research on corporeal physiology, in which the contrast between models often became reason for cooperation or contamination. In the seventeenth century, the Aristotelian-Galenic tradition was still very relevant, but there were many overlapping lines of research, from Cartesian iatromechanics to Van Helmont's iatrochemistry (based on the consideration of chemical phenomena linked to fermentation and the production of gases), from Glisson's *fibrillaire* approach to studies like Robert Hooke's "micrographia". As usual, seen from closer up, the picture becomes much less clear. On the topic, cf. Grmek 1990, Duchesneau 1998, Clericuzio 2000, Nunziante 2011.
5 On the distinction/confusion between the two frames of intelligibility, see Sellars 1959, p. 154; p. 162. On the part played by the evolutionary theories in overcoming the manifest image, cf. Sellars 1963, p. 17.

serl's in the *Crisis of European Sciences*,⁶ we can however contextualize it better by restricting its field of application. There is a semantic dimension, in fact, characteristic of early Modernity, in which such dichotomous tension displays all its strength, namely the lexical dimension that refers to the world of organisms, organizations and "organic mechanisms".

As it is known, the word "organism" does not exist in Ancient Greek, and it was also very little used in Medieval Latin (Cheung 2006, p. 321).⁷ The most striking thing about its first occurrences in the seventeenth century is that the noun "organism" was uncountable, not countable, since it denoted a "mass or complex of things that have the property of being organized" (Pasini 2011, p. 1219). In other words, it referred to a "principle of order" (Cheung 2006, p. 395), and did not mention individuals; this would happen later, towards the end of the eighteenth century. In this case one might say, using Sellars's analytical tool, that the pragmatics of the term is entirely internal to the event-law lexicon: the organism designates a procedural form, a way of organizing, and not a substantial individual. It does not mean "living substance", rather it is the "lifeless disposition of parts": it is that mode of organization necessary to the unfolding of the vital effects of the soul (Cheung 2006, pp. 324–325). To use the language of our proto-Modern ancestors it is an "organic mechanism". There is in fact something more that is needed for there to be life, something that cooperates with the organism: it takes a soul, a plastic nature, a dominant monad. In other words, there is the need of natures endowed with a certain form (be it in the entelechial-Aristotelian sense brought to light by Leibniz, or in the Platonic sense of spiritual principle highlighted by More and Cudworth): that is to say, something that serves as the principle of metaphysical unity.⁸ All this naturally represents the other side of the coin, since it refers to the thing-nature framework and it is only by assuming this more specific perspective that we can see why organisms

---

6 See Husserl 1970, pp. 21–59. On the relevance of Husserl's analysis on Sellars, cf. Hampe 2010.
7 In the first text acknowledged by scholars (tenth-eleventh centuries), an unknown author describes an alchemic method for distilling fluids, and the Greek term "*organismòs*" refers to "an apparatus in which liquids are distilled" (see Cheung 2006, p. 321). In the second text known to us (eleventh-twelfth centuries), the word only occurs in the plural and indicates a "polyphony" of human voices; so it is related to the idea of a melodic harmony, though the text in question refers to a disharmonic tune sung by human voices (see Cheung 2006, p. 321).
8 Ralph Cudworth (1617–1688) and Henry More (1614–1687) were leading figures among the so-called Cambridge Platonists. They believed that there was an echo of divine Reason in the human soul. They assumed the presence in nature of intelligible forms they refer to as "plastic natures" (Cudworth) or "hylarchic principles" (More). Cf. Duchesneau 1998, pp. 149–181.

went on to become the prototype of what we call living beings (as in Kant, for instance).[9]

The peculiar feature of the lexicon of organisms therefore lies in the fact that it establishes a sort of correlation "between metaphysics and physiology" (Pasini 2011, p. 1231). It provides a way to hold together the procedural organization of the body's physiology as well as the metaphysical dimension of its individual form. It is a lexicon that serves as an interface between the legality of the body and the normativity of (living) individuals.

The tension then also reveals its effects on the epistemic level. To understand *scientifically* how the body-machine works, we must refer to the laws governing animal physiology (in his controversy against Stahl, Leibniz argued that the physician must take care of the body, not of the soul).[10] On the other hand, to understand *metaphysically* what a living substance is, we must refer to its intimate normative structure (the dominant monad as the generating principle of an individuality).[11] This consideration raises another issue, which is probably the most decisive of all: in the Cartesian-mechanistic age (and particularly for Leibniz) the world of proper natural objects – in the Aristotelian sense of "things of which the final cause coincides with the formal, and the what-it-is-for and the what-it-is are one" – is a world of machines (Pasini 2011, p. 1229). The expression "organic mechanism", that to us sounds like an oxymoron, was pleonastic before the seventeenth century, since "mechanical" and "organic" simply meant the same thing (Pasini 2011, p. 1217). Nature is essentially a machine ("*horologium Dei*" – Leibniz, A II 1, pp. 22–23), precisely in the sense of a mechanical device whose position is defined by the realm of the law:

> everything must happen in the bodies in such a way that it is possible to explain it distinctly from the very nature of the bodies, that is, from the size, the figure and the laws of motion: this is what I call "mechanical" (*Animadversiones*, p. 68).

The laws of motion explain the mechanical dimension of bodies, and the latter should not be perceived in a metaphorical sense because it is not that bodies *resemble* machines, *they are* machines. The technical-artefactual model becomes

---

[9] In Kant's late works, we find both types of occurrence, with "organism" used to mean a "principle of order", and as a generic name for "individuals". See Cheung 2006, p. 331.
[10] Georg Ernst Stahl (1659–1734) was a German chemist and physician. He published an important work entitled *Theoria medica vera* (Halle 1708) in which he somehow defended a peculiar form of animism. The book prompted a famous controversy with Leibniz (see Nunziante 2011).
[11] The use of the adverbs "scientifically" and "metaphysically" has a rhetorical (though not arbitrary) function in this case.

the epistemological model that encompasses, in a broad sense, the very concept of "natural". But there is more than this, because Leibniz himself gives a further twist to this mechanical ontology, making it almost burst from within. In the multifaceted world of machines (everything around and within us is a machine), some machines in fact have a particular shape. Some are so perfectly machined that they steadfastly remain the same. What Leibniz baptizes with the new name of "natural machines" have the primary feature of always exhibiting an identical form of organization, right down to their smallest details (such an organization is infinite in the sense that it constantly and identically repeats itself). As Deleuze pointed out, "organism" in this context means "machinery" in the sense of something that is perpetually "machined" and it is only these particular types of "natural" machine that will be marked as "living" from then on (Deleuze 1993, p. 8).

In this peculiar dimension *laws* and *norms* meet and merge together to some degree. The main feature, here again, is the organization, or the principle of manifest order in the arrangement of the bodily parts, which can therefore only be understood on the grounds of the laws governing corporeal physiology. This organization, however, differs from that of an artificial machine, which can stop at some point, when it ceases to be organized (when a cog in mechanical gearing breaks off and no longer fits its companion part). The living always remains organized because its organization is the expression of a different normativity that is more than just mechanical and more than physical, because it has to do with information codes that Leibniz sees at the root of an individual living being. The reference is to the world of monads, of primitive entelechies and of those particular types of monad that "dominate" the body-machine, making it "one" (Leibniz 1989, p. 177/GP II, p. 252). The reference here is to a metaphysical normativity that constitutes the generating principle of a "series". But, before coming to this point, it may be better to rewind the tape and analyze some of the general features of the nature-machine paradigm of the Cartesian age.

## 3 Cartesian Mechanical Embryology and Its Limits

The very concept of "nature" in early-Modern Western tradition is thus a mechanically conceived notion. It is not simply modeled on the pattern of artificial machines, since nature is the prototype of any good mechanical functioning: nature is the *real, well-functioning* machine.

The "nature-machine" paradigm is not without problems, however. It works well in describing the mechanical phenomena related to motion and the collision of bodies, but seems limited when applied to the broad array of biological phenomena. Animal physiology was explained in mechanical terms, but this approach was not always very successful and the field of medicine, for instance, provided an extraordinary terrain for comparing theoretical approaches and different practical solutions.

Generally speaking, the biological events relating to animal generation, self-regulation, growth, and deterioration, were those proving most resistant to a mechanical model of causal explanation (Duchesneau 1998, pp. 45–46). The methodological goal of the Cartesian program was actually quite clear: explaining a physical process means specifying the efficient causes that generated it, and this involved providing an adequate representation of the mechanism that, under such circumstances, prompts the occurrence of the physical event in question (Duchesneau 1998, p. 46).

This very idea of proof as a kind of a priori deduction starting from the general level of mechanical principles met with some unexpected difficulties, however, when it came to biological events. In the case of medicine, for example, practical solutions came first and a general theory was formulated afterwards; this meant that the theory was developed retrospectively rather than deductively.

The idea of a universal mechanism dominated the corporeal physiology of animals, as we can see from this meaningful passage from Descartes:

> [In conclusion], I would like you to reflect [...] on how all the functions that I have attributed to this machine, such as the digestion of food, the beating of the heart and arteries, the nutrition and growth of the members [...], how these functions follow completely naturally in this machine solely from the disposition of the organs, no more nor less than those of a clock or other automaton from its counterweights and wheels, then it is not necessary to conceive on this account any other vegetative soul, nor sensitive one, nor any other principle of motion and life, than its blood and animal spirits, agitated by the heat of the continually burning fire in the heart, and which is of the same nature as those fires found in inanimate bodies. (AT XI, *Le monde*, pp. 201–202)

The elements described in the above passage can be found in the human body, as in any other animal. We do indeed have a machine that works like a clock, and our movements can be explained as a kind of "automaton". Yet, all this does not seem to be enough because what is really difficult to explain in the corporeal machine is its self-movement (AT XI, p. 120), which is hard to deduce from anything else – unless such a machine has been made directly by the hands of God (AT XI, p. 120).

In short, the model was that of analogical deduction. A certain vital event was considered (such as digestion, breathing, or homeostasis), any kind of finality or intervention by the soul (as claimed in the previous Aristotelian tradition) was ruled out, and after that an analogy was sought with some known mechanical models. This is what Harvey did, for instance, when he described the blood circulation in the vessels as a closed hydraulic circuit:

> The heart, consequently, is the beginning of life; the sun of the microcosm, even as the sun in turn might well be designated the heart of the world; for it is the heart by whose virtue and pulse the blood is moved, perfected, and made nutrient, and is preserved from corruption and coagulation; it is the household divinity which, discharging its function, nourishes, cherishes, quickens the whole body, and is indeed the foundation of life, the source of all action. (Harvey 1928, ch. VIII, p. 42)

In this passage we have a dizzying transfusion of meanings: mechanical principles are continuously interwoven with metaphors and analogies – to such a degree that each term explains and is explained by the other. The heart is like the sun, which in turn is the heart of the world, and so on.

There was still something not quite right, however, and it was embryology that eventually brought researchers back to square one. In fact, the theoretical problem posed in the case of embryology sounded more or less like this: how can organization be produced from what is not at all organized (the bare extended matter deprived of inner qualities), and how can such a concept be justified by virtue of mechanical categories alone? The theoretical impasse was also partly due to the very notion of knowledge that, in the meantime, had become dominant. For a broad consensus, in fact, "knowing" was referred to as the ability to "re-produce" something, in the strict sense of manufacturing it (we know something insofar as we are capable of reproducing it with our own hands). As a consequence, scientists were driven towards anthropomorphic interpretations of vital apparatuses (what scholars call "technological anthropomorphism"), since they explained them (the functioning of the heart, the kidneys, or homeostasis) by drawing analogies on a horizon of concepts borrowed from a different field of application, namely from the human world of mechanics (Duchesneau 1998, p. 83).

Therefore, on the one hand, it was theoretically forbidden for modern science to explain natural phenomena by appealing to anthropomorphic or zoomorphic concepts, as the pre-Modern natural scientists had done, reducing nature to mere anthropic projections. But on the other hand, in ordinary practice, ample use was made of theoretical models based on such finalistic-mechanical models since the handicraft lexicon *is* an anthropic lexicon (Jonas 1966, p. 10; pp. 109–110).

The emergence of this so-called "handicraft-lexicon" had to tackle two further basic difficulties. First, there was the problem of the designer: if nature is essentially a machine, some entity – namely, God – must have designed it. Second, there is the teleological problem of the purposiveness of nature, since everything produced by a designer stems from an intentional plan, but the idea of attributing intentional goals to nature is highly problematic.

As Georges Canguilhem pointed out, a first assumption of the Cartesian theory of life was that, prior to nature, there was God, the great creator of life. A second assumption concerned the everlasting bond established between "life" and "functioning machines" (Canguilhem 1965, p. 112). The "nature-machine" theory implied the acceptance of a kind of axiomatic system of the sort: life exists because God has decided so in creating nature (first assumption); and it has an organized form of a mechanical type (second assumption). Taking this ontological-epistemological framework for granted, which involved accepting some profound theological as well as teleological suppositions, modern scientists attempted to produce theoretical models compliant with this framework.

As a result, the "nature-machine" model suffered from an internal stress that is extremely interesting to analyze. Admittedly, there is a sense in which the metaphor of nature as a universal clock was used to indicate that nature was autonomous in relation to God (clocks are a sort of self-propelled machine, and therefore relatively independent). This humanization of nature did not really remove God from the universe, however; it just made His presence consistent with an epistemology centered on the above-mentioned handicraft-lexicon (we only have true knowledge of what we are able to produce by ourselves). In fact, images of God as the "architect", "artisan" or "watchmaker" of the universe became very popular in the culture of the time. As Leibniz wrote in a brilliant passage of the *Discourse on Metaphysics* (1686):

> It is appropriate to make this remark in order to reconcile those who hope to explain mechanically the formation of the first tissue of an animal and the whole machinery of its parts, with those who account for this same structure using final causes. Both ways are good and both can be useful, not only for admiring the skill of the Great Worker, but also for discovering something useful in physics and in medicine. And the authors who follow these different routes should not malign each other [...] It would be best to join together both considerations, for if it is permitted to use a humble comparison, I recognize and praise the skill of a worker not only by showing his designs in making the parts of his machine, but also by explaining the instruments he used in making each part, especially when these instruments are simple and cleverly contrived. *And God is a skillful enough artisan* to produce a machine which is a thousand times more ingenious than that of our body, while using only some simple fluids explicitly concocted in such a way that only the ordinary laws of nature are required to arrange them in the right way to produce so admirable an

> effect; but it is also true that this would not happen at all unless God were the author of nature. (A VI 4 B, pp. 1564–65 – *italics in the text*)

In some ways, God was humanized, while the vocabulary of production was deified (Husserl, 1970, p. 66). All this posed a hard problem that would be revealed not so much in the Modern age, but after Darwin, because the legacy of a handicraft-lexicon semantically referring to the plan of a divine intentionality was bound to remain highly problematic, since it was unacceptable to the post-Darwinian naturalists. In the Early Modern Age notions like "life", "machine", "organism" were in fact all part of a same semantic set, which was basically consistent with the presence of a Great Craftsman of the universe. Such concepts were doomed to become controversial much later on when, along with the arrival of philosophical naturalism, there came the problem of a design without a designer, and of purposes without intentions (Kitcher 1993; Ayala 2007).

## 4 Leibnitian Organisms: Percpetive Super-Machines

Among the philosophers of early-Modern times, G.W. Leibniz (1646–1716) was perhaps the one who paid the greatest attention to the issue of a philosophical understanding of life. Leibniz was probably also the author who led the Cartesian mechanistic paradigm towards a new and unexpected normative turn. Although "organism" was not a core word in his philosophical vocabulary, his reflection on the "machines of nature" was destined to become prototypical of the semantic history of the term.

Generally speaking, it could be said that Leibniz saw the notions of "order" and "organization" as coming prior to the concept of machine, intending the former approximately as a "mutual relationship of the parts" (A VI 4 B, p. 1320). In a natural universe with machines scattered all around, there are two basic types of organization: a *finite* one, typical of the machines that, from now on, will be called "artificial machines"; and an *infinite* one, belonging to the so-called "machines of nature", which will be denoted as belonging to living substances. A "finite" organization is unable to replicate itself, as in the case of the parts in the mechanisms of a clock, whereas the capacity to perpetually replicate their internal order is peculiar to living substances.

> We must then know that the machines of nature have a truly infinite number of organs, and are so well supplied and so resistant to all accidents that it is not possible to destroy them. A natural machine still remains a machine in its least parts, and moreover, it always re-

mains the same machine that it has been, being merely transformed through the different enfolding it undergoes, sometimes extended, sometimes compressed and concentrated, as it were, when it is thought to have perished. (GP IV, p. 482, p. 142)

If we replace "machine" in the above text with "organization", the quoted passage works even better: there is a natural organization that remains organized even in its smallest ingredients. This is the infinite dimension of nature we were referring to: an organization remains steadfastly organized, and persists in preserving the form of its relational identity.

As a consequence, the machines of nature display a greater degree of complexity than artificial ones, as the latter are only supported by a mechanical type of organization. Paradoxical as it may seem, the problem of this latter kind of machine is that, under certain circumstances, the arrangement of its components ceases to be "mechanical" (i.e., functionally organized), and so Leibniz writes in the § 64 of the *Monadology:*

> A machine constructed by man's art is not a machine in each of its parts. For example, the tooth of a brass wheel has parts or fragments which, for us, are no longer artificial things, and no longer have any marks to indicate the machine for whose use the wheel was intended. But natural machines, that is, living bodies, are still machines in their least parts, to infinity. That is the difference between divine art and our art. (Leibniz 1989, p. 221)

Leibniz's unspoken target is the notion of "self-organization" because this marks the dividing line between natural and artificial, and it is highly relevant that such a concept represents an internal development of mechanical categories, to such a degree that he stretched the very program of Modern mechanics beyond its epistemological limits. In fact, the concept of organization exhibits two different levels. On the one hand, it simply indicates something that is already "organized" (in the sense of something that *has been designed* by someone else). In other words, it is a process that in some respects has already been accomplished by an external designer, and it is no longer capable of renewing itself. On the other hand, organization indicates a self-arranging capacity, in the sense of a process directed towards self-preservation through a perpetual re-arrangement of its own structure.

This last remark leads us to clarify another key feature of the living substance: "Corpus viventis est machina sese sustentans et sibi similem producens." (A VI 4 A, p. 568)

The living body is a self-governed system that aims to assure the self-preservation of its own internal organization, and is eventually able to produce new replicating systems. The phenomenon of life accordingly entails a tendency towards self-preservation. In this regard, it is important to emphasize that what is properly maintained in living beings is the form of their organization, a kind of relational identity, that comes before the atomistic model comprising the system as the sum of (already formed) different parts. In the case of living systems, in fact, it is the relationship of the parts to the whole that remains unchanged, and that is therefore ontologically primary.

This also explains why, technically speaking, the concept of order represents a "simple primitive term" that cannot be further broken down into more basic elements, nor can it be deduced from something else. Using Leibniz's own terminology, the concept of order is a "purely integral" term (*terminus integralis* – A VI 4 A, p. 741): it enjoys a logical-ontological priority because it depends neither on the properties of the parts, nor on the physical properties of matter. It mainly concerns an internal system of relationships, which has the advantage of keeping itself invariant with respect to the shifting of the parts.

In a brief fragment entitled *De machina animata*, Leibniz writes: "By no-one can a body that is perfectly similar to the human body be produced, unless someone is able to preserve the order of the division to infinity." (A VI 4 B, p. 1801) We need to focus on the words he uses here: a living body can be produced only by preserving the order of the division to infinity. It is characteristic of Leibniz's philosophy to connect the mathematical notion of *infinity* to the logical one of *identity*, as well as to the metaphysical notion of *individual*. A notion is infinite, Leibniz says, when "the same reason always exists" (A VI 6, p. 154). The "true infinite" (A VI 6, p. 158), he adds, is the absolute "which is anterior to all composition, and is not formed by the additions of parts" (A VI 6, p. 154).[12] So we can paraphrase the above text from *De machina animata* in the following sense: to produce a living body it is necessary to produce an ongoing self-replicating order that keeps itself steadily identical, and it is only if this strong requirement is fulfilled that the resulting body will be identical to a living one.

Turning to another extremely meaningful passage:

> Corpus vivens est Automaton sui perpetuativum ex naturae instituto, itaque includit nutritionem et facultatem propagativam, sed generaliter vivens est Automaton (seu sponte agens) cum principio unitatis, seu substantia automata. (A VI 4 A, p. 633)

---

[12] Here is another meaningful passage: "Let us take a straight line and prolong it until it is double the length of the first. Now it is clear that the second line, being perfectly similar to the first, may itself be doubled in order to have a third, which is still similar to the preceding; and the same ratio still holding, it is never possible to stop the process; thus the line may be prolonged to infinity, so that the consideration of the infinite arises from that of similarity or from the same ratio, and its origin is the same of that of universal and necessary truths" (A VI 6, p. 158).

A possible translation runs more or less as follows: the living body is a self-moving (*automaton*) and self-preserving machine organized by nature, which thus includes nutrition and the capacity to spread itself. In more general terms, a living being is a self-moving machine that is such by virtue of a principle of unity: the living being is an automatic substance. Perhaps the sense of what we said at the beginning of our analysis has now become more perspicuous: Leibniz seems to bring the mechanism of an entire era to its extreme consequences, because it is as though he were aiming to expand the notion of machine indefinitely, letting the concept of order spring from the very inside of an infinite mechanical logic.

Now here is the point to develop. So far, a kind of phenomenological description of natural machines has been produced, since we have just referred to the observable macroscopic features of continuously self-replicating organic structures. But the question could also be posed more rigorously in the following terms: by virtue of which internal property is the living body actually able to maintain itself? What makes the self-preservation of an organism possible? Such questions bring out another key feature of Leibniz's biological thought, which refers to the Aristotelian tradition of forms and to that "principle of unity" encountered in a previous quotation. Living substances are spontaneously self-organizing machines because they do indeed have an internal principle of unity that effectively makes them "substantial" (i.e., the dominant monad). To come straight to the point, living beings are *perceptive kinds of units*. "Perceptionis gratia sunt organa sensuum; procurandae perceptionis sive actionis gratia sunt organa Motus." (Leibniz 1996, p. 212) The translation here could be read as follows: sensory organs are finalized to perceptions, whereas the aim of organs of movement is to attain new perceptions, in the sense that their purpose is to let corporeal machines exhibit new perceptions (perception and appetite being the peculiar kinds of action performed by living substances).

This passage is important and deserves some further supporting comments. In a somewhat parallel passage, Leibniz observes that "we use the external senses as a blind man uses his stick" (GP VI, p. 499), by which he means that only perceptions enable the animal to orient itself in the world. Though it might seem just a technicality, it is important to emphasize that Leibniz is setting up a conceptual distinction between "motion" and "action".[13] The *motion* of the animal's body has to do with the proper functioning of its organs and the neurophysiology of its corporeal status. It therefore concerns the proper functioning of its heart, muscles, tendons, and everything that can be described in terms of me-

---

13 The distinction was first made by Leibniz in the dialogue *Pacidius Philaleti* (1676). See A VI 3, pp. 528–571 and particularly p. 571.

chanical events (according to the Cartesian tradition). The *action* of the animal, on the other hand, has to do with a completely different lexicon referring to the world of forms contained in the substantial core of the animal's individuality, namely with a lexicon inherited from the Aristotelian tradition.

Although these two distinct levels of explanation work simultaneously and in parallel (bodily motions are explained by bodily motions, and actions by actions), Leibniz nonetheless seems to establish a hierarchy. In a sense, he says that what really matters to the animal's self-preservation primarily concerns its capacity to impress the shape of a representation on the surrounding environment, so he bestows metaphysical priority on the representative dimension of perceptions.

> Now, it is not conceivable how a perception can begin naturally, no more than matter: because whatever kind of machine one can imagine, one will conceive in it nothing but collision of bodies, size, shapes, motions and things which we understand to be very different from perception – which therefore cannot start naturally and neither will have an end. (GP III, p. 344–45).[14]

So there we have it, the distinction between "motions" and "perceptions" (or actions). The mechanism of the animal body's physiology produces shocks, balancing movements, calibrations and adjustments, but perception involves more than just a physical collision. Indeed, in accordance with certain movements of the body some specific perceptions will also be tracked, but it is not epistemologically correct, in Leibniz's view at least, to explain such events by connecting them in a causal way.

Consider the case of the relationship between mind and brain. To us, today, it seems like an old-fashioned metaphysical thinking, but Leibniz's argument is that no-one can hope to exhibit perceptions (in the sense of experiencing them, as from a first-person perspective) simply by recording different neurophysiological tracks.

> Moreover, we must confess that *perception*, and what depends upon it, is *inexplicable in terms of mechanical reasons*, that is through shapes, size, and motions. If we imagine a machine whose structure makes it think, sense, and have perceptions, we could conceive it enlarged, keeping the same proportions, so that we could enter into it, as one enters a mill. Assuming that, when inspecting its interior, we will find only parts that push one another, and we will never find anything to explain a perception. (Leibniz 1989, p. 215 – Monad. § 17)

---

[14] See also GP III, pp. 340–41 for a rather similar passage.

Whether we like it or not, this is the path Leibniz takes and, coming back to the main point, it follows that the animal's self-preservation can be described both in mechanical terms, by referring to the functioning of its corporeal physiology (using the lexicon of motions governed by rules), or by appealing to the internal dimension of its perceptions (using the normative, finalistic-oriented lexicon of actions). In both cases, we are speaking of perfectly well-suited considerations, taking for granted the general harmony of the universe and the consequent validity of the pre-established psycho-physical parallelism.

According to Leibniz, however, the most enigmatic and decisive factor for the preservation of life is represented by the very nature of perception, because here the question becomes: where does the source of the capacity to transform a disrupted multiplicity into a well-connected unity lie? It is precisely in this key unifying element that the secret of the representing act lies, and this explains why perceptions cannot be explained by mere mechanical reasons, and why – at the same time – they are decisive for the nature of living beings. Perception makes living substances capable of homogeneous, unified behavioral responses, since it is always the whole animal machine that reacts to any single environmental stimulus. According to Leibniz, this happens because it is only the perceptual dimension that actually makes the animal body "one". So what kind of properties do perceptions have? What makes them so essential to animal so much so that Leibniz, in several passages, feels the need to say that wherever there is life, there is also perception, and vice versa?

To answer these questions, we need to switch our lexicon, abandoning the mechanical framework of causal explanations. Perception belongs to the realm of actions and Leibniz sees it as the simplest of our cognitive activities, shared by every sort of living being.[15] Perception essentially has to do with an organizing and ordering capacity because, if "order" means "mutual relationship of the parts", then perceptions represent the most basic units of order since they turn the manifold environmental inputs into a single representative output. The most straightforward definition of perception is therefore: "The passing state which involves and represents a multitude in the unity or in the simple substance is nothing other than what one calls *perception*." (Leibniz 1989, p. 214 – Monad. § 14)

The keyword here is "multitude in the unity": a multiplicity of sensory impressions is expressed by the unity of a single passing perceptual representation,

---

[15] As we said, Leibniz establishes a difference between motions and actions. Concerning the latter, and following the Aristotelian tradition, he further distinguishes between acts and actions. See Piro 2002, pp. 55–66; Nunziante 2011, pp. 179–185.

the order coming not from the outside, but from within the simple nature of a substance. By virtue of that, the animal gains a propensity to accomplish an action, since due to the representative content conveyed by the perception (i.e., the specific degree of information displayed by it) it begins to move and to orientate itself in the world. Perception thus gives unitary form to what would otherwise remain a confused background noise. In other words, it transforms the world outside of the animal into a well-ordered environment of viable interactions. The muscles, the heart, and all the necessary motions we mentioned before are concomitant elements of this process, but without such a perceptual implementation the environment would remain silent and the outside of the animal would simply remain an un-decoded wall.

The most extraordinary element of this representational process therefore lies in its formal structure, which does not causally depend on the physiological structure of the animal (within this Aristotelian lexicon, "form" and "matter" are not mutually reducible terms). Leibniz carefully distinguishes the representational content (what is *represented*) from the representative act in itself (i.e., the *representing* act of giving shape to a certain content). He is not very generous with details in this regard, but we can imagine the act of perception as establishing a system of relations that enables the acquisition of every minimal environmental input by ordering it within a framework of differences and comparisons. The core idea is that such a system of relations stands for the identity of the perceived item by substituting the tokened occurrence of the input with an analogical framework of symbolic references made available by the perceiving substance. Going along with this line of reasoning, and using Leibniz's technical lexicon, perceiving thus means *expressing* something with something else. Animal machines do not have any immediate access to the world, but their representation of it denotes the use of a symbolic resource.

Perception therefore contains within itself something of an "ideal" nature, since it already involves a kind of symbolic activity, by means of which every incoming token is displayed as an environmental type. Simply put, the primary skill of what we call "living" lies, to Leibniz's mind, in a kind of categorizing capacity. Living substances transform single tokens of experience into general patterns of representation, and this is only possible by virtue of an innate expressive aptitude:

> That is said to express a thing in which there are relations [*habitudines*] which correspond to the relations of the things expressed. But there are various kinds of expression; for example, the model of a machine expresses the machine itself, the projective delineation on a plane expresses a solid, speech expresses thoughts and truths, characters express numbers, and an algebraic equation expresses a circle or some other figure. What is common to all these expressions is that we can pass from a consideration of the relations in the ex-

pression to a knowledge of the corresponding properties of the thing expressed. Hence it is clearly not necessary for that which expresses to be similar to the thing expressed, if only a certain analogy is maintained between the relations. (A VI 4 B, p. 1370).[16]

The concept of expression refers to an analogical processing device: certain "occurrences" (*habitudines*) are placed in relation to other occurrences. It indicates a symbolic capacity of putting something in the place of something else ("characters express numbers", in the above quotation) and, in a sense, it is representative as well as innate because, from the mere consideration of that which expresses, we can come to know the corresponding properties of that which is expressed. In their representations, living beings are therefore not dealing directly with things in themselves, but with a system of normatively related symbols. Environmental inputs are placed in an expressive relationship with the established content of a corresponding representation, and this happens according to an internal rule-governed code: "Series est multitudo cum ordinis regula" (A VI 4, p. 1426).

Here is the normative-metaphysical source of the aforementioned "multitude in the unity". The formation of an organic representing individuality is closely related both to the physiological laws governing the corporeal machine ("organism" as uncountable) and to the presence of an internal *regula ordinis*, i.e., a principle of activity capable of generating norm-governed patterns of representations. The code for the order coincides with the metaphysical role of the dominant monad: it is only when the organism dimension (i.e., the corporeal arrangement of the parts) is combined with the unity dimension of the "actuating monad" that the organic aggregate is actually living. In this case, instead of a body mass imbued with decentralized forms of organization, we have a "corporeal substance, which the dominant monad in the machine makes one" (Leibniz 1989, p. 177/GP II, p. 252).

These considerations bring us back to the root of the word organism mentioned at the beginning of our analysis. As we recall, in the Early Modern period, "organism" did not denote a single entity, but rather a form of organization. This organization had a mechanical-phenomenological side consisting of an infinite replication of structures (the natural machine that steadily replicates its same mechanical order), but this process of mechanical replications governed by laws has also revealed a more internal normative-metaphysical dimension: the *regula ordinis* instantiated by the dominant monad, capable of generating sequences of perceptions governed by rules, thereby bestowing a representing individual form on the corporeal machine.

---

**16** Other interesting definitions of "expression" can be found in A VI 6, p. 131 and GP II, p. 112.

Something like an environment can only appear to an already well-ordered representing individual substance. Living beings are "mirror of the universe", Leibniz says, in the sense that they can represent the mechanical order of the universe by virtue of their exhibiting the internal *regula ordinis* of their dominant monad. The laws of bodily movement and the activities of the soul are intrinsically coordinated in such a way that "the soul is configured as the essential representative element of the body, and the body as the essential instrument of the soul" (*Animadversiones*, p. 32). Accordingly, "*harmonia*" has the last word in his system because it summarizes the mutual relationship between the mechanical order of the efficient causes (that can be grasped with the aid of mathematical principles) and the formal order of the final causes (that are "bound to metaphysical rules"). The concepts of "organism" and of "life" respectively summarize this state of affairs. By itself, the organism is not alive, it denotes a lifeless disposition of parts. The dominant monad brings life to an aggregate, but can never figure as a disincarnate ingredient of the world ("omnem mentem est organicam" – A VI 3, p. 394).[17] Life is an activity that can only be expressed mechanically (as action in motion), since it is a metaphysical principle of order that presupposes the complementary and integrated presence of a further mechanical dimension. That is why the word *harmonia* is so important to our understanding of the form of Leibniz's philosophical system: it indicates the necessary coexistence of two different and yet mutually integrating orders of legality. The self-normativity of individuals is part and parcel of the rational (mechanical) lawfulness of things.

## 5 Conclusions

The word organism enshrines different traditions, as it merges together the late-Aristotelian legacy and the lexicon of modern mechanics. Reference to Leibniz has shown us that this was not a mere juxtaposition of different epistemic-ontological models as might be inferred from Sellars's distinction between the event-law lexicon and the thing-nature framework. It entails their profound integration – even if the dominant model was that of mechanical epistemology, at least initially. This is an absolutely pivotal point: for Leibniz, the right approach to the topic of physical legality was that of modern mechanics. In some respects, Leibniz unreservedly embraces the primacy of the "scientific image of the world", and his cosmological conception is radically mechanistic in every re-

---

[17] Cf. Nunziante 2002, pp. 84–85.

spect. The "normative turn" of his thought comes about from within this mechanical paradigm, and can in no way be configured as an external addition to it.[18] Leibniz retrieves the lexicon of substantial forms, deepening his studies on dynamics and discovering the invisible world of forces. It is from this perspective that, little by little, he arrives at the conviction that the principles of mechanics alone are not enough to explain the complex phenomena of bodily organization.

All the analyses conducted to date have done nothing but gravitate around the very concept of "organization", since the great merit of Leibniz lies in his addressing the distinction between "organization" and "self-organization". It is precisely around this distinction that the fusion between the lexicon of the Modern philosophers and the late-Aristotelian tradition eventually came about. "Organization" has to do with the domains of law, of events, of macroscopic and microscopic physics. The idea that there are "bodies within bodies to infinity" (A VI 2, p. 241) is one of Leibniz's earliest beliefs, and stems from researches that were conducted using the microscope.[19] Organization is therefore an uncountable noun that brings out the mathematical dimension of bodies like fractal structures that replicate themselves to infinity.

"Self-organization", on the other hand, is a term that Leibniz does not use directly because it rather designates a conceptual framework that came to be defined by his original reflection on the issue of organized bodies. Self-organization refers to the realm of a finalistic-oriented normativity. It is a dimension, the core element of which is what we might be tempted to call biological individuality – as opposed to physical legality – were it not for the fact that the term "biology" is totally inappropriate in this context because there was still no properly deployed "science of life" in Leibniz's time. But it is by this route that organism will later become a countable noun.

The appeal to substantial forms was motivated by Leibniz having identified the need to find a principle of unity that could explain that vast array of phenomena that today we would define as biological, and that in his time were called "vital". Leibniz is aware of the fact that this principle of unity (the dominant monad that gives form to individuality) must be introduced as a primitive (metaphysical) concept to account for the manifest features of those things that we call living, be they plants, animals or human beings. The passages in which

---

[18] I take up the expression "normative turn" from O'Shea 2007, so as to underscore the peculiar characteristic of Sellars's thinking. *Mutatis mutandis*, I think something similar can be applied to Leibniz too.

[19] On the importance of the invention of the microscope in early-Modern philosophy, see Wilson 1995.

Leibniz explains that perceptions and representations "cannot start naturally" (GP III, pp. 340–41; 344–45) are very relevant: in this context, the word "nature" and the adjective "natural" are declined in an eminently mechanical sense, according to the Modern dimension of the event-law lexicon. Yet, although they are not compatible with the mechanical model, the metaphysical principles of unity, endowed with representative form, must be included as a supplementary integration of such model, just as the lexicon of "persons", according to Sellars, must integrate the conceptual resources of the scientific image of the world.

The concept of life, which is therefore different from the concept of organization, holds together – it has indeed been construed to hold together – the two poles of organization and self-organization: the "organism dimension" and the "unity dimension". It is a concept built on the integration of the two poles, representing their perfect fusion – in Leibniz's intentions, at least.

But this fusion was short-lived because – with the weakening of the Aristotelian legacy in the centuries that followed, and with the eruption of Darwinian evolutionary theories on the scene – the balance was disrupted, and the semantics of life and organisms was renewed by cutting away parts of the material with which it had been built. And the topic of normative factors that "do not begin naturally" became part of the so-called "placement problem" typical of every form of naturalism (De Caro and Macarthur 2010, pp. 1–17; Price 2011, pp. 187–189).

## Abbreviations

| | |
|---|---|
| AT | Descartes, René (1897–1913): Oeuvres, Publiées par Charles Adam & Paul Tannery, Paris: Léopold Cerf Imprimeur Éditeur. seconda edizione a cura di B. Rochot – P. Costabel – J. Beaude – A. Gabbey, 11 voll., Paris 1964–74. |
| A | Leibniz, Gottfried Wilhelm (1923-): Sämtliche Schriften und Briefe, hrsg. von der Deutschen Akademie der Wissenschaften, Darmstadt. [Followed by the number of the series, the volume (and possibly tome), the page number]. |
| GP | Leibniz, Gottfried Wilhelm (1875–90): Die philosophischen Schriften von Gottfried Wilhelm Leibniz, VII Bde, hrsg. von C.I. Gerhardt, Berlin, Weidmann (unveränd. Nachdruck Hildesheim, Olms 1960–61). |
| Animadversiones | Leibniz, Gottfried Wilhelm (2011): Animadversiones in G.E. Stahlii Theoriam Medicam. In: Antonio-M. Nunziante (Ed.): Gottfried Wilhelm Leibniz. Obiezioni contro la teoria medica di Georg Ernst Stahl. Sui concetti di vita, anima, organismo. Macerata: Quodlibet, pp. 23–121. |

# References

Ayala, Francisco J. (2007): "Darwin's Greatest Discovery: Design Without Designer". In: John C. Avise, Francisco J. (Eds.): *In the Light of Evolution: Volume I: Adaptation and Complex Design.* Washington (DC): National Academies Press (US); Available from: https://www.ncbi.nlm.nih.gov/books/NBK254313/

Canguilhem, Georges (1965): *La connaissance de la vie. Deuxième édition revue et augmentée.* Paris: Librairie philosophique J. Vrin.

Cheung, Tobias (2006): "From the organism of a body to the body of an organism: occurrence and meaning of the word 'organism' from the seventeenth to the nineteenth centuries". In *The British Journal for the History of Science* 39(3), pp. 319–339.

Clericuzio, Antonio (2000): *Elements, Principles and Corpuscles. A Study of Atomism and Chemistry in the Seventeenth Century.* Dordrecht-Boston-London: Kluwer.

De Caro, Mario and Macarthur David (Eds.) (2010): *Naturalism and Normativity.* New York: Columbia University Press.

Deleuze, Gilles (1993): *The Fold. Leibniz and the Baroque.* Trans. by Tom Conley, London: The Atholone Press.

Duchesneau, François (1998): *Les modèles du vivant de Descartes à Leibniz.* Paris: Vrin.

Grmek, Mirko Drazen (1990): *La première révolution biologique. Réflexions sur la physiologie et la médecine du XVIIe siècle.* Paris: Éditions Payot.

Hampe, Michael (2010). "Science, Philosophy, and the History of Knowledge: Husserl's Conception of a Life-World and Sellars's Manifest and Scientific Images". In: David Hyder/Hans-Jörg Rheinberger (Eds.): *Science and the Life-World.* Stanford (CA): Stanford University Press, pp. 150–163.

Harvey, William (1928): *Exercitatio anatomica de motu cordis et sanguinis in animalibus.* With an English translation and annotations by Chauncey D. Leake. Springfield, Baltimore: Charles C. Thomas.

Heidegger, Martin (1976): "Vom Wesen und Begriff der *physis.* Aristoteles, Physik B, 1". In: Martin Heidegger, *Gesamtausgabe,* Band 9, *Wegmarken,* Frankfurt am Main: Vittorio Klostermann, pp. 239–301.

Husserl, Edmund (1970): *The Crisis of European Sciences and Transcendental Phenomenology.* Evanston: Northwestern University Press.

Jonas, Hans (1966): *The Phenomenon of Life. Toward a Philosophical Biology.* New York: Harper & Row.

Kitcher, Philip (1993): "Function and Design". In: "Midwest Studies in Philosophy" 18(1), pp. 379–397.

Leibniz, Gottfried Wilhelm (1989): *Philosophical Essays.* Ed. and trans. by Roger Ariew/Daniel Garber. Indianapolis, Cambridge: Hackett Publishing Company.

Leibniz, Gottfried Wilhelm (1996): *De scribendis novis medicinae elementis.* Enrico Pasini (ed.). In: E. Pasini, *Corpo e funzioni cognitive in Leibniz.* Milano: Franco Angeli, pp. 212–217.

McDowell, John (1996): *Mind and World: with a new introduction.* Cambridge, London: Harvard University Press.

Nunziante, Antonio-M. (2002): *Organismo come Armonia. La genesi del concetto di organismo vivente in G.W. Leibniz.* Trento: Pubblicazioni di Verifiche.

Nunziante, Antonio-M. (2011): "Vita e organismo tra filosofia e medicina: le ragioni di una polemica". In: *Gottfried Wilhelm Leibniz. Obiezioni contro la teoria medica di Georg Ernst Stahl. Sui concetti di anima, vita, organismo*. Macerata: Quodlibet, pp. 125–186.

O'Shea James R. (2007): *Wilfrid Sellars. Naturalism with a Normative Turn*. Cambridge (UK), Malden (USA): Polity.

Pasini, Enrico (2011): "Both Mechanistic and Teleological. The Genesis of Leibniz's Concept of Organism, with Special Regard to his 'Du rapport general de toutes choses'. In: Hubertus Busche (Ed.): *Departure for Modern Europe. A Handbook of Early Modern Philosophy (1400–1700)*. Hamburg: Felix Meiner Verlag, pp. 1216–1235.

Piro, Francesco (2002): *Spontaneità e ragion sufficiente. Determinismo e filosofia dell'azione in Leibniz*. Roma: Edizioni di storia e letteratura.

Price, Huw (2011): *Naturalism Without Mirrors*. Oxford, New York: Oxford University Press.

Sellars, Wilfrid (1949): "Aristotelian Philosophies of Mind". In: Roy Wood Sellars/V.J. McGill/Marvin Farber (Eds.): *Philosophy for the Future. The Quest Modern Materialism*. New York: The MacMillan Company, pp. 544–570.

Sellars, Wilfrid (1959). "Meditations Leibnitziennes". In. Wilfrid Sellars (Ed.): *Philosophical Perspectives*. Springfield, Illinois: Charles Thomas Publisher, pp. 153–181.

Sellars, Wilfrid (1963). "Philosophy and the Scientific Image of Man". In: Wilfrid Sellars (Ed.): *Science, Perception and Reality*. London, New York: Routledge & Kegan Paul, pp. 1–40.

Stahl, Georg Ernst (1708). *Theoria medica vera. Physiologiam et pathologiam tanquam doctrinae mediace partes vere contemplativas, e naturae et artis veris fundamentis, Intaminata ratione, et inconcussa Experientia sistens*. Halae: litteris et impensis Orphanotrophei.

Wilson, Catherine (1995): *The Invisible World. Early Modern Philosophy and the Invention of the Microscope*. Princeton: Princeton University Press.

Hugh Desmond, Philippe Huneman
# The Ontology of Organismic Agency: A Kantian Approach

**Abstract:** Biologists explain organisms' behavior not only as having been programmed by genes and shaped by natural selection, but also as the result of an organism's agency: the capacity to react to environmental changes in goal-driven ways. The use of such 'agential explanations' reopens old questions about how justified it is to ascribe agency to entities like bacteria or plants that obviously lack rationality and even a nervous system. Is organismic agency genuinely 'real' or is it just a useful fiction? In this paper we focus on two questions: whether agential explanations are to be interpreted ontically, and whether they can be reduced to non-agential explanations (thereby dispensing with agency). The Kantian approach we identify interprets agential explanations non-ontically, yet holds agency to be indispensable. Attributing agency to organisms is not to be taken literally in the way we attribute physical properties such as mass or acceleration, but nor is it a mere heuristic or predictive tool. Rather, it is an inevitable consequence of our own rational capacity: as long as we are rational agents ourselves, we cannot avoid seeing agency in organisms.

## Introduction

Stags lock antlers to gain access to mates. Arctic poppies rotate and track the sun in order to maximize solar exposure. Bacteria swim up a sucrose gradient in order to get better access to the source of sucrose. When biologists explain organisms' behavior by referring to their goals in this way, then they are using what we in this paper will call *agency explanations*. Such explanations make sense of organisms' behavior *as if* they were agents with goals.

Despite its philosophical pedigree (going back in some form to Aristotle), the problem of organismic agency was neglected in much of twentieth century philosophy of biology and mainstream evolutionary theory, which was dominated by the *artifact approach* to organisms. The organism was understood to be a collection of functional traits, designed by natural selection in much the same way Paley's watch was designed by an intentional creator. Thus, each pattern of apparently purposive behavior was understood to be a functional trait that is pur-

---

**Hugh Desmond,** University of Antwerp and KU Leuven
**Philippe Huneman,** CNRS / Université Paris I Sorbonne

https://doi.org/10.1515/9783110604665-003

posive in name only ('teleonomy', cf. Pittendrigh 1958; Ernst Mayr 1961). In philosophy of biology this view was enshrined by the 'selected effects' account of function, where all biological functions can be explained by a process of natural selection (Wright 1973, 1976; Millikan 1984; Neander 1991).

In recent decades a more robust approach to organismic agency, which we will call the *agential approach*, has become increasingly influential. Organisms are agents with goals and purposes that interact with their environments, and their behavior can only be understood with reference to the goals of organisms as wholes rather than as mere collections of parts (e.g., genes or traits). This focus on whole organisms (following Bateson 2005) is linked with several ongoing developments in evolutionary biology, most notably the so-called Extended Synthesis (e.g. Müller 2017).

This motivates taking a new look at the long-standing question whether or not organismic agency is 'real', in the same way that the wings of a bird are, or the claws of a bear. After all, agency ascriptions to organisms have long been suspected of being mere metaphors and fictions of the human mind – an anthropomorphic projection even – rather than an accurate description of the mind-independent world. We call this question the question of whether to adopt an *ontic view* of agential explanations (cf. Salmon 1989). In an ontic view of agential explanation, agential explanations explain because they refer to an element of the ontology of the world (i.e., agency) which is responsible for the explanandum (i.e., organismic behavior) – just in the same way that causal-mechanical explanations explain because they single out the actual mechanism that causes the explanandum phenomenon (Craver 2014).

Our approach in this paper will be to overlay this question with a distinct but closely related one: whether agential explanations ultimately can be reduced to non-agential explanations (a worry raised in e.g., Lewens 2007). For instance, once one asks the question why organisms have such-and-such purposes and not others in the first place, the agential approach rapidly becomes inadequate. Why do stags want access to mates in the first place? The most plausible explanation would seem to involve a selection explanation, along the lines of 'those stags who tended to not engage in sexual competition did not get to transmit their genes to the next generation'. Thus, the genesis of organismic purposes is explained through a process of natural selection. However, does this also imply that purposeful behavior can be explained without reference to organismic purpose or agency, without loss of explanatory power? We call this question the question of *explanatory dispensability*. Agency is (explanatorily) dispensable if and only if an agential explanation can be replaced by an explanation void of any reference to organismic agency or purposes, without any loss of explanatory power.

Even though there are four possible combinations of answers to these two questions, most contemporary thinking about agential explanation focuses on two. The first is the non-ontic (epistemic) view of agential explanation where agency is dispensable. Its main representative today is what we call the 'Neo-Fisherian option', which holds that agency is invoked in explaining behavior for purely heuristic reasons – in particular, as shorthand for other types of explanation (especially selectionist explanation). This option has been especially widely adopted in behavioral ecology where, following Grafen's 'maximizing agent analogy' (Grafen 1984), organisms' behavior is analyzed as maximizing inclusive fitness (Grafen 2006).[1] The Neo-Fisherian option can be traced back to Fisher's fundamental theorem of natural selection, which states that, under the influence of natural selection, populations of organisms have a tendency to increase fitness (equal to the population's genetic variance in fitness: cf. Fisher 1930, chapter 2). In this way, agential explanations could be adequately replaced by explanations that do not refer to organismic agency and purposes (but only to natural selection), and organismic agency therefore is not a mind-independent causal power in the way, for instance, natural selection is assumed to be.

The second option, decidedly less mainstream but increasingly defended, combines indispensability with an ontic view of agential explanation. We call this the 'Neo-Aristotelian option', since it expands fundamental ontology to include organismic purposes. Different versions of this option have been developed in recent years: most prominently by Walsh (Walsh 2012, 2015), but Moreno and Mossio's analysis of biological autonomy also follows the Neo-Aristotelian option (Moreno and Mossio 2015), as does Varela's notion of autopoiesis (Varela 1979).

In this paper we seek to identify a third option[2] which we call the 'Kantian option' regarding agential explanations: (1) the concept of organismic agency is indispensable to scientific explanation and (2) agential explanations are to be conceived non-ontically[3]. In particular, viewing organisms as agents with pur-

---

[1] Inclusive fitness is a fitness measure that includes the (expectation of the) number of kin offspring (so a sterile individual could have a high inclusive fitness if its relatives had many offspring), mitigated by the degree of genetic relatedness between relatives; for this latter reason, the maximizing analogy forms a bridge between behavioral ecology and population genetics.
[2] The fourth option – where agency is explanatorily dispensable and yet considered robustly real – is possible but does not strike us as particularly compelling. After all, if a concept is dispensable, Ockham's razor directs us to discard it from our ontology.
[3] To what extent 'non-ontic' should be interpreted as 'epistemic' sensu Salmon (1989) is a rather complicated question which we discuss at the end of section 5.

poses is a "demand of reason": it is necessary given our rational nature. This means that attributing agency is not a consequence of our limited computational capacity, or of our contingent evolved nature that causes us to detect agency falsely (cf. the so-called 'agency detection' cognitive modules: Atran 2002; Barrett 2000). Yet at the same time, it is a mistake to believe that agency is a natural regularity or causal process, belonging to the 'furniture' of the world in the same sense as physical processes. In this way we will suggest how one can obtain the robust explanatory indispensability desired by (some) Neo-Aristotelians (i.e., agency is not just a heuristic) without the ontological price that Neo-Fisherians would be loath to pay.

The paper is structured as follows: in the first section we give a broad introduction to organisms and the major streams in biological thought, written for non-specialists (i.e., philosophers outside the philosophy of biology). In the second section we define with more precision what an agential explanation is and contrast it with functional explanations. In the third we discuss various attempts to replace agential explanations with non-agential explanations, and argue that – despite widespread hopes – once one looks at the details, one cannot but conclude that attempts to make agency dispensable, even today, remain aspirational rather than clearly successful. In the fourth section we discuss Kant's original approach to teleology in the natural world and show how it can be the basis for our Kantian approach to agency. In the final section we show how the Kantian approach entails viewing agential explanations as a 'demand of reason'.

# 1 Artifacts and Agents

Much of mainstream twentieth-century evolutionary biology operated within the framework of what is called 'the Modern Synthesis', a term coined by Julian Huxley (Huxley [1942] 1974). The Modern Synthesis was forged in the 1930s and 1940s by Ronald Fisher, Sewall Wright, Theodosius Dobzhansky, and John Haldane, among others, and is often described as the synthesis of Mendelian genetics and Darwin's theory of natural selection. It was very much focused on how allele (different versions of the same gene) frequencies change over time in response to evolutionary forces, such as natural selection, mutation, drift, or migration.

Organisms were essentially analyzed as epiphenomena arising from changes in underlying allele frequencies. In the words of Huxley ([1942] 1974), they were viewed as "bundles of adaptations" where each adaptive trait was shaped by natural selection in response to environmental demands – just as artifacts are designed and put together, piece by piece, by an artisan. Such a view of organisms has never been unanimously accepted, even among the major architects of the Modern Synthesis (cf. Mayr 1982; Simpson 1944, 1953), but the view has nonetheless been the dominant one, and has been popularized in the work of Richard Dawkins (Dawkins 1976). Dawkins introduced a dichotomy between replicators (alleles) and interactors (organisms), with the consequence that organisms are mere tools in a never-ending arms race between genes, with genes the genuine actors in evolutionary history. Even apparently goal-directed organismic behaviors, such as beavers building dams, are expressions ('extended phenotypes') of the underlying genotype (Dawkins 1982). In sum, while it may *seem* that an organism undertakes behavior to further its own goals (e. g., secure food, fend off predators, etc.), it does so actually for the benefit of the genes, which get to replicate when the organism does well. In this way, the theoretical resources of the Modern Synthesis were used to support a philosophical view of agency as dispensable and fictional.

The metaphor of Paley's watch, which dominated in the early days of the Modern Synthesis,[4] was supplemented after the 1960s with analogies borrowed from computer science. Organismic behavior was often described as *programmed*, starting with influential papers by Mayr (1961) and Jacob and Monod (1961):

> The purposive action of an individual, *insofar as it is based on the properties of its genetic code*, therefore is no more nor less purposive than the actions of a computer that has been programmed to respond appropriately to various inputs. (Mayr 1961, 1504, our emphasis)

So even if the behavior of an organism may seem goal-directed, that is only because its genetic code has been 'programmed' by natural selection to direct the organism to react in certain ways to certain inputs, and in other ways to other inputs. Organisms are no more goal-directed than computers are.

Despite the metaphors of 'design' and 'program,' it is important to note that even biologists operating squarely within the Modern Synthesis were well aware of the limits of the metaphors. In the quote above, Mayr qualified the programming analogy with "insofar as it is based on the properties of its genetic code." Mayr is not claiming that individual organisms behave exactly like pre-programmed computers, only that some aspects of their behavior are determined by environmental inputs in the way that a computer program responds to user inputs. Similarly, in *The Extended Phenotype*, Dawkins devotes a whole chapter to debunking the view that genes determine all aspects of organismic behavior, a view he calls the 'myth' of genetic determinism.

---

[4] Cf. Lewens 2005 for an in-depth discussion of the artifact metaphor.

The limitations of the artifact metaphor are built into one of the very foundations of the Modern Synthesis: the analysis of phenotypic variance as proposed by Fisher (Fisher 1919). This analysis states that, in general, only a part of the variation of phenotypes in a population is explained by a corresponding variation in genotype. The rest is variation in environment (impacting how the organism develops), or variation in how genotype and environment correlate (cf. e.g. Hamilton 2009).

Thus, no practicing biologist holds that organismic behavior (or phenotype) is entirely determined by a genetic program[5], for the very simple reason that the *environment* is the second element that goes into determining phenotype.

The role of the environment points to limitations in speaking about the adaptive 'design' of organisms. A genotype may be designed for a particular type of environment, i.e., there may be a particular 'normal' environment in which the bulk of the selection for that genotype occurred. In that normal environment, the genotype develops into an adaptive phenotype. However, in reality, environments are highly heterogeneous, so in a population of identical genotypes, only a fraction will develop in the 'normal' environment. Other environmental inputs – inputs that differ from the normal environment – cause the organism to diverge from its 'designed' phenotype. In this way, while theoretical resources in the Modern Synthesis lend some support to the artifact metaphor, the same resources point also to the metaphor's limitations.

Moreover, the role of environment in organismic behavior (and phenotype more generally) also provides a direct motivation for the agential approach. To see this in more detail, consider the phenomenon of phenotypic plasticity. A trait is 'plastic' (in the context of quantitative genetics[6]) when the underlying genotype can develop into different phenotypes solely due to environmental variation. The degree of plasticity of a trait is represented by the term $V_E$ in the equation above.[7] Plasticity, defined in this way, is an incredibly basic phenomenon: it simply refers to how different environments cause genotypes to develop into different phenotypes. At its most basic, it can refer to phenomena that are the result of physical or chemical (rather than properly biological) processes, such as the stunting in the growth of a plant in response to poor nutrition. There are few if

---

[5] Whether organismic behavior can be entirely explained by natural selection is a more difficult question, since the environment also can be influenced by natural selection through niche construction. We discuss this in Section 3.

[6] There is also cell plasticity, referring to the multiple dispositions of a totipotent cell in developmental theory. This is not relevant here.

[7] The term describes how different genotypes are correlated with different degrees of plasticity; or in other words, how different genotypes react differently to environmental novelty.

any organisms that lack some form of phenotypic plasticity in some of their traits.

The phenomenon of plasticity was not considered to be of any special significance until the work of Bradshaw (Bradshaw 1965); before him, the phenotypic variation due to environmental perturbation was often viewed as noise. Bradshaw showed that plasticity in a trait can be adaptive to heterogeneous environments. If an organism can vary a trait in response to changes in its environment so as to be able to adopt a more adaptive phenotype, such an organism can be at a selective advantage in variable environments, compared to an organism without that capacity. In particular, Bradshaw distinguished between four types of environmental heterogeneity where plasticity can be adaptive[8] (Bradshaw 1965, 21): (1) when the environment changes on a time-scale that is equal to or shorter than generation time; (2) when the environment varies over very short spatial scales; (3) when the magnitude of environmental variation is very large; (4) when it is beneficial to maintain a stable phenotype in a population while maintaining genetic diversity.

Adaptive scenario (4) shows how maintaining stable phenotypes in the face of environmental change is also a form of plasticity. When it becomes inscribed into developmental pathways, it has been termed 'canalization' (Waddington 1940); moreover, plastic maintenance of stable phenotypes is hypothesized to precede genetic accommodation where the phenotype is produced by genetically determined developmental pathways (West-Eberhard 1989). Finally, some degree of canalization in organismic traits is nearly ubiquitous, since thermodynamic fluctuations in the molecular bases of genes would be detrimental and then counter-selected if they were significantly affecting the development of phenotypes.

Adaptive scenarios (1)-(3) refer to organismic behavior that is often thought of as (apparently) agential. For instance, in response to chemical cues emitted by sea slugs, bryozoans will develop spines to defend themselves (Godfrey-Smith 1996). Such forms of plasticity open up parallels with cognition, and not surprisingly, theorists and philosophers concerned with the evolution of cognition often take the evolution of adaptive phenotypic plasticity to be a model (van Duijn, Keijzer, and Franken 2006; Lyon 2017; Calvo Garzón and Keijzer 2011; Sterelny 2000; Caporael, Griesemer, and Wimsatt 2013; Godfrey-Smith 1996). Organisms exhibit a whole range of cognitive, or at least apparently cog-

---

[8] See also Nicoglou (2015) for the history of Bradshaw's study, and Desmond (2018) for a more detailed discussion of the role of temporal and spatial scale.

nitive[9], behaviors: they sense changes in the environment, are able to process this information and select a response from a repertoire of responses. Far from being a late-stage development in evolutionary history, we see these types of behaviors in bacteria, which can undertake evasive action upon detecting predators (Pérez et al. 2016), or swim to a food source upon detecting sucrose gradients (Auletta 2013).

In sum, the impact of the environment on phenotype shows – via the phenomenon of phenotypic plasticity – how it is not entirely adequate to view organisms simply as artifacts. Moreover, it motivates a definition of organismic agency as the capacity to respond to changes in the environment in such a way as to further organismic purposes. Organismic agency thus understood is a much broader concept than the agency traditionally ascribed to human, rational subjects, which is typically characterized by means of some mental state, like an intention (cf. Schlosser 2015). The approach to organismic agency in the biological sciences, by contrast, blackboxes whatever cognitive processing may or may not be going on. In this sense organismic agency is best understood as an ecological property (cf. Walsh 2015), namely, a property of the interaction between organism and environment. We will now discuss agency and agential explanations in more systematic detail.

## 2 Agential Explanations

### Definition of Agency

For the purposes of this paper, we will operate with the following minimal working definition of agency:

A system is an agent if and only if (1) it possesses a certain purpose P, where P is a particular state of the system, (2) it maximizes the realization of P in response to environmental change, and (3) the system itself is a cause of the realization of P.

While we view this definition as being continuous with established work in this area (cf. Moreno and Mossio 2015, 92–93), it may be helpful to explain the various elements involved in the definition. Condition (1) models the purpose of a system as a particular state. For organisms, purposes may refer to developmental states, physiological states, or behavioral states. Condition (2) specifies that

---

[9] The application of the term cognition, as well as other terms such as communication or memory, to organisms such as bacteria remains a controversial point. See discussion in Lyon 2015.

goal-directedness is to be interpreted as a maximization or optimization. This equation of purposefulness with some type of optimization is common across the sciences. Finally, condition (3) is intended to exclude clear non-agent systems, even where a process of maximization is occurring, such as the marble rolling down into the middle of the bowl (minimizing gravitational potential energy). Here the marble is not considered a cause of its own maximization behavior. The same is true of more complex physical systems, such as Bénard convection cells, which are patterns of heat flow that appear spontaneously when the temperature gradient is large enough. Such structures may maintain their organization even in the face of perturbation in their environment, such as movement of the container walls (Manneville 2006); nonetheless, they are widely considered not to be agents (Moreno and Mossio 2015). By contrast, an organism that modifies its phenotype in order to be more adaptive to a new environment is considered to be a cause of the modification of its own phenotype.

While 'self-causation' can function as a label to distinguish agents from complex physical systems, it remains controversial as to what precisely self-causation means and how the boundary should be drawn (or, how blurry the boundary is). For instance, it has been (controversially) argued that self-propelling oil droplets are agents (Hanczyc and Ikegami 2010). Consequently, many rival accounts of self-causation have been given, pointing to various factors such as internal organization, or control of environmental constraints (Moreno and Mossio 2015; Barandiaran, Di Paolo, and Rohde 2009; Skewes and Hooker 2009; Shani 2013; Burge 2009; Horibe, Hanczyc, and Ikegami 2011).[10] For the purposes of this paper we do not take a stance on how self-causation should be analyzed; what will be of importance is how it should be interpreted (i.e., whether it refers to an ontic causal process, or is a convenient heuristic).

### Definition of Agential Explanation

With this operational definition of agency in place, we can introduce 'agential explanations' as scientific explanations that explain in virtue of reference to a system's agency:

---

[10] The literature on naturalized agency is interdisciplinary to a high degree, with contributors coming from backgrounds ranging from biology or nonlinear physics to artificial intelligence, robotics, or cybernetics. A systematization of all the various contributions and approaches is still lacking.

> Explanandum: In response to environmental change $E1 \to E2$, the system undergoes the change $S1 \to S2$.
> Explanans: (1) The system has purpose $P$,
> (2) $S2$ maximizes the realization of purpose $P$ in environment $E2$,
> (3) the system itself is a cause of the realization of $P$.

An agential explanation is a type of 'extremal explanation', where the explanandum is explained as some extremal state of affairs maximizing some scalar variable $w$,[11] given certain conditions (i.e., the purpose of the system). As we will discuss later, an important class of extremal explanations, commonly used in physics, explains the explanandum as a mathematical consequence of the structural set-up of the system $S$ (this typically involves various parameters $pi$).[12] By contrast, an agential explanation involves a reference to 'self-causation', where the realization of the purpose is 'caused' by the system $S$ itself.

## Contrast with Functional Explanation

When it comes to the use of teleological language in biology, the philosophy of science has been overwhelmingly focused on functional statements and functional explanations, e.g., "the heartbeat in vertebrates has the function of circulating blood through the organism" (Hempel 1959). Insofar as a behavior is simply an organismic trait, can one not just say that the purpose is the 'function' of purposeful behavior – thus reducing agential explanations to special cases of functional explanations?

This is not quite correct. Ascribing agency to an organism involves a different type of teleological statement than ascribing a function. The main difference between functional and agential explanations is that functional explanations attribute a purpose (function) to a *trait* of an organism, whereas agential explanations attribute a purpose to the *whole organism*. Functions are attributed to traits of organisms, whereas agency is attributed to the organisms themselves.

Nonetheless, agential and functional explanations can interact in subtle ways. For instance, a case could be made that philosophical accounts of functional explanations often presuppose it is possible to ascribe purposes to the

---

[11] Some of the most frequently used variables include potential energy, entropy, free energy, fitness, utility.
[12] See also Birch (2012) for a compatible account of what he calls 'agent-talk' in terms of robustness and stable states.

whole organism. So, for example, the causal role account of functions (roughly) holds that a function is what contributes to some 'capacity' of a larger complex system that contains it (Cummins 1975); however, how should such a 'capacity' be analyzed if not as a property of the system as a whole? Similarly, the recent organizational account (Mossio, Saborido, and Moreno 2009) uses organism-level goals that can be used to ground trait-level functions.[13]

Potentially, a similar point could be made about the selected effects account, which holds that a function is what explains why some structure was selected for in the past (Wright 1973, 1976; Millikan 1984; Neander 1991). The selected effects account presupposes there was some 'normal environment' in the evolutionary past, and while this seems like a good presupposition for structures like the heart or lungs, it is much less clear what the 'normal environment' of certain types of animal behavior should be. Since this point relates to the dispensability of agency, we will come back to this line of thought in the next section.

**Agential Explanations in Social and Cognitive Sciences**

In principle, agential explanations, as defined above, can also be extended to rational agents, where the purpose is defined as value or a general utility measure. Such explanations, commonplace in economics, often (and controversially) assume that economic actors are utility-maximizing agents (which is of course unrealistic, cf. Tversky and Kahneman 1974). There is a deep parallelism here between economics and behavioral ecology, noticed by Maynard-Smith in his seminal book on evolutionary game theory when he says that selection is to fitness what rationality is to economics, both being about maximization (whether of utility or fitness). This parallel also underlies formal approaches to behavioral ecology (Grafen 1984, 2014), where organisms maximize fitness in the same way rational agents maximize utility.

But the parallelism between economics and evolutionary biology goes deeper than fitness- or utility-maximization. The apparently irrational behavior diagnosed by research following Kahneman and Tverky's seminal insight on biases can be accounted for when one takes an ecological perspective. Here, considering that human agents have been shaped by evolution, and that their decision-making modules or protocols evolved in environments where information was partial and decision time was very short (due to predators, competitors etc.), then crude cognitive biases that yield a utility-enhancing solution most of the

---

**13** For a critique, see Huneman (2019).

time would have been selected. This is what Gigerenzer calls 'ecological rationality' (Gigerenzer 2000), which gives rise to a bounded rationality, which in turn refers to how apparently irrational biases can originate as heuristics that actually are, on average, utility-maximizing given constraints (limited information and time). Thus, adopting an evolutionary viewpoint allows many instances of apparently irrational behavior to be analyzed as (boundedly) rational. All these parallels between economics and evolution by natural selection give rise to a notion of 'agency' that has recently been systematically explored (in Okasha 2018).

## 3 The Ontology and Dispensability of Agency

### Should Agential Explanation Be Viewed Ontically?

With this systematization in place, we can now consider in some more detail the question whether the explanatory relation is to be interpreted as ontic or merely epistemic, i.e., whether an agential explanation explains in virtue of referring to an element of the ontology of the world.[14] In an agential explanation, the system itself is said to be the 'cause' of its own behavior; but what does 'causation' mean in this context?

A first safe observation is that the explanatory relation in agential explanations does not explain by simply referring to a mechanism, or to any process of causal production for that matter. There are clearly some causal processes causing the system's change of state $S1 \to S2$ (e.g., neurological processes causing behavioral change), but an agential explanation, at least as stated above, does not explicitly refer to such causal processes. It explains the behavior in terms of a purpose, and a condition linking that purpose to concrete conditions in reality (i.e., $S2$). In this sense, an agential explanation cannot be viewed ontically in the same way as a causal mechanical one (Craver 2014).

A second, relatively safe observation is that if the realization of $P$ is ultimately a mathematical consequence of the structural set-up of a system, then there is little reason to invoke 'purposes' and 'agents' as part of the ontology. For instance, the minimization of potential energy is a mathematically deductive consequence from the forces impinging on it as it rolls down the hill: there is no need to invoke some 'self-causation' of the rolling ball. So, if one takes 'self-cau-

---

[14] As shorthand, one can refer to this issue as the 'ontology of agency', but just for the sake of clarity we emphasize that our approach to this issue in this paper is not directly metaphysical, but is indirect, through analyzing how agential explanations should be interpreted.

sation' to be shorthand for a pattern of behavior that is a non-causal (whether mathematical or structural) consequence of the causal set-up of the system (e.g., approach to an attractor state), then ultimately agential explanation is a non-causal explanation that identifies structural (or mathematical) consequences of how causal powers interact.

The Neo-Fisherian option largely follows this route, where organismic behavior as analyzed along the lines of the 'maximizing agent analogy', where organisms behave in such a way that maximizes their inclusive fitness. The underlying assumption is called the "phenotypic gambit" (Grafen 1984, 2014), which holds that the choice of a phenotype by the organism mirrors the allele dynamics that underlie evolution. In this way, natural selection is taken to design organisms so that they make decisions similar to what, as it were, natural selection would do if it were making the decision.

So, it would seem that an ontic interpretation of agential explanation requires a causal interpretation of agential explanation. This is indeed suggested by the inclusion of 'self-causation' in the definition of agential explanation, although the challenge for the Neo-Aristotelian option is then to specify how self-causation should be interpreted.

While Walsh is unambiguous that agents should be included in an expanded ontology (especially Walsh 2015: 211 ff.)[15], it is in his account of natural purposes that we can see how this is fleshed out. Walsh describes natural purposes as "counterfactually robust difference makers" where purpose and means are related by invariance relations (Walsh 2015: 198) in much the same way that cause and effect are related by invariance relations in Woodward's interventionist account of causation (Woodward 2003). Explaining a behavior as purposive involves identifying the disposition of "conducing" (Walsh 2015: 199), analogous to how mechanistic explanation *(sensu* Glennan 2002 or Machamer et al. 2000*)* involves identifying the dispositions of pushing or pulling (Walsh 2015: 198). Even though Walsh does not describe such relations as 'causal'[16], the account clearly involves some notion of causal difference-making where, moreover, explanations involving natural purposes are interpreted ontically.

---

[15] In particular, it is implied if one adopts a Gibsonian view of the environment as a set of *affordances* proper to a species or subspecies of organism (Gibson 1979 [2014]). These affordances refer to the potential actions of an organism in an environment (e.g. running, jumping, eating, sleeping, etc.) that are jointly determined by the environment and the purposes of an agent. Moreover, these affordances in turn dispose the agent to act in certain ways.

[16] In fact, at one point he emphasizes that teleological explanations are not a species of causal explanations (Walsh 2015: 196). However, here we read him as having in mind a concept of causal production.

Beyond a reluctance to expand fundamental ontology beyond what is strictly necessary, we would like to point to two reasons for being dissatisfied with the pure Neo-Aristotelian option. First, while we argued that an ontic interpretation of agential explanation should attribute some causal reality to agency, when this entails broadening the concept of causation, it becomes less clear what precisely is gained by an expanded Neo-Aristotelian ontology. Can robust patterns of counterfactual dependence be objectively judged to be causal or non-causal? This is notoriously dependent on the concept of causation one uses: once the concept is broadened enough, then any counterfactual or even counterpossible proposition will appear as causal (Huneman 2010). After all, if the agent with purpose $P$ were modified to an agent with purpose $P'$, then the observed behavior would be (much) less likely, and this is sufficient to count as a causal relation for some accounts of causation. This raises the question: if agential explanation ultimately boils down to patterns of counterfactual relations, why does it matter if one interprets agential explanation ontically?

Second, pragmatic factors complicate the ontic interpretation of agential explanation. An agential explanation entails some counterfactual relation between explanans (agent, purpose $P$) and explanandum (behavior); however, both a particular explanandum behavior as well as the agent's purpose $P$ can be described at finer and coarser grains. Depending on the granularity with which explanandum/explanans is described, the causal character of the corresponding explanatory relation changes (for an argument, see Desmond 2019). If agential explanation is to be viewed ontically, and thus as explanatory in virtue of picking out some self-causing capacity of an organism, one would not want this causal character to disappear merely due to pragmatic factors, such as the grain at which explanandum/explanans is described.

### Can Agency Be Dispensed with?

While we do not pretend to have given any direct argument against the ontic interpretation of agential explanation, we do hope to have clarified how the ontic interpretation will lead to a host of problems, some of which are perhaps insuperable. However, we now want to turn our attention to the other side of the coin: explanatory dispensability. The strongest argument in favor of a non-ontic conception of agential explanation is that agency is dispensable (this is

Ockham's razor).¹⁷ We will illustrate in this section how it is misguided to believe that organismic agency has been dispensed with in science. Whether science may be able to dispense with agency in the future is a whole other question. What we wish to show is a more limited point: given a pessimistic meta-induction on attempts to dispense with agency, there are at least good grounds to believe that science will *not* be able to dispense with agency. This will lay the ground for the Kantian approach, which uniquely combines two positions: agency is indispensable (for scientific reasons), but agential explanations should not be conceived ontically.

Many phenomena clearly do not ask for agential explanations. If a tree branch cracks and falls to the ground during a storm, and we seek to explain the change in the tree's state, we spontaneously tend not to appeal to any type of 'agency' of the tree. A property of the tree as a whole could be explanatorily relevant – for example, a disposition such as brittleness could be referred to in order to explain why oaks tend to crack more than willows during storms. Nonetheless, we tend not to explain this tree 'behavior' in terms of the purposes of the tree. Rather, given certain forces created by the wind, and perhaps given certain structural properties of the tree, the outcome of the branch cracking was determined. No agency is involved.

Even if extremal explanations were to be used, no agency would be required. Classical mechanics provides a perfect example of how extremal explanations exist alongside causal-mechanical explanations. Newtonian analyses of the behavior of masses, in terms of a mechanistic account of the continued action of local forces, can always be rephrased with the Principle of Least Action (through the Hamiltonian or Lagrangian formalism), which abstracts away from a great number of degrees of freedom in a system, and instead ascribes a certain scalar (i.e., the 'action' $S$) to a system. The behavior of the system is then the behavior that maximizes or minimizes the action (cf. Coopersmith 2017).¹⁸

When a system has an extremely large number of degrees of freedom (~$10^{23}$), a different type of extremal explanation is needed, but even here the explanation remains non-agential. Consider a generic thermodynamic phenomenon, such as

---

**17** Going in the other direction, from indispensability to a realist interpretation of a concept, is more controversial but has of course been widely explored since Quine and Putnam.

**18** Historically, such non-agential extremal explanations are exactly what Leibniz had in mind, when he stated that each mechanistic explanation, given in terms of the differential equations governing the trajectories of the parts, could be reformulated in terms of final causes (*Discours de métaphysique* § 13 (Leibniz, 1890)). As a further aside, even though such final causes were dispensable for Leibniz, explanations involving them were to be preferred because they were the most conducive to theology and were compatible with God's moral maxims.

the flow of heat from hot to cold. In statistical-mechanical analysis, the molecules in a gas or liquid fluctuate randomly, but after some time, it is likely that the faster-moving molecules will not remain bunched up in one area of the container (i.e., the 'hot' area), but will spread out over the whole container, either by diffusion or by transferring momentum to slower-moving molecules through collisions.

This type of explanation, first introduced by Boltzmann, is non-causal in the sense that it relies only on principles of combinatorics together with some boundary conditions. A uniform temperature is a vastly more likely outcome than any other since it corresponds to a much greater number of possible microstates, or ways of distributing molecular speeds among the molecules in the container. Erwin Schrödinger aptly named this type of extremal explanation, the 'method of the most probable distribution' (Schrödinger [1946] 2013).

This type of extremal explanation has been widely applied to more complex systems, including open systems that are far from thermodynamic equilibrium ('dissipative systems'). Ilya Prigogine, a pioneer in this field, proposed the principle of minimal entropy production: i.e., systems in far-from-equilibrium conditions organize themselves so as to minimize the increase of entropy (Prigogine 1947). However, universal extremal principles that govern the behavior of all dissipative systems have not been found. For instance, the principle of maximal entropy production has also been proposed (Paltridge 1979). It remains unclear to what extent these extremal principles are instances of the method of the most probable distribution – or whether they bring goals and purposes to the table that cannot be explained through statistics alone.

When we look at more recent applications of statistical physics, gradual progress can be discerned. For instance, an upper bound on the rate of bacterial replication has been proposed (England 2013). However, this remains a research program, and while there is not yet any clear reason why the program cannot continue to make gradual progress, the prospect of reducing organismic behavior to statistical physics remains remote.

When one departs from the reductive rigor of statistical physics, then axiomatic thermodynamic extremal principles can seemingly be used to explain animal behavior (and human behavior in particular). The work of Karl Friston, which has enjoyed success in theoretical neuroscience, is an example of this approach. Here animal behavior is analyzed as minimizing free energy – intuitively, this means that organisms minimize the quantity of 'surprise' in their environment (Friston 2010). However, free energy minimization is taken as axiomatic and is not given a deeper derivation in the way Boltzmann had done for entropy maximization in the context of equilibrium thermodynamics. In this way, it seems that the concept of 'organismic purpose' (in this case, the purpose of minimizing free energy) cannot be dispensed with within Friston's framework.

We have discussed two types of non-agential extremal explanations – causal mechanical ones, and non-causal statistical ones – and argued that both fail to dispense with organismic agency. We would now like to consider in more detail what is perhaps the most serious contender for dispensing with organismic agency, namely selectionist explanations. To what extent selectionist explanations reduce to non-causal statistical ones remains controversial. Some have argued that they do: evolution caused by fitness differences is structurally identical to, for instance, the differential growth rates of bank accounts with different interest rates (Matthen and Ariew 2002; Walsh, Lewens, and Ariew 2002).

Regardless of the interpretation of natural selection, it is clear how it should be combined with causal-mechanical explanation so as to seem to dispense with agency. Consider the behavior of chemotaxis, where bacteria swim up sucrose gradients. An agential explanation of this behavior would refer to the purpose of the bacteria to get nutrition. However, one could attempt to explain chemotaxis by referring to how, given a certain environmental input into the mechanism of chemotaxis, a certain output (swimming behavior) is to be expected. And why is the mechanism of chemotaxis set up in this particular way (connecting these inputs with those outputs)? Here the selectionist explanation comes in: those bacteria that came up with the mechanism of chemotaxis were able to take distance from competitors and maximize their access to resources (Wei et al. 2011). This in turn allowed them to reproduce more, eventually crowding out the bacteria incapable of chemotaxis. There is no need to reference agency here.

Can agential explanation be reduced to selectionist explanation in this way? We will not take a stand on whether it can or not ; however, we would like to argue that this issue is considerably more complicated than the simple selectionist-mechanical explanation above suggests. A selectionist explanation may be adequate for chemotaxis, but it is far from clear that this can be generalized to organismic behavior in general, especially concerning cases where organisms produce adaptive behavior even in novel environments.

To see this, recall that a selectionist explanation requires a homogeneous selective environment (Brandon 1990), which means that selection pressures must be relatively uniform across the environment. So, if an organism is exposed to a 'novel' selective environment, this means that it is exposed to selection pressures that the organism's ancestors never encountered. One of two scenarios then presents itself. The first is that the fitness-maximizing analogy breaks down: the organism sticks to its behavior that was previously adaptive, but maladaptive in the new environment. The second is that the fitness-maximizing analogy holds, and the organism chooses a new behavior that maximizes its (inclusive) fitness. However, in this case, referring merely to a selectively-determined function cannot explain why the new behavior was chosen. In other words, a mere selectionist explanation is not adequate.

To give this line of thought some more systematic detail: assume that organism $O$'s behavior $B$ was selected for in selective environment $E$. Furthermore, $B$ is the output of some heritable function $F$. What if the environment shifts to some radically different $E^*$? Different organisms will behave differently, depending on $F$. Some will continue producing $B$ regardless; others will be sensitive to cues in the environment, and the function $F$ will produce as output $B^*$ instead of $B$ (this is behavioral plasticity). If $B^*$ turns out to be adaptive to $E^*$, is this not a lucky coincidence assuming that $O$'s ancestors never encountered the selection pressures in $E^*$? In other words, if $F$ is designed for $E$, is it not a lucky coincidence that $F$ should also produce adaptive behavior in a radically different environment? This is how organismic purposes and agency can be introduced, to provide a better explanation of the production of adaptive behavior in novel environments.

Of course, plasticity itself can be selected for (Bradshaw 1965), and this is where the problem gets complex and interesting. So, if $F$ underlies a plastic trait that is modulated to produce adaptive behavior in $E^*$, the selectionist could respond that the appropriate explanation is not an agential explanation, but rather that $E^*$ and $E$ are simply not two different selective environments. They may be different *physical* environments, but they are instances of the same selective environment – for instance, they may be similar instantiations of the same pattern of heterogeneity, such as possessing the same varying cues.

In this way, the question of whether agential explanations can be reduced to selectionist explanations opens up to a large and fundamental problem of what selective environments are and how they should be delineated. For this reason, we do not wish to take a stand on whether agential explanations can be reduced to selective explanations. A safer conclusion we would like to draw is this: it is currently unclear whether agential explanations can be reduced to selectionist explanations, and therefore we should not assume that the theory of natural selection easily dispenses with agency. We should take seriously the option that agency may be indispensable.

# 4 The Kantian Approach to Purposiveness

Kant's work on teleology can offer an interesting perspective in that he considered a closely related problem – apparently incompatible ways of viewing biological organisms – but resolved it in a way that cuts across the dichotomy that pairs the indispensability of agency with an ontic view of agential explanation. Most interest in the Kantian perspective on teleology has focused on developmental phenomena.[19] A passage that is often quoted as particularly relevant is the following where Kant introduces the term 'self-organization':

> In such a product of nature each part is conceived as if it exists only *through* all the others, thus as if existing *for the sake of the others* and *on account of* the whole, i.e., as an instrument (organ), which is, however, not sufficient (for it could also be an instrument of art, and thus represented as possible at all only as a purpose); rather it must be thought of as an organ that *produces* the other parts (consequently each produces the others reciprocally), which cannot be the case in any instrument of art, but only of nature, which provides all the matter for instruments (even those of art): only then and on that account can such a product, as an *organized* and *self-organizing* being, be called a *natural purpose*. (Original emphasis; translation slightly modified. Kant [1790] 2001, 274; 5:374)

What Kant is arguing here is that organisms are not simply machines (e.g., artifacts), where each part may be designed to contribute to the whole, but where some external source (e.g., the artisan) is the cause of the production and maintenance of each part of the machine. Thus, for instance, the minute hand of a watch is produced by the artisan and not by any other part of the watch. By contrast, the various anatomical and physiological traits of an organism are produced by processes internal to the organism. Organisms are thus not to be judged as machines: an essential property of organisms is that the parts also cause the production and maintenance of the other parts, as we see in the ontogenesis.

In this way, the passage in which Kant introduces the notion of self-organization is most directly relevant to issues concerning the development of organisms; not surprisingly this is where the connection between contemporary biology and Kant's thought has most often been made (see also Huneman 2017). Here however, we would like to draw out more explicitly the implications of this for organismic behavior and organismic agency.[20] In particular, we will look in more detail at Kant's general idea of purposiveness, and at his general

---

**19** For a discussion of these different perspectives, and the relevance of the Kantian approach to contemporary debates in evolutionary biology, see Huneman 2017. See also references to Kant in Varela 1979, Kauffman 1993.

**20** To a certain extent, the division between development and behavior is artificial. Development typically refers to morphological changes (cell differentiation, growth, etc.) that are relatively irreversible and slow in comparison to physiological changes (metabolism) or behavioral changes (movement through space). Some explicitly distinguish between development and behavior (e.g. Burge 2009); by contrast, most behavioral ecologists consider any trait (for instance a tree growing small vs. large leaves) as a 'behavior'.

treatment of the antinomy of teleological judgment, which concerns the apparent clash between 'mechanistic' and 'teleological' approaches to the organism.

## 4.1 The Antinomy of Teleological Judgment

In his *Critique of the Power of Judgment,* Kant posits the following two conflicting maxims concerning 'generation' (a contemporary close-equivalent: development) and 'mechanical laws' (namely, laws that govern the way parts yield wholes – see McLaughlin 1990):

*Thesis:* All generation of material things is possible in accordance with merely mechanical laws.

*Antithesis:* Some generation of such things is not possible in accordance with merely mechanical laws. (Kant [1790] 2001, 258–259; 5:387).

In particular, Kant had biological organisms in mind as possible entities that are not generated merely according to mechanical laws. This thesis-antithesis pair is simply a contradiction, leading to mutually incompatible views with no prospect of reconciliation.

Kant's first step, then, is to make explicit that such pronouncements about the nature of reality are actually *judgments* that are necessarily relative to our *cognition* of reality. Hence, he proposes the following thesis-antithesis pair:

> The *first maxim* of the power of judgement is the *thesis:* All generation of material things and their forms must be judged as possible in accordance with merely mechanical laws. The *second maxim* is the *antithesis:* Some products of material nature cannot be judged as possible according to merely mechanical laws (judging them requires an entirely different law of causality, namely that of final causes). (Kant [1790] 2001, 258–259; 5:387)

This is the antinomy of teleological judgment. The motivation underlying the antithesis draws on the idea that mechanical laws do not seem to adequately account for the organization that can be found in biological organisms. In particular, Kant writes:

Nature, considered as a mere mechanism, could have formed itself in a thousand different ways without hitting precisely upon the unity (KU, AA, V: 360).

The mechanical laws do not privilege any particular organization over another; hence, if the organization were to be explained with merely mechanical laws, the organization of organisms could only be judged to be the result of *chance* (see Huneman 2006).

## 4.2 Contingency and Kant's Concept of Purposiveness

In this way, Kant is relying on a concept of purposiveness that can be described as the 'lawfulness of the contingent as such' (First Introduction to the *Critique of Judgement*, Kant [1790] 2001, 20; 20:217). An initial illustration of the concept can be given in the context of development. For instance, if one were only to take the mechanical laws of nature into account, the fact that the development of a chicken leads to a chicken appears to be contingent – once the initial and boundary conditions are sufficiently changed, it might develop into a monster. However, the laws themselves cannot explain this wide divergence in outcome, since in both cases the same laws apply. One must introduce the idea that the development is the development *of a chicken*, and therefore is oriented towards this goal. Thus, such an idea brings some necessity into a process that is, with regard to nature itself (i.e., the mechanical laws of nature), contingent. The same goes for the functions of organisms: whether or not an elephant's lungs breathe seems highly contingent if one only takes into account the laws of nature, but appears as necessary when we introduce the idea that breathing is the function of the lung. This entails invoking the idea of a functioning organism. Thus, biological functions and embryogenic development instantiate the same epistemic pattern (Huneman 2006).

Kant's theory of purposiveness is intended to reflect this epistemic fact. To introduce it, he gives a famous example: what if one were to come across a regular hexagon drawn in the sand (Kant [1790] 2001 §62)?[21] This, says Kant, can only be understood as an instance of purposiveness, because if we do not posit a concept ('regular hexagon') that is 'at the basis of' (i.e., guided) its production, we cannot understand why it is drawn in the sand. In other words, while the laws of nature can lead to the appearance of all sorts of figures in the sand, the specific kind of figure we see is not privileged by those laws (i.e., it is not any more probable than any other kind of figure). When we see a physical instance of a regular hexagon then, and we judge that it fits the concept of a 'regular hexagon', there is no indication in the laws of nature as to why a regular hexagon should be produced rather than another one. Hence, we reasonably assume that the concept 'regular hexagon' was at the basis of its production – namely, someone thought of this concept and has drawn the hexagon – and thereby

---

[21] Note that Kant chose an example from mathematics as part of his overall strategy to decouple the notion of purposiveness he intends to capture from the usual scheme of craftsmanship, fabrication, etc.

the contingent figure we see on the sand features some lawlikeness (since it has been drawn according to some rule).

To put the argument in a more contemporary idiom, consider the following. Among the set of all possible hexagons, the size of the subset of regular hexagons (i.e., with equal sides) is extremely small (measure = 0). Hence, given that we observe a regular hexagon, and that the probability that a process governed only by mechanical laws of motion would cause a regular hexagon to appear is 0, appealing to the presence of a concept at the ground of the production allows for a (much) better explanation of the appearance of the hexagon.

This same line of reasoning can be applied to organisms. In Kant's *Unique Argument for a proof of God's existence*, the first major text in which he deals with life and finality, he considers the traditional example of the eye, describing the example in the following way: in an eye there are many parts, each following different and mutually independent laws, and yet, the parts function not only in such a way that the eye can see, but if we were to even slightly change the structure or behavior of one of the parts, the eye as a whole would no longer achieve sight. Similarly, in the third *Critique* Kant gives the example of the bird whose different anatomical parts seem to be organized in very specific ways in order to enable flight: "the structure of a bird, the hollowness of its bones, the placement of its wings for movement and of its tail for steering, etc." (Kant [1790] 2001, 233; 5:360). And yet, it remains possible also to view an organism as a clump of dead matter, obeying the laws of mechanics. The price to pay for the latter possibility is that there is no longer any answer to the question of why those parts are so *contrived* – to use a word that will become crucial for Darwin – to allow flight.[22]

The notion of purposiveness as elaborated by Kant is closely related to extremal explanations. Consider the evolution of the camera eye, and its dependence on the laws of nature (and causal-mechanical processes). Assume we can vary the laws of nature (and causal-mechanical processes) by manipulating a parameter vector $\{(A_i)\}$, and let some scalar variable $W$ represent the functional

---

[22] Note that the judgment that there is a causal relation between two objects or events (like two billiard balls colliding) is a constitutive use of reason (understanding), whereas judging according to mechanical laws is a regulative use of reason, even though mechanical laws are clearly closely related to causality as an *a priori* principle. But, as said before, mechanism is about the relation between parts and wholes – knowing wholes from the parts – while causation is about the succession of events or facts. Disentangling how precisely Kant understood the relation between causality and mechanical laws is the subject of some debate in Kant scholarship. See, for example, Allison 2001.

value of the eye (for instance, the representational accuracy[23] of its images). Further assume that a particular set of parameter values ($A1, A2... An$) maps onto the extreme value $W0$ for $W$ – namely, the best functionality or representational validity – and let $W0$ also be the precise value that $W$ assumes in empirical nature. Yet, it would seem highly improbable that the parameter vector should attain the exact ($A1, A2... An$) among all possible values of $\{(Ai)\}$; correspondingly, functional sight seems highly improbable. Referring to the concept of 'sight' as a concept that somehow guides the fixation of the parameter values ($A1, A2... An$) allows one to ascribe a kind of necessity (or at least a much higher probability) to ($A1, A2... An$). Thus, interpreting $W0$ as the purpose of 'sight', allows for the explanation of why sight-enabling structures emerged.

So, when Kant holds here that the fact that ($A_1, A_2... A_n$) obtains is not explainable except if one thinks of a concept ('sight') at its ground, this account is perfectly analyzable as positing an extremal explanation which explains the explanandum as the extremal value (in turn corresponding to functional sight) of a mathematical function. Thus, the reason why the vector ($A_1, A_2... A_n$) – otherwise wholly contingent – is the one that we find in nature is that the vector realizes some extremum. Moreover, the concept of vision ultimately appeals to the idea of a functioning organism that is able to survive (e. g. catch prey, avoid predators, track motions and light in its environment, etc.). In this way, the reference to the concept of vision introduces lawlikeness into the contingent unity of mechanical laws involved in the design of the eye.

The lawfulness of this contingent unity, i.e., the notion of purposiveness, is for Kant only a *regulative* and not a *constitutive* concept or principle. Regulative and constitutive principles refer to two uses of reason. While not going into too much detail,[24] the latter refer to the synthetic *a priori* principles (causality, permanence, reciprocal action, etc.) that ground any science of nature, and when events or facts can be subsumed under such principles, they can be considered as 'objective'. By contrast, in the regulative use of reason, the principles inform our cognition of the objects and allow for knowledge of objects, but do not posit anything as objective. An instance of the "regulative use" of ideas of reason, described in the *Dialectic* or the *Critique of Pure Reason*, involves prescribing the idea of the "synthesis of all conditions" to the world. This allows us to require new conditions for the conditioned events, empirical laws, forces or facts we have found. Nonetheless, we cannot posit as objective the *whole* of conditions –

---

[23] See Burge (2009) on veridicality as a norm for representational systems. Burge's notion of norm is here accounted for in terms of extremal value.
[24] For an introduction see Huneman 2007.

which Kant calls the "unconditioned" and which can refer to, for instance, the whole world, or God (which in turn refers to what Kant calls "Ideas of reason"[25]). Likewise, the idea that each individual belongs to a species that in turn belongs to a higher order class (family, genus, etc.) is not an objective fact, but a regulative principle of our knowledge, without which we would be unable to cognize an ordered world.

The regulative principles that allow for biology (since, at least as stated in the third *Critique*, the constitutive principles of judgment lead to viewing the contingent as simply contingent and lawful) are precisely the lawlikeness of the contingent as purposive: this kind of lawlikeness implies, as we said, the idea of a functional or developing organism. Moreover, from the moment the reference to such totality – namely, the organism – is introduced, a new level of necessity is brought into a set of facts and events that would otherwise appear wholly contingent. This then allows these facts to be studied in a scientific manner: biologists will ask which mechanisms fulfill this or that function, or what processes lead to the formation of such and such an organ and then the whole organism. Inversely, any such scientific enquiry already assumes the lawlikeness of the contingent.

## 5 Organismic Agency and the Demand of Reason

We will now depart from a description of Kant's framework, and draw out the (Kant*ian*) implications of the concept of purposiveness for the central issue of this paper, namely the ontology and dispensability of agency. In particular, we will argue against three ways of viewing agency: first the position that agency is a mere projection (non-ontic, dispensable); second that it refers to an element of objective nature (ontic, indispensable); finally, we will also contrast the Kantian view with the position that agency is a mere heuristic, but that it is indispensable given our evolved nature and limited computational capacity. This is a view where agential explanation is viewed as non-ontic, and agency as dispensable for cognition of reality, but indispensable for the *human* cognition of reality.

First, can purposiveness be seen as a projection of the human mind onto the natural world? In this view, goals and functions are in fact anthropomorphic projections onto the world (e. g., Lewens 2007: 544–5). Such projections may serve some purpose as heuristics, but they do not reveal anything objectively real

---

25 On this notion see Allison 2001, Grier 1995.

about the world and are entirely dispensable: they are to be replaced by mechanistic or law-based explanations whenever the latter become available.

In response, recall from the previous section how, in a Kantian framework, the question whether or not agential explanation should be viewed ontically is bound up with the distinction between 'regulative' and 'constitutive' principles. Only the latter give rise to ontic explanations; nonetheless, that does not mean that purposiveness is merely a projection of the human mind onto the natural world. Granted, purposiveness does not constitute nature as such and is therefore not objective in the same way that laws of nature are. However, they are not a 'projection' in the sense that it is an optional way for a cognizing subject to see the world. Once biological items are the object of a quest for knowledge, there is no alternative to purposiveness for the faculty of knowledge.

This can be emphasized by referring to one last element in Kant's work, namely how the structure of the faculty of knowledge is 'finite':

> Absolutely no human reason (or even any *finite reason that is similar to ours in quality, no matter how much it exceeds it in degree*) can ever hope to understand the generation of even a little blade of grass from merely mechanical causes. (our emphasis, Kant [1790] 2001, 279; 5:410)

Our reason is 'finite' because it cannot derive intuitions from concepts, and therefore, the particular from the universal.[26] An 'infinite' reason, by contrast, would not be limited in this way. However, Kant does not posit that such an infinite reason actually exists, or for that matter, is even possible under some counterfactual scenario. Instead, it is a mere idea that orients philosophical enquiry into knowledge, or if you will, a thought experiment aimed at clarifying what reason is. This distinction between finite and infinite reason can be connected to two modes of understanding: discursive and intuitive understanding. Intuitive understanding (which, like infinite reason, is a mere idea that orients the philosophical enquiry about knowledge) would be able to cognize the particular instances of concept X at the same time it cognizes the (universal) concept of X. By contrast, discursive understanding must go through 'mediations' in order to arrive at the particular. Simple acts of observation can be such mediations (to check whether anything corresponds to the concept X).

The concept of purposiveness is also such a mediation, since it allows reason to proceed from the universal laws of nature to particular organisms. A living being can, in general, be analyzed by means of mechanistic laws, e.g., the uni-

---

[26] Concepts allow us access to the universal, while intuitions provide us access to the particular.

versal laws governing the dynamics of each part. However, here only a very specific combination of the part-level processes results in a living being (think of the various laws involved in the building of the eye, mentioned above). Hence, the finite reason has to shift to the level of the "lawlikeness of the contingent as such", namely, to assume the regulative principle of purposiveness by introducing the reference to the whole organism. This "idea of the whole," he says in § 65, is only a principle of cognition, not of production.

This finiteness of reason leads to what Kant in other passages describes as a 'demand' of reason for the 'unconditioned'. We previously described it as the 'synthesis of all conditions', but in a more contemporary idiom, it could also be described as the following:

> The demand for the unconditioned is essentially a demand for ultimate explanation, and links up with the rational prescription to secure systematic unity and completeness of knowledge. Reason, in short, is in the business of ultimately accounting for all things. (...) the demand for the unconditioned is inherent in the very nature of our reason, [and] is unavoidable and indispensably necessary... (Grier 2018)

Kant thus takes this demand of reason to deliver a kind of impossibility result for the possibility of a non-purposive explanation of organismic development (and by extension, the same could be said of organismic agential-like behavior).

In this way, in contrast to interpreting agential explanations as involving anthropomorphic projection, for the 'Kantian option' there is no alternative to explaining organismic behavior as agential. Moreover, seeing an organism as an agent is even a precondition (the transcendental ground) to being able to make a projection onto a natural system. For instance, if in some agential explanation, a repertoire of actions is projected onto a living organism, this presupposes seeing an organism as an agent. Assuming agency makes ascribing empirical methodology and even (behavioral) property to organisms possible. This is how the 'indispensability' implied by the Kantian option should be understood.

Others have taken the 'blade of grass' passage cited above as support for an ontic view of agential explanation, where "organisms are subjects having purposes according to values encountered in the making of their living" (Weber and Varela 2002, 102). But an ontic interpretation of agential explanation – where organisms are (objectively) subjects with (objective) purposes – clashes with Kant's overarching transcendental framework, since only constitutive principles can ground ontic explanations. Given that regulative principles such as purposiveness are a consequence of the finite nature of reason, they are not empirically discoverable facts, but are instead presupposed in any epistemic strategy for searching empirical truths. This shows how the Kantian option implies a non-ontic view of agential explanation.

Does the Kantian option imply an epistemic view of agential explanation? We take 'epistemic' here to refer to expectability *sensu* Salmon (Salmon 1989), where an explanation explains in virtue of showing the explanandum as expected (i.e., with high probability). In this sense, the Kantian option does certainly interpret agential explanations as showing how the explanandum is to be expected; however, much also depends on how 'expectability' is interpreted. Consider the subjective interpretation[27], where expectability is analyzed as dependent on the amount of information available to the subject; as the information changes, so does the expectability. This is what is presupposed if one views agential explanations as arising from bounded rationality, where agency is ascribed as a heuristic or computational shortcut given time and/or information constraints. The Kantian option is not 'epistemic' in this way: it does not refer to properties of what could be called 'evolved human nature' but rather to a fundamental structure of reason itself. Any finite reason, even if it would be as computationally powerful as the largest supercomputer, would not be able to understand organismic purposes only in terms of causal mechanisms. Even if our empirical nature were very different – for instance, if we had evolved very different cognitive heuristics for understanding the world – as long as we are endowed with a finite reason then we would still employ teleological concepts such as agency. Thus, agential explanations are non-ontic in the sense that agency as a concept ultimately can be traced back to a fundamental structure of reason (and not a structure of the objective world, nor to a quantity of information about the world available to a subject).

In sum, the Kantian approach suggests that agential explanations are to be viewed as non-ontic explanations but in which agency is indispensable. Viewing organisms as agents is a heuristic – it allows organisms to be identified as wholes in the first place (cf. Breitenbach 2008), and thereby allows a research program about the mechanisms of functions and development – but it is not merely a heuristic: it is unavoidable for a *finitely rational understanding* of nature. Agential explanations may be predictive tools – they may accurately summarize complex patterns of behavior and allow us to predict how organisms will respond to environmental inputs – but they are more than mere predictive tools, because if they were merely predictive tools, agential explanations would be replaceable by an explanation that integrates a mass of complex causal detail. Even though the latter may be predictively equivalent or even superior to an agential explanation, it does not afford *understanding* to rational beings.

---

**27** An objective interpretation seeks to analyze expectability (and probability) in terms of objective structures, and thus leads to a variation on the ontic view of agential explanation.

# 6 Conclusion: Organismic Agency and the Demand of Reason

The shift in contemporary biology towards the agential approach motivates paying closer philosophical attention to agential explanations. Yet agential explanations are still today interpreted along the lines of a dichotomy between ontic/indispensable or non-ontic/dispensable, even though both options are ultimately unsatisfactory. In this paper we elucidated the Kantian option, where viewing organisms as agents is a demand of reason, and thus indispensable to our cognition of reality, but yet where agency is not added to the 'furniture' or basic ontology of the world.

This implies that agential explanations are *unavoidable* given our rational nature. This goes further than merely stating that agential and non-agential explanations are complementary. While it is of course possible also to view organisms as combinations of mechanisms, scientists, as rational beings, have no choice but to use agential explanations as well. Agency is thus not simply an investigative heuristic or a predictive tool that can be dispensed with once our scientific knowledge is sufficiently advanced, like a ladder that is climbed only then to be kicked away. Seeing agency in the natural world is not like a form of superstition that can be dispelled by the onward march of scientific reason; it is inherent to reason itself and is therefore not a ladder that can ever be kicked away.

# References

Allison, Henry E. (2001): *Kant's Theory of Taste: A Reading of the Critique of Aesthetic Judgment.* Cambridge, UK: Cambridge University Press.

Atran, Scott (2002): *In Gods We Trust: The Evolutionary Landscape of Religion.* Oxford, UK: Oxford University Press.

Auletta, Gennaro (2013): "Information and Metabolism in Bacterial Chemotaxis." In: *Entropy* 15, pp. 311–326.

Barandiaran, Xabier E./ Di Paolo, Ezequiel/ Rohde, Marieke (2009): "Defining Agency: Individuality, Normativity, Asymmetry, and Spatio-Temporality in Action." In: *Adaptive Behavior* 17, pp. 367–386.

Barrett, Justin L. (2000): "Exploring the Natural Foundations of Religion." In: *Trends in Cognitive Sciences* 4, pp. 29–34.

Bateson, Patrick (2005): "The Return of the Whole Organism." *Journal of Biosciences* 30 (1), pp. 31–39. https://doi.org/10.1007/BF0270514.

Birch, Jonathan (2012): "Robust Processes and Teleological Language." In: *European Journal for Philosophy of Science* 2, pp. 299–312.

Bradshaw, Anthony D. (1965): "Evolutionary Significance of Phenotypic Plasticity in Plants." In: *Advances in Genetics* 13, pp. 115–155.
Breitenbach, Angela (2008): "Two Views on Nature: A Solution to Kant's Antinomy of Mechanism and Teleology." In: *British Journal for the History of Philosophy* 16, pp. 351–369.
Burge, Tyler (2009): "Primitive Agency and Natural Norms*." In: *Philosophy and Phenomenological Research* 79, pp. 251–278.
Brandon, Robert N. (1990):wei *Adaptation and Environment.* Princeton University Press.
Calvo Garzón, Paco/Keijzer, Fred (2011): "Plants: Adaptive Behavior, Root-Brains, and Minimal Cognition." *Adaptive Behavior* 19, pp. 155–171.
Caporael, Linnda R./Griesemer, James R./Wimsatt, William C. (2013): *Developing Scaffolds in Evolution, Culture, and Cognition.* Cambridge, MA: MIT Press.
Coopersmith, Jennifer. (2017): *The lazy universe: an introduction to the principle of least action.* Oxford, UK: Oxford University Press.
Craver, Carl F. (2014): "The Ontic Account of Scientific Explanation." In: Kaiser, Marie I./Scholz, Oliver R./Plenge, Daniel/Hüttemann, Andreas (Eds.): *Explanation in the Special Sciences*, Dordrecht: Springer Netherlands, pp. 27–52.
Cummins, Robert (1975): "Functional Analysis." *The Journal of Philosophy* 72 (20): 741–765.
Dawkins, Richard (1976): *The Selfish Gene.* Oxford, UK: Oxford University Press.
Dawkins, Richard (1982): *The Extended Phenotype: The Long Reach of the Gene.* Oxford, UK: Oxford University Press.
Desmond, Hugh. (2018). Natural selection, plasticity, and the rationale for largest-scale trends. In: *Studies in History and Philosophy of Science Part C: Studies in History and Philosophy of Biological and Biomedical Sciences, 68–69*, 25–33.
Desmond, Hugh. (2019). Shades of Grey: Granularity, Pragmatics, and Non-Causal Explanation. In: *Perspectives on Science.*
Duijn, Marc van/Keijzer, Fred/Franken, Daan (2006): "Principles of Minimal Cognition: Casting Cognition as Sensorimotor Coordination." In: *Adaptive Behavior* 14, pp. 157–170.
England, Jeremy L. (2013): "Statistical Physics of Self-Replication." In: *The Journal of Chemical Physics* 139, 121923.
Fisher, Ronald A. (1919): "The Correlation between Relatives on the Supposition of Mendelian Inheritance." In: *Transactions of the Royal Society of Edinburgh* 52, pp. 399–433.
Fisher, Ronald A. (1930). *The Genetical Theory of Natural Selection.* Oxford, UK: Oxford University Press.
Friston, Karl. (2010): "The Free-Energy Principle: A Unified Brain Theory?" In: *Nature Reviews Neuroscience* 11, pp. 127–138.
Gibson, James J. ([1979] 2014). *The Ecological Approach to Visual Perception.* New York and London: Psychology Press.
Gigerenzer, Gerd (2000): *Adaptive Thinking: Rationality in the Real World.* Oxford, UK: Oxford University Press.
Glennan, Stuart (2002): "Rethinking Mechanistic Explanation." In: *Philosophy of Science,* 69 pp. 342–353.
Godfrey-Smith, Peter (1996): *Complexity and the Function of Mind in Nature.* Cambridge, UK: Cambridge University Press.

Grafen, Alan (1984): "Natural Selection, Kin Selection and Group Selection." In: Krebs, John Richard/Davies, Nicholas Barry (Eds.): *Behavioural Ecology*. Oxford, UK: Blackwell, pp. 62–84.

Grafen, Alan (2006): Optimization of inclusive fitness. In: *Journal of Theoretical Biology* 238, pp. 541–563.

Grafen, Alan (2014): "The Formal Darwinism Project in Outline." *Biology & Philosophy* 29, pp. 155–174.

Grier, Michelle (1995): "Kant's Rejection of Rational Theology." In: *Proceedings of the Eighth International Kant Congress* 2, pp. 641–650.

Grier, Michelle (2018): "Kant's Critique of Metaphysics." In: Zalta, Edward N. (Ed.): *The Stanford Encyclopedia of Philosophy*. https://plato.stanford.edu/archives/sum2018/entries/kant-metaphysics/.

Hamilton, Matthew B. (2009): *Population Genetics*. Hoboken, NJ: Wiley-Blackwell.

Hanczyc, Martin M./Ikegami, Takashi (2010): "Chemical Basis for Minimal Cognition." In: *Artificial Life* 16, pp. 233–243.

Hempel, Carl G. (1959): "The Logic of Functional Analysis." In: Gross, Llewellyn (Ed.): *Symposium on Sociological Theory*. New York: Harper and Row, pp. 271–87.

Horibe, Naoto/Hanczyc, Martin M./Ikegami, Takashi (2011): "Mode Switching and Collective Behavior in Chemical Oil Droplets." In: *Entropy* 13, pp. 709–19.

Huneman, Philippe (2006): "Naturalising Purpose: From Comparative Anatomy to the 'Adventure of Reason.'" In: *Studies in History and Philosophy of Science Part C: Studies in History and Philosophy of Biological and Biomedical Sciences* 37, pp. 649–674.

Huneman, Philippe (2007): "Reflexive judgement and Wolffian embryology: Kant's shift between the first and the third *Critique*." In: Huneman, Philippe (Ed.): *Understanding purpose? Kant and the philosophy of biology*. Rochester: University of Rochester Press, pp. 75–100.

Huneman, Philippe (2010): "Topological Explanations and Robustness in Biological Sciences." In: *Synthese* 177 (2), pp. 213–45. https://doi.org/10.1007/s11229-010-9842-z.

Huneman, Philippe (2017): "Kant's Concept of Organism Revisited: A Framework for a Possible Synthesis between Developmentalism and Adaptationism?" In: *The Monist*, 100, pp. 373–390.

Huneman, Philippe (2019): "Revisiting Darwinian Teleology." In: *Studies in History and Philosophy of Science Part C: Studies in History and Philosophy of Biological and Biomedical Sciences*, 76: 101188.

Huxley, Julian ([1942] 1974): *Evolution: The Modern Synthesis*. London, UK: Allen and Unwin.

Jacob, François/Monod, Jacques. (1961). "Genetic regulatory mechanisms in the synthesis of proteins." In: *Journal of Molecular Biology* 3, pp. 318–356.

Kant, Immanuel ([1790] 2001): *Critique of the Power of Judgment*. Cambridge, UK: Cambridge University Press.

Kauffman, Stuart A. (1993): *The Origins of Order: Self-Organization and Selection in Evolution*. Oxford, UK: Oxford University Press.

Leibniz, Gottfried W. (1890): "Discours de métaphysique". In: *Philosophische Schriften* Band IV, Carl Immanuel Gerhardt, (Ed.), pp. 427–465.

Lewens, Tim (2005): *Organisms and Artifacts: Design in Nature and Elsewhere*. Cambridge, MA: MIT Press.

Lewens, Tim (2007): "Function." In: Matthen, Mohan/Stephens, Christopher (Eds.): *Handbook of the Philosophy of Science*. Amsterdam: Elsevier, pp. 525–47.
Lyon, Pamela (2015): "The Cognitive Cell: Bacterial Behavior Reconsidered." In: *Frontiers in Microbiology* 6, article 264.
Lyon, Pamela (2017): "Environmental Complexity, Adaptability and Bacterial Cognition: Godfrey-Smith's Hypothesis under the Microscope." In: *Biology & Philosophy* 32, pp. 443–65.
Machamer, Peter/Darden, Lindsay/Craver, Carl. F. (2000): "Thinking about Mechanisms." In: *Philosophy of Science* 67, pp. 1–25.
Manneville, Paul (2006): "Rayleigh-Bénard Convection: Thirty Years of Experimental, Theoretical, and Modeling Work." In: Mutabazi, Innocent/Wesfreid, José Eduardo/Guyon, Etienne (Eds.): *Dynamics of Spatio-Temporal Cellular Structures*. New York, NY: Springer, pp. 41–65
Matthen, Mohan/Ariew, André (2002): "Two Ways of Thinking About Fitness and Natural Selection:" In: *Journal of Philosophy* 99, pp. 55–83.
Mayr, Ernst (1961): "Cause and Effect in Biology." In: *Science* 134, pp. 1501–6.
Mayr, Ernst (1982): *The Growth of Biological Thought: Diversity, Evolution, and Inheritance*. Cambridge, MA: The Belknap Press of Harvard University Press.
McLaughlin, Peter (1990): *Kant's Critique of Teleology in Biological Explanation: Antinomy and Teleology*. Lewiston: E. Mellen Press.
Millikan, Ruth Garrett (1984): *Language, Thought, and Other Biological Categories: New Foundations for Realism*. Cambridge, MA: MIT Press.
Moreno, Alvaro/Mossio, Matteo (2015): *Biological Autonomy*. Dordrecht: Springer.
Mossio, Mossio/Saborido, Christian/Moreno, Alvaro (2009): "An Organizational Account of Biological Functions." In: *The British Journal for the Philosophy of Science* 60, pp. 813–841.
Müller, G. B. (2017). "Why an extended evolutionary synthesis is necessary." In: *Interface Focus* 7, 20170015.
Neander, Karen (1991): "Functions as Selected Effects: The Conceptual Analyst's Defense." In: *Philosophy of Science* 58, pp. 168–84.
Nicoglou, Antonine (2015): "The evolution of phenotypic plasticity: Genealogy of a debate in genetics." In: *Studies in History and Philosophy of Science Part C: Studies in History and Philosophy of Biological and Biomedical Sciences*, 50, pp. 67–76.
Okasha, Samir (2018): *Agents and Goals in Evolution*. Oxford, UK: Oxford University Press.
Paltridge, Garth W. (1979): "Climate and Thermodynamic Systems of Maximum Dissipation." In: *Nature* 279, pp. 630–31.
Pérez, Juana/Moraleda-Muñoz, Aurelio/Marcos-Torres, Francisco Javier/Muñoz-Dorado, José (2016): "Bacterial Predation: 75 Years and Counting!" In: *Environmental Microbiology* 18, pp. 766–79.
Pittendrigh, Colin S. (1958): "Adaptation, Natural Selection, and Behavior." In: Roe, Anne/Simpson, George Gaylord (Eds.): *Behavior and Evolution*. New Haven: Yale University Press, pp. 360–416.
Prigogine, Ilya (1947): *Étude thermodynamique des phénomènes irréversibles*. Liège: Desoer.
Salmon, W. C. (1989). *Four Decades of Scientific Explanation*. Pittsburgh, PA.: University of Pittsburgh Press.

Schlosser, Markus (2015): "Agency." In: Zalta, Edward N. (Ed.): *The Stanford Encyclopedia of Philosophy*. https://plato.stanford.edu/archives/fall2015/entries/agency/.

Schrödinger, Erwin. ([1946] 2013): *Statistical Thermodynamics*. New York, NY: Dover Publications.

Shani, Itay (2013): "Setting the Bar for Cognitive Agency: Or, How Minimally Autonomous Can an Autonomous Agent Be?" In: *New Ideas in Psychology* 31, pp. 151–65.

Simpson, George Gaylord (1944): *Tempo and mode in evolution*. New York, NY: Columbia University Press.

Simpson, George Gaylord (1953): *The Major Features of Evolution*. New York, NY: Columbia University Press.

Skewes, Joshua C./Hooker, Cliff A. (2009): "Bio-Agency and the Problem of Action." In: *Biology & Philosophy* 24, pp. 283–300.

Sterelny, Kim (2000): *The Evolution of Agency and Other Essays*. Cambridge, UK: Cambridge University Press.

Tversky, Amos/Kahneman, Daniel (1974): "Judgment under Uncertainty: Heuristics and Biases." In: *Science* 185, pp. 1124–31.

Varela, Francisco J. (1979): *Principles of Biological Autonomy*. Amsterdam: North Holland.

Waddington, Conrad Hal (1942): "Canalization of Development and the Inheritance of Acquired Characters." In: *Nature* 3811, pp. 563–65.

Walsh, Denis (2012): "Mechanism and Purpose: A Case for Natural Teleology." In: *Studies in History and Philosophy of Biological and Biomedical Sciences* 43, pp. 173–81.

Walsh, Denis (2015): *Organisms, Agency, and Evolution*. Cambridge, UK: Cambridge University Press. https://doi.org/10.1017/CBO9781316402719.

Walsh, Denis/Lewens, Tim/Ariew, André (2002): "The Trials of Life: Natural Selection and Random Drift." *Philosophy of Science* 69, pp. 429–46.

Weber, Andreas/Varela, Francisco J. (2002): "Life after Kant: Natural Purposes and the Autopoietic Foundations of Biological Individuality." In: *Phenomenology and the Cognitive Sciences* 1, pp. 97–125.

Wei, Yan, Xiaolin Wang, Jingfang Liu, Ilya Nememan, Amoolya H. Singh, Howie Weiss, and Bruce R. Levin (2011): "The Population Dynamics of Bacteria in Physically Structured Habitats and the Adaptive Virtue of Random Motility." In: *Proceedings of the National Academy of Sciences* 108, pp. 4047–52.

West-Eberhard, Mary Jane (1989): "Phenotypic Plasticity and the Origins of Diversity." *Annual Review of Ecology and Systematics* 2, pp. 249–78.

Woodward, James (2003): *Making Things Happen: A Theory of Causal Explanation*. Oxford, UK: Oxford University Press.

Wright, Larry (1973): "Functions." In: *The Philosophical Review* 82, pp. 139–68.

Wright, Larry (1976): *Teleological Explanations*. Berkeley, CA: University of California Press.

Michela Bordignon
# Teleology, Backward Causation and Contradiction. Hegel's Dialectical Account of Organic Nature

**Abstract:** This paper is aimed at analysing Hegel's dialectical account of living organisms. Generally speaking, in Hegel's thought "dialectic" is a process of self-determination that necessarily involves the integration of some kind of otherness. In the case of life, the self-determining process of living organisms articulates itself on the basis of internal purposiveness, which represents the core of the structure of living beings. The article will be divided in four parts: 1) in the first part, I will analyse Kant's notion of internal purposiveness, which will work as conceptual background to investigate Hegel's theory of teleology; 2) in the second part, I will explore Hegel's theory of teleology, its circular structure and its paradoxical implications in the *Science of Logic*; 3) in the third part, I will expound upon the way Hegel's dialectic of internal teleology is incorporated in his account of organic life in the Philosophy of Nature; 4) in the fourth part, I will investigate Hegel's treatment of animal organisms, with a specific focus on the assimilation process, which is the point of Hegel's analysis where the dialectical character of organism comes fully to light.

> Quando eu passo
> Perto das flores
> Quase elas dizem assim:
> Vai que amanhã enfeitaremos o seu fim
>
> (Nelson Cavaquinho. *Eu e as flores*)

# 1 Introduction

In *Fragment of a System*, that is one of the unpublished essays of the youth years, Hegel presents one of his first and most significant efforts to account for the specific structure of life: "it [life] must also be considered as capable of entering into relation with what is excluded from it, as capable of losing its individuality or being linked with what has been excluded" (Hegel, 2014, p. 341; p. 310). That is why "life is the union of union and nonunion" (Hegel, 2014, pp. 343–344;

---

**Michela Bordignon,** Federal University of ABC

https://doi.org/10.1515/9783110604665-004

p. 312). This text, which was written in 1800, discloses the main intuition that orients Hegel's mature conception of the structure of life, that is the dialectical relationship of both inclusion and exclusion between living being as a whole and what is other then itself. The rational account of this insight is properly displayed in Hegel's mature writings and it is this rational account and its philosopical implications that I aim to expound upon in this article.

I will start from a brief presentation of the main conceptual source of Hegel's dialectical theory of organism, that is Kant's theory of internal purposiveness. I will show how Hegel's critical analysis of Kant's insights is the basis for the elaboration of his own dialectical theory of teleology in the *Science of Logic* and of the way this theory is unfolded in his account of organic life in the *Philosophy of Nature*. In the last part of the article I will focus my attention on the second moment Hegel's treatment of animal organism – the assimilation process –, where the dialectical character of Hegel's theory comes fully to light.

## 2 The Kantian Notion of Organism

Kant's analysis of the organisation of living beings in the second part of *The Critique of the Power of Judgment* represents a crucial conceptual source for Hegel's conception of life.

In Kant's view, the organisation of living beings cannot be accounted for through a mere mechanistic pattern of causality.[1] Something more than efficient causality is needed.[2] Kant's strategy is to refer to a model of causality that looks back at the notion of final cause in a way that, at the same time, makes explicit the limits of this conceptual pattern of causality. More specifically, Kant's explanation of life is based on a teleological account of the relationship between the different parts of the organism and the organism as a whole.[3]

---

[1] In Kant's explication of living beings, the introduction of purposiveness is needed in so far as the relation of efficient causality constitutes a series that always goes from cause to effect, whereas living organisms are understandable only if they are conceived according to a causal relation where each member of the series is cause as well as effect. This kind of causality can be accounted for only through purposiveness (cf. Chiereghin 1990, p. 137). Cf. also Menegoni 2008, p. 128–132.

[2] Cf. Dahlstrom 1998, p. 169.

[3] On Kant's treatment of teleology cf. McFarland, 1970, Düsing 1986b, McLaughlin 1990, Ginsborg 2001, Kreines 2005, Steigerwald 2006, Quarfood 2006, Ginsborg 2006, Breitenbach 2006, Huneman 2007, Cohen 2007, Goy 2008, Fisher 2008, Breitenbach 2009, Beisbart 2009, Illetterati 2014b, Goy 2014.

Such an account obviously implies the risk of presupposing an extra-naturalistic and thus extra-scientific principle of explanation of living beings. Every teleological explanation involves a dynamic of backward causation: something coming afterwards – the end – is the cause of something coming backwards – the object through which the end is realised. This is problematic because we normally conceive of causality as efficient causality, which develops in a linear dynamic. Instead, final causality is constituted by a structure that curves and closes the linearity of efficient causality in itself, being characterised by a circular structure:

> A causal nexus can also be conceived in accordance with a concept of reason (of ends), which, if considered as a series, would carry with it descending as well as ascending dependency, in which the thing which is on the one hand designated as an effect nevertheless deserves, in ascent, the name of a cause of the same thing of which it is the effect. [...] Such a causal connection is called that of final causes (*nexus finalis*) (Kant 1908, § 65, p. 372; p. 244).

The dynamic of final causality is retroactive and circular and thus it is paradoxical: an effect cannot be the cause of its cause.[4] In the context of human activity, this incoherence is normally solved through the anticipation of the end of an object in the intention of the producer of the object. This strategy is not applicable in the context of living beings, because it would imply the presence of a designer that produces living beings according to an end, and this would represent an extra-naturalistic element undermining the scientific character of the account of organisms.

Kant's account of internal purposiveness in the organisation of living beings is meant to solve this problem. This is Kant's starting point: there is no rational justification for the presence of final causality in living organisms. Nor can the evidence of our experience of the presence of ends in nature justify the claim that there actually are ends in nature. Nevertheless, this does not prevent us from using final causality, but we need to look for a concept of purpose that does not imply the presence of a designer who anticipates the end to be realised. This is why Kant distinguishes between external and internal purposiveness:

1) External purposiveness takes place when an entity is oriented towards something else's utility: "one thing in nature serves another as the means to an end" (Kant 1908, § 82, p. 425; p. 293). Purposiveness, in this case, is "contingent in the thing itself to which it is ascribed" (Kant 1908, § 63, p. 368; p. 240).

---

[4] On the circular and paradoxical character of this kind of causality, see Barbagallo 2012, p. 252; Garelli 2008, pp. XXXII–XXXIII.

2) Internal purposiveness takes place when a single thing is simultaneously a "cause and effect of itself" (Kant 1908, § 64, p. 370; p. 243), that is to say, when the end of an object is the realisation of the object itself.

On the basis of this distinction, Kant rehabilitates the notion of purposiveness and saves it from the criticism of teleology, which characterised the birth of modern science. According to this criticism, humans usually interpret the natural world in the same way they interpret artefacts. This criticism is effective against the teleological approach built on external purposiveness, but not against the one based on internal purposiveness. A natural end conceived as a cause and effect of itself does not imply the presence of a designer that is external to the natural item and that anticipates the end that the item is meant to realise. Therefore, the inner purpose of a natural item is nothing separated from it, because it is linked to its way of being, namely, it is the realisation of what the item really is. The specificity of living beings as natural ends [als Naturzweck] is the relation between their parts and the whole as if each one was mutually cause and effect of the existence and the forms of all the others.

Nevertheless, according to Kant, we can use final causality only if we do that by analogy with the only way we can think of such a kind of causality, that is, in terms of technical-practical purposiveness:

> Nevertheless, teleological judging is rightly drawn into our research into nature, at least problematically, but only in order to bring it under principles of observation and research in *analogy* with causality according to ends, without presuming thereby to *explain* it (Kant 1908, § 61, p. 360; p. 234).

According to the analogical use, teleological judgment does not have a determining value with respect to the organisation of living beings, that is to say, it cannot shed light on the ontological status of what is living. Rather, it is only a reflective judgment, and thus it is employed "for guiding research into objects of this kind" (Kant 1908, § 65, p. 375; p. 247).

Therefore, Kant presents the possibility of a teleology based upon inner purposiveness in the consideration of living organism, albeit mainly with a regulative and heuristic function. This point is of crucial importance in the Kantian explanation of organisms, and it is also the point on which Hegel's criticism of Kant's view is based.

## 3 The Notion of Teleology in Hegel and Kant

In the teleology section of the *Science of Logic*,[5] Hegel offers a critical analysis of Kant's theory of teleology and purposiveness.[6] First of all, Hegel, as well as Kant, associates the perception of a purposiveness with the presence of a designer: "Where *purposiveness* is discerned, an *intelligence* [*Verstand*] is assumed as its author, and for the end we therefore demand the concept's [*des Begriffes*] own free existence" (Hegel 1981, p. 154; p. 734).[7] As in Kant's analysis, this presence implies the risk of going beyond the boundaries of scientific enquiry:

> The more the teleological principle was linked with the concept of an *extramundane* intelligence, and to that extent was favoured by piety, the more it seemed to depart from the true investigation of nature, which aims at cognizing the properties of nature not as extraneous, but as *immanent determinatenesses* and accepts only such cognition as a valid *comprehension* (Hegel 1981, p. 155; p. 735).

Secondly, Hegel contrasts teleology and mechanism, associating the first with the notion of final causality, and the second with the notion of efficient causality. Whereas purposiveness is the basis of an essential unity of the different determinacies of the object that it determines (their principle of unity affects their content), mechanism leaves the determinacies of the object as external and indifferent (their principle of unity does not affect their content).[8]

---

5 Cf. Manser 1986.
6 Hegel's analysis of Kant's theory of teleology starts with *Faith and Knowledge* (1802), especially in the pages dedicated to reflecting judgment. Then, in the *Phenomenology of Spirit* (1807), at the beginning of the section on reason, Hegel investigates the teleological relation. Hegel focuses his attention on teleology also in the *Science of Logic* (1816) and then in the corresponding paragraphs of the logic of the *Encyclopedia Logic* (§§ 153–160 in the 1817 *Enciclopedia*; §§ 204–212 in the 1830 *Enciclopedia*). For a detailed analysis of Hegel's reception of Kant's theory of teleology, cf. Findlay 1971, Wilkins 1974, Greene 1976, Verra 1981, Wohlfahrt 1981, Menegoni 1989, Baum 1990, Chiereghin, 1990, Düsing 1990, Giacché 1990, Souche-Dagues 1990; DeVries 1991, Stanguennec 1990, Lugarini 1992, Dahlstrom 1998, Pierini 2006, Garelli 2015.
7 I have elected to change Miller's translation of 'Begriff' from 'Notion' to 'concept'.
8 "In mechanism they become so through the *mere form of necessity*, their *content* being indifferent; for they are supposed to remain external, and it is only understanding as such that is supposed to find satisfaction in cognizing its own connective principle, abstract identity. In teleology, on the contrary, the content becomes important, for teleology presupposes a concept, something *absolutely determined* [*ein an sich und für sich bestimmtes*] and therefore self-determining, and so has made a distinction between the *relation* of the differences and their reciprocal determinateness, that is the *form*, and the *unity that is reflected into itself, a unity that is determined in and for itself* and therefore *a content*" (Hegel 1981, p. 156; p. 736).

Thirdly, Hegel recalls the importance of Kant's theory on the distinction between external and internal purposiveness in the exploration of the structure of living beings. At the same time, he criticises Kant's theory of the reflecting value ascribed to teleological judgment in nature:

> The concept, as end, is of course an *objective judgment*, in which one determination, the *subject*, namely the concrete concept, is self-determined, while the other is not merely a predicate but external objectivity. But the end relation is not for that reason a *reflective judgment* that considers external objects only according to a unity, *as though* an intelligence had given this unity *for the convenience of our cognitive faculty*; on the contrary it is the absolute truth that judges *objectively* and determines the external objectivity absolutely (Hegel 1981, p. 159; p. 739).

Hegel conceives of purpose as a concept that determines external objectivity. Thus, the problem of Kant's notion of internal purposiveness is that it is ascribed only a reflecting value, that is to say, it is conceived as if it was placed only in an intelligence that gave purposes for the convenience of our faculty of cognition. It is a purposiveness conceived according to the paradigm of utility and then it is reducible to external purposiveness.[9]

Therefore, since internal purposiveness in Kant's account is only a regulative and explicative principle, it is separated from objectivity, and it is only a finite, relative and subjective purposiveness, that always presupposes an intelligence as its source:

> Its finitude consists in its being confronted by an *objective*, mechanical and chemical *world* to which its activity relates itself as to something *already there*; its self-determining activity is thus, in its identity, immediately *external to itself* and as much reflection outwards as reflection-into-itself. To this extent end still has a genuinely *extramundane* existence – to the extent, namely, that it is confronted by this objectivity, just as the latter on the other hand confronts it as a mechanical and chemical whole not yet determined and pervaded by the end (Hegel 1981, p. 161; p. 742).

If the purpose is one that realises itself independently from an extramundane existence, it has to be an objective purpose, that "is in its own self the urge to realize itself" (Hegel 1981, p. 162; p. 742) or that "the teleological process is the *translation* of the concept that has a distinct concrete existence as concept into objectivity" (Hegel 1981, p. 167; p. 747). Therefore, purpose is a concept that realises itself in objectivity.

---

[9] Cf. Düsing 1990, p. 150.

Notably, in this context, the word 'concept' is not meant in the traditional way, namely as an idea placed in someone's intelligence. Rather, concept is an objective structure endowed with the impulse to its own realisation.[10] This concept, as purpose, is at the same time the principle guiding the teleological process and the result of the process itself.[11] This takes us back to the problem of backward causation that we encountered in Kant's analysis of teleology and that is pointed out by Hegel himself:

> It can therefore be said of the teleological activity that in it the end is the beginning, the consequence the ground, the effect the cause, that it is a becoming of what has become, that in it only that what already exists comes into existence, and so on (Hegel 1981, p. 167; p. 748).

Purposiveness, as objective and determining purposiveness, is internal purposiveness. Its realisation implies a self-determining process that always involves some kind of backward causational dynamic and, thus, a contradiction: something coming afterwards – the effect – is the cause of something coming backwards – or is the process that accomplishes it.

As we saw, Kant's strategy is to escape the two horns of the problem implied by the use of final causality, that is to say, the contradiction of backward causation or the dissolution of this contradiction through the reference to an extra-naturalistic element anticipating the end to be realised. Kant's way out of these two alternatives is the notion of an internal purposiveness that is ascribed only as a regulative value in scientific enquiry.

Hegel shows that this strategy fails to effectively account for a proper notion of internal purposiveness. The reflecting value of teleological judgment in nature and the regulative role that internal purposiveness plays in scientific enquiry implicitly leads internal purposiveness back to the conceptual pattern of external purposiveness. Hegel's strategy instead is not to escape the backward causation problem, but to recognise that inner purposiveness, if it is meant to be determining and objective, necessarily implies backward causation and its contradictory consequences:

> Since the concept here in the sphere of objectivity, where its determinateness has the form of *indifferent externality*, is in reciprocal action with itself, the exposition of its movement here becomes doubly difficult and involved [*verwickelt*], because this movement is itself doubled and a first is always a second also. In the concept taken for itself, that is in its subjectivity, its difference from itself appears as an *immediate* identical totality on its own ac-

---

10 Cf. Illetterati 1995a, p. 234.
11 Cf. Zebina 2015, p. 100.

count; but since its determinateness here is indifferent externality, its identity with itself in this externality is also immediately again self-repulsion, so that what is determined as external and indifferent to the identity is the identity itself; and the identity as identity, as reflected into itself, is rather its other. Only by keeping this firmly in mind can we grasp the objective return of the concept into itself, that is, the true objectification of the concept (Hegel 1981, p. 171; p. 752).

Purpose – "the concept" – determines itself in something that is indifferent to and other than itself, namely in indifferent externality. Therefore, it needs to negate itself, in order to realise its identity in this other. This self-negation corresponds to what Hegel defines as "self-repulsion" of the purpose. What is external and indifferent to the first immediate identity of the purpose becomes identical with this identity in the process of realisation of the purpose itself. The first immediate identity of the purpose becomes a concrete identity precisely through this process of realisation in its other. In this sense, Hegel claims that the determinateness of the purpose "now possesses within itself the *determinateness of externality*, and its simple unity is consequently the unity that repels itself from itself and in so doing maintains itself. End therefore is the subjective concept as an essential effort and urge to posit itself externally" (Hegel 1981, p. 160; p. 740).

The contradictory implications of the backward causation dynamic embodied in the self-repelling move of the purpose are not made explicit in the treatment of teleology in the *Science of Logic*. On this point, Hegel is much clearer in the *Encyclopedia Logic:*

> As this contradiction of its identity with itself opposite the negation and the opposition posited in it, it [purpose] is itself the sublating, the *activity* of so negating the opposition that it posits it as identical with itself. This is the *process of realizing the purpose* in which, by rendering itself something other than its subjectivity and objectifying itself, it has sublated the difference of both, has joined *itself* together *only with itself* and has *preserved* itself (Hegel 1992, § 204, p. 209; p. 276).

The contradiction Hegel refers to in the *Encyclopedia Logic* is precisely the contradiction implied by the backward causation dynamic at the basis of internal purposiveness, where the purpose realised at the end in objectivity is, at the same time, what guides the very process of realisation of the purpose itself. The best and clearest example of this kind of contradiction takes place, as Hegel points out, in living organisms:

> Need and drive are the examples of purpose lying closest at hand. They are the *felt* contradiction that takes place *within* the living subject itself and they enter into the activity of negating this negation that is still mere subjectivity. The *satisfaction* produces the peace be-

tween the subject and object, in that the objective dimension standing *over there* in the still on hand contradiction (to the need) is equally sublated with respect to this. Its one-sidedness, through the unification with the subjective dimension (Hegel 1992, § 204 R, p. 210; p. 227).

The analysis of Kant's theory of teleology and of Hegel's critical analysis provided the conceptual background for shedding light on Hegel's account of organism in his *Philosophy of Nature*, which will be the focus of the next section of the article. I will give particular attention to the assimilation process and to the way internal teleology orients it. In this process, in fact, the way organism maintains itself as living in the relation to its other – external environment – will allow me make explicit the relation of both exclusion and integration of otherness in the organism itself, and thus the dialectical character of Hegel's account of living beings.

# 4 Organic Physics. Teleology and Philosophy of Nature

Assimilation is analysed in the third part of Hegel's *Philosophy of Nature*, which is organic physics. Internal purposiveness assumes a constitutive function in this part of Hegel's philosophical system. Already in the introductory section of organic physics, Hegel notes that

> Mechanism constituted the first part of the philosophy of nature, chemistry constituted the apex of the second part, and teleology constitutes this third part (see § 194 Add. 2). Life is a means, but for this concept, not for another; it is perpetually reproducing its infinite form. Kant had already determined living existence as constituting its own end. There is change here, but it is only present on behalf of the concept, for it is merely the otherness of the concept that changes. It is only in the absolute negativity of this negation of that which is negative, that the concept can remain in communion with itself (Hegel 1970, § 337, p. 339; p. 11).[12]

Said another way, "[...] life has its contrary within itself however, and in itself it is a rounded totality [*abgerundete Totalität*], or its *own end* [*Selbstzweck*]" (Hegel

---

**12** Similarly, in the *Science of Logic*, Hegel states that "since the concept is immanent in it, the *purposiveness* of the living being is to be grasped as *inner*; the concept is in it as determinate concept, distinct from its externality, and in its distinguishing, pervading the externality and remaining identical with itself" (Hegel 1981, p. 184; pp. 766).

1970, § 338 Z, p. 339; p. 11 (trad. mod.)).[13] Teleological orientation distinguishes what is alive from what is not.[14] The end – which is internal to the organism and which organises and directs its development – is what turns the organism itself into a totality where each part is simultaneously the end and means of all the other parts, and which thus establishes an immanent relationship between them. Hence, the organism is an integrated whole. What is not alive, on the contrary, is articulated not on the basis of an inner purpose, but as something composed of parts which are not properly members of a whole, because each part maintains a kind of self-subsistence.

Inner purposiveness then characterises the whole realm of living beings, namely geological nature, vegetable and animal organisms. In the mineral kingdom, however, inner purposiveness exists only as a precondition, that is to say, as a condition of possibility for the development of life.[15]

Vegetable organisms are also characterised by a teleological dynamic that is, however, not completely internal. Actually, the development of plants depends on light, which is their source of energy and warmth, namely their principle of determination. Light is not immanent to them, since it comes from a source that stands outside plants themselves. Only in animal organisms is the purpose really internal and so they can properly be said to be able to determine themselves. In other words, animals are the kind of organisms where life actually comes into existence. Unlike minerals and plants, animals have the "sun in itself", or, they have the principle of their determination within themselves:

> In the animal, light has found itself, for the animal checks its relationship with an other. The animal is the self which is for the self, it is the existent unity of differences, and pervades their distinctness. The plant's tendency towards being-for-self gives rise to the plant and the bud, which are two independent individuals, and are not of an ideal nature. Animal being consists of these two posited in unity. The animal organism is therefore this duplication of subjectivity, in which difference no longer exists as it does in the plant, but in which

---

13 Hegel describes life as "the concept that, distinguished from its objectivity, simple within itself, pervades its objectivity and, as its own end, possesses its means in the objectivity and posit the latter as its means, yet is immanent in this means and is therein the realized end that is identical with itself" (Hegel 1981, p. 177; p. 760). Cf. also Hegel 1981, p. 187; p. 769.
14 On Hegel's treatment of the notion of life cf. Düsing 1986a, d'Hondt 1986, Vieillard-Baron 2004, Arndt, Cruysberghs, Przylebski, Fischbach 2006, Spieker 2009, Sell 2014, Ng 2018.
15 "There are accordingly flashes of life in the earth and the elements surrounding the earth which do not develop into lasting individuality, or the even longer lasting life of the species, but represent something more than chemical mechanical materiality, having perhaps a spontaneity of origin and a brief self-maintenance of which inert matter is incapable" (Findlay 1984, p. 95).

only the unity of this duplication attains existence. True subjective unity exists in the animal (Hegel 1970, § 350 Z, p. 430; p. 102).

Animal organism is a subjectivity, namely an integrated unity which is able to determine itself in its other. Another way of saying this is that animals are organisms that are able to realise their own nature through the immanent relationship with the external environment.[16] The subjectivity of plants, instead, is only formal: since plants don't have their centre in themselves, their parts maintain a certain independence which is not present in the world of animal organisms.[17] Therefore, the difference between animals' concrete subjectivity and plants' formal subjectivity depends on the different way in which animals determine themselves in relation to the external environment. For example, in the assimilation process, the animal is able to choose the natural product from which it feeds and it can interrupt the assimilation when it does not need it anymore. Moreover, it transforms the assimilated material into its own subjectivity through the process of digestion. In this way, the animal's relationship with the external environment is mediated, whereas the plants' relationship with nature around them is more simple and immediate. Plants cannot choose what they assimilate to feed themselves, nor can they stop the process of assimilation, because they directly assume what the environment offers to them, namely water, air and light. These features are related to another difference between animals and plants: animals have sensibility, they are able to perceive external stimulations and they can feel what they need from the external environment. Moreover, animals are able to determine themselves also in space, since they can move within it on the basis of their needs (food, water, rest, etc.). On the contrary, the position and the movement of plants is determined by external forces, such as light and gravity.[18] Therefore, the animals' independence from the external environment is higher than the plants' one. This is testified also by their capacity to maintain bodily temperature and being able to emit sounds with which they

---

[16] "Organic individuality exists as *subjectivity* in so far as the externality proper to shape is *idealized* into members, and in its process outwards, the organism preserves within itself the unity of selfhood" (Hegel 1970, § 350, p. 430; p. 102). "Even when the organism's activity, starting from the *need* it is experiencing, moves away from its singularity and towards what is other, it always realizes itself. This means that the animal, in its inward activity, has a movement that, in moving outward, always has in itself its objective and its centre. This makes it a *subject*" (Illetterati 2014a, p. 157).

[17] Plants are unaccomplished subjectivity, that is a totality that is not yet able to involve determinate difference in itself. Cf. (Illetterati 1995a, p. 167).

[18] Self-movement represents a crucial moment of rupture of the immediate unity of the living organism with external environment (cf. Hösle 1988, p. 325).

can express themselves. In addition, animals reproduce themselves through sexual relations, that is, through the relation with another animal individual of the same species, whereas the plants' reproductive process does not imply any relation with another individual, but occurs within the same single organism. Finally, the animals' process of formation is quite distinct from that of plants. The seed already contains the preformed structure of the organism: the plants' development is only a growth, since they don't go through the process of self-differentiation of internal organs that we find in animals.[19] While in plants the beginning and end of the process of formation are quite the same, in animals the beginning and end of the development are different, and at the same time they are also identical. In fact, the structure of the animal organism at the end of its formation is constituted by nothing different from the principle that guides and orients the first steps of its development. Therefore, this very development has a circular dynamic that is not fully transparent in the process of the formation of plants.[20] This circular dynamic is highlighted by Hegel, for example, in the following passages:

> Living existence has *being*, and preserves itself only as this reproductiveness, not as mere being [*Seiendes*]. It has being only because it turns itself into what it is. It is a pre-existent end, and is itself merely result (Hegel 1970, § 352, p. 107).
> As an intro-reflected unity of various singularities, the animal exists as a spontaneously self-producing end, and is a movement which returns into its *particular* individuality (Hegel 1970, § 351 Z, p. 107).
> As animal life is its own product and purpose [*Selbstzweck*], it is simultaneously both end and means. The end is a determination which is of an ideal nature, and which is already present as an antecedent. As the realizing activity which then occurs has therefore to conform to the determination which is present, it brings forth nothing new, so that the realization is to an equal extent a return-into-self. The accomplished end has the same content as that which is already present in the activation [...]. The organization itself is to an equal extent both end and means, for it consists of nothing which is subsistent. As the viscera and the members in general are in a state of reciprocal activity, they are always posited as being of an ideal nature, and as each member is a centre which produces itself at the expense of all the others, it has existence only by means of the process. In other words, that which is sublated and reduced to means, is itself end and product (Hegel 1970, § 352 Z, p. 108).

---

**19** Nevertheless, according to Hegel, plants have parts that can transform into other parts of the same individual, and so their structure is constituted of what can be called 'indifferent' parts, such that "when a branch or twig is broken off and layered, it puts forth roots and constitutes a whole plant" (Hegel 1970, § 345 Z, p. 383; p. 56).
**20** On the difference between animal and vegetal organisms cf. Hösle 1987, Bach 2004.

The living being is not something given, since it develops itself and its development as oriented by an inner purpose. Thus, this self-determination has a teleological character where the end of the formation process is also its guiding principle. It should now be clear that this teleological process does not have the linear structure of efficient causality, because it incorporates the circular structure of final causality that I have expounded upon in the first part of the article. Such circular structure corresponds to the one characterising backward causation.

To analyse this structure in detail, we must then look at the specific dialectic that characterises the way the animal organism determines itself. The dialectic of the animal organism is constituted of three moments: a) the formation (shape), that is the animal's relation to itself; b) the assimilation process, that is the animal's relation to inorganic nature; and c) the generic process, that is, the animal's relation to another animal organism. I will focus on the second moment of this dialectic, which is one in which the dynamics of internal purpose and its contradictory consequences are more transparent.

## 5 The Assimilation Process. Teleology and Contradiction

The assimilation process is that through which "the organism must [...] posit the subjectivity of externality, appropriate it, and identify it with its own self" (Hegel 1970, § 357 Z1, p. 136).[21] In this process there is the living individual on the one hand, and the external environment on the other. External environment is given as pre-existent and, consequently, it is independent from the organism. At the same time, however, the assimilation process, that is, the organism's tendency to appropriate the external environment, is nothing but the negation of this very independence. In this sense, the assimilation process implies the contradiction of the inorganic. This contradiction arises on the basis of the tension between the independence of the environment from the organism on the one hand and its constitutive relation to it on the other.[22]

---

**21** According to Breidbach: "Hegel definiert den Organismus, wie er sich im Tier realisiert, als ein geschlossenes sich in sich setzendes Ganzes, das die Materie einbildet (assimilier), das das Assimilierte und sich in dieser Assimilation dann auch selbst formiert und so in eine eigene Prozeßbestimmtheit (sein Selbst) überführt" (Breidbach 2004, p. 219).
**22** "Organic being is orientated towards externality to the same extent that it is internally strung in opposition to it, and this consequently gives rise to the contradiction of this relationship, in

This contradiction has its roots in another contradiction, which stands at the basis of the assimilation process, that is, the contradiction of need [*Bedürfnis*]. In needs, the relation of both exclusion and integration of Hegel's dialectical account of organism fully comes to light. In this dialectic and in its contradictory structure, inner purposiveness has a crucial function. I will show how this occurs in this last part of this article.

In need, the animal has its other in itself since it feels it in itself as the other that it needs, as something that is negated and that it tends to appropriate. Therefore, in the first place, need is characterised by a negative structure: "these determinations [*impulse, instinct, need*] [...] are negations posited as contained within the affirmation of the subject itself" (Hegel 1970, § 359 Z, pp. 141–142).[23] The other of the organism – its negation – is not indeterminate, because it is the specific other that the organism needs:

> The animal can be stimulated only by means of *its* inorganic nature, because for the animal, the only opposite is its *own*. The animal does not recognize the other in general, for each animal recognizes its *own* other, which is precisely an essential moment of the special nature of each (Hegel 1970, § 361 Z, p. 147).

The negation at the basis of need is thus a determinate negation, and it has a constitutive function with respect to the way in which the organism articulates itself.

In the second place, this negation is not simply the other of the organism, but it is an other which stands in the organism itself. This is shown by the pain felt by the organism. As Mure points out, "that the tension of unsatisfied want is painful is the clearest indication that the wanted object lies within the subject" (Mure, 1950, p. 659).[24] The other stands in the organism in its radical

---

which two independent beings come forth in opposition to each other, while at the same time externality has to be sublated" (Hegel 1970, § 357 Z, p. 136).

**23** Hegel also claims that "as the animal is a true self, which is for itself and has attained to individuality, it establishes its separateness and particularity. It detaches itself from the universal substance of the Earth, which has an external determinate being for it. For the animal, the externality which has not come under the domination of its self, is a negative and indifferent being" (Hegel 1970, § 351 Z, p. 106).

**24** Düsing writes: "Das lebendige Individuum ist damit selbst 'absoluter *Widerspruch*'; es findet sich als negativ bestimmt, und es negiert jene Welt und ist dadurch positiv identisch mit sich. Dieser Widerspruch ist in ihm gegenwärtig als grundlegender '*Schmerz*', der alles Lebendige durchzieht. In diesem Schmerz gründen das Bedürfnis, in dem das Lebewesen sein Negiertsein, seinen Mangel fühlt, und der Trieb, der auf die Negation der äußeren Welt und die Gewinnung der positiven Identität mit sich geht" (Düsing 1986a, p. 286).

otherness, because it is in it as something negated. In need, the other of the organism is found in the organism as the presence of the absence.

In the third place, the negative, that is what is needed, is not only found in the organism itself, but it is also perceived as negative by the animal itself as what it needs to appropriate:

> Only a living existence is aware of *deficiency*, for it alone in nature is the *concept*, which is the unity *of itself* and *its specific antithesis*. Where there is a *limit* [*Schranke*], it is a negation, but only *for a third term*, an external comparative. However, the *limit* constitutes deficiency only in so far as the *contradiction* which is present in *one term* to the same extent as it is in the *being beyond it*, is as such immanent, and is posited within this term (Hegel 1970, § 359 A, p. 141).[25]

The other that the animal needs is not only a limitation, a defect, something missing that is perceived as such only from someone external to the organism itself. The animal itself perceives this need, this negation, and the necessity of overcoming this inner lack within itself.[26] In order to do that, the organism has to appropriate this negative and turn it into a positive, into something of its own:

> The *process* [...] begins with the self's internal diremption, the awareness of externality as the *negation* of the subject. The subject is, at the same time, positive self-relatedness, the *self-certainty* of which is opposed to this negation of itself. In other words, the process begins with the awareness *of deficiency*, and the *drive* to overcome it (Hegel 1970, § 359, p. 141).

At this point of Hegel's analysis, the dialectical character of the organism becomes transparent. In the assimilation process, otherness is, at the same time, excluded and included in the organism itself, and this occurs due to the inner purpose orienting the assimilation process.[27] As Illetterati points out, "also

---

**25** In another passage of the text, Hegel writes the following: "it is the subject's feeling of dependence which is primary here; the subject feels that it is not self-contained, and that another negative being is necessary to it, not contingent. This is the unpleasant feeling of need [*Bedürfnis*]. The deficiency in a stool which has three legs is in us. In life, the deficiency is in life itself however, although to an equal extent it is also sublated, for life is aware of the limit [*Schranke*] as a deficiency" (Hegel 1970, § 359 Z, pp. 490–491).
**26** The capacity of the organism to feel itself and what it needs plays a crucial function in the process or self-preservation of the organism itself. Cf. Klotz 2010, p. 235.
**27** "The organism is an end in itself, because it assimilates alien material which fuels the metabolism through which it keeps itself alive. In this sense, the organism represents an *autoteleological structure*, i.e., a *self-regulating system*" (Brinkmann 1996, p. 141). Cf. also Ferrini 2009, p. 77.

when the activity of the organism is wholly oriented towards what is outside, it always tends to realize itself as individual" (Illetterati 1995a, p. 174).[28] In this sense, the end of the organism is the unity of itself with the other that it needs, since this unity is what allows the organism to realise itself, that is to say, it allows the development and the preservation of its bodily structure. This unity is the principle that orients the assimilation process:

> The need is specific, and its determinateness is a moment of its universal concept [...]. The drive is the activity of overcoming the deficiency of such a determinateness, i.e., of overcoming its form, which is initially merely subjective. In that the content of the determinateness is primary, and in that it maintains itself in the activity in which it is merely carried into effect, the drive constitutes purpose (§ 204), and confined solely to living existence, instinct. This deficiency in form is inner stimulation, the specific determinateness of whose content appears at the same time as an animal's relation to the particular individualization of the various spheres of nature (Hegel 1970, § 360, p. 145).

Thus, the assimilation process is not reducible to the simple mechanical or chemical appropriation of external objectivity by animal organism. It does not correspond to the simple and immediate transposition of external objectivity within the organism itself. Rather, the material which is assimilated is transformed by the organism through digestion, which is the mediation necessary in order for the organism to turn the material into something of its own, that is to say, into something able to serve its inner purpose, which is the development and preservation of its bodily structure.

The reference to Kant's notion of inner purposiveness in Hegel's account of need is not then accidental:

> The fundamental determination of living existence is that it is to be regarded as acting purposively. This has been grasped by Aristotle, but has been almost forgotten in more recent times. Kant revived the concept in his own way however, with the doctrine of the inner purposiveness of living existence, which implies that this existence is to be regarded as an end in itself. The main sources of the difficulty here, are that the relation implied by purpose is usually imagined to be external, and that purpose is generally thought to exist only in a conscious manner. Instinct is purposive activity operating in an unconscious manner (Hegel 1970, § 360 A, p. 145).

Hegel's main point is that the animal organism is an end in itself exactly on the basis of the Kantian model of inner purposiveness. Nevertheless, Hegel's theory of inner purposiveness is distinct from Kant's. There is no space for the analogy

---

**28** This self-relation based on inner purposiveness is what allows the self-preservation of the structure of the organism in the relation to external environment. Cf. Marques 2016, p. 123.

with the purposiveness characterising the technical-practical sphere of artefacts, i.e., a purposiveness that is thought to exist in a conscious manner. The revolution of Hegel's model of teleology is the possibility to think of an unconscious model of finalistic activity.[29] To this extent, Hegel does not adopt the strategy of anticipation of the end that would solve the problem of the circular and paradoxical structure of backward causality:

> One must not think of the purposive action of nature as self-conscious understanding. Each step involved in the consideration of nature is out of the question unless the end has been perceived. The end is precisely that which is predetermined and active, and which preserves itself by assimilating the other to which it relates itself (Hegel 1970, § 365 Z, p. 167).[30]

The assimilation process does not have the linear structure of mechanical processes.[31] On the contrary, assimilation has the circular structure where the end of the process – its effect, that is the unity of the organism with what it needs – is the very cause of the process itself.[32] At the beginning of the process, what is needed is a negative that is already constitutively present in the organism itself as what the organism is lacking. This presence of the otherness in the organism is precisely what pushes the organism itself to assimilate what it needs in order to realise and make explicit its unity with this otherness, which, at the beginning of the process, was only implicit.

Using Hegel's terms, we could say that need is not only a first abstract negation. In need, the organism is not simply negated. Need is also what Hegel calls 'second negation', since it is a negation through which the living organism realises itself. At the beginning of the assimilation process the animal is united with the otherness that it needs in so far as this otherness is posited in it as negated. The result of the assimilation process is the concrete realisation of this unity: it is

---

[29] Cf. Findlay 1984, p. 98.
[30] Marmasse points out that: "Selon l'auteur de l'*Encyclopédie*, la finalité interne de l'organisme ne renvoie pas à l'esprit mais peut être comprise à partir de seules ressources de la nature. Pour lui, l'auto-médiation ne requiert pas une activité spirituelle [...] mais simplement le retour à soi de l'être naturel, de telle sorte qu'el en vienne à constituer son propre principe et à échapper à tout assujettissement à l'égard de ce qui est extérieur à son corps. [...] Chez Hegel, le télos qui caractérise l'organisme est purement naturel" (Marmasse, 2008, p. 291).
[31] "Assimilation itself is the enveloping of the externality within the unity of the subject. Since the animal is a subject, a simple negativity, the nature of this assimilation can be neither mechanical nor chemical, for in these processes, the substances, as well as the conditions and the activity, *remain external* to one another, and lack an absolute and living unity" (Hegel 1970, § 363, p. 151).
[32] Cf. Zebina, 2010, p. 160.

a unity with external objectivity that, in so far as it is something other than organism itself, is excluded from the organism; but, at the same time, it is an otherness which is properly interiorised and integrated in the structure of the organism through the process of digestion.

In this way, need is the presence of the absence, but it is also that which moves the animal toward its other, in order to assimilate it, that is to say, it is that through which animals turn the presence of the absence into a concrete presence. In the process raised by need, the animal organism concretely integrates in itself the other than itself. The organism and its other are one and the same thing:

> The organism [...] separates itself from the one-sided subjectivity of its *hostility* towards the object, and so becomes *explicitly* what it is implicitly. It therefore becomes the non-neutral identity of its concept and its reality, and so finds the end and product of its activity to be the already established beginning and origin of its being. It is thus that *satisfaction conforms to reason*; the process which enters into external differentiation turns into the process of the organism with itself, and the result is not the mere production of a means, but the bringing forth of an end, the unity of the self (Hegel 1970, § 365 A, p. 155).

The contradiction of need in the assimilation process is the structure of a specific tension, that is the one between the organism initially oriented by inner purposiveness towards the appropriation of exteriority on the one hand and the organism that realises this appropriation and the unity with exteriority on the other. The first one is the organism separated from the external objectivity that it needs. The second one is the organism that has assimilated external objectivity and thus has realised its inner purpose, namely the development and preservation of its bodily structure, and this occurs precisely through the concrete unity with this objectivity. The two organisms are distinct, but they are also the same. They are distinct in so far as the structure of the animal organism before and after the assimilation process is different. They are the same in so far as the organism that performs assimilation is identical with itself in this process of self-determination, which is a process of self-realisation: "From pain begin the *need* and the *impulse* that constitute the transition by which the individual, in being for itself the negation of itself, also becomes for itself identity – an identity which only is as the negation of that negation" (Hegel 1981, p. 685). Simply said, in the assimilation process, the organism is identical with itself in its own process of self-differentiation.

Thus, in the assimilation process, the organism realises its concrete identity with itself, which is gained through a constitutive process of self-differentiation. This identity through difference represents the core of Hegel's dialectical account of the animal organism, where the external objectivity is – for the organism –

negated, excluded, needed, but it is also interiorised and integrated by the living organism. This dialectic is oriented by an inner purposiveness that is the main feature of what is living, and its core is represented by the contradiction marking the organisation of the animal organism. A living being is thus a being that remains identical with itself in its process of self-differentiation, that is to say, it is a being which is able to determine itself. This is why Hegel claims that the organism is a 'subject': "The *subject* is a term such as this, which is able to contain and *support* its own contradiction" (Hegel 1970, § 359 A, p. 141).

Because of this dialectic, living organisms are not individuals that are simply, immediately and abstractly identical with themselves. Abstract identity, which excludes the presence of otherness, far from preserving the inner structure of the living organism, necessarily implies its death. The organism needs the external environment in order to develop and maintain its internal articulation. Hegel's dialectical account of living beings is meant to shed light on precisely this point. The contradiction of need, drive, instinct, and the way in which the organism feels what it needs, therefore, far from being the point where the self-determining process of organism loses its consistence, is the guiding principle of its self-determination, self-movement, and of its self-realisation. This contradiction stands at the basis of the constitution of everything which is living: "its self-determination has therefore the form of objective externality, and since it is at the same time self-identical, it is the absolute *contradiction*" (Hegel 1981, p. 684).[33]

# 6 Conclusions

The contradiction that Hegel's dialectic refers to is not the symptom of a mistake in the account of the organism. Rather, as we have seen, it corresponds to the constitutive structure underlying Hegel's dialectical account of the organism itself. This structure is implied by the immanent relation with the external environment in which the living organism is necessarily involved and by the teleological dynamic that orients this very relation. Teleology is necessary in order to account for the realm of living beings, which is not reducible to the linear character of mechanic and chemical laws. The application of these laws is not mistaken, but it provides only a partial and one-sided explication of the organisation characterising living beings.

---

**33** "Den ‚Widerspruch' versteht die sp.-d.L. [= spekulativ-dialektische Logik, A.S.] nicht im Sinne einer allgemeinen Aufhebung des Widerspruchsverbots […], sondern als die ontologische Bestimmung der negativen Einheit entgegengesetzter Begriffsmomente, die, gegenüber der abstrakten Identität, das Prinzip aller Bewegung und Lebendigkeit bildet" (Jaeschke 1980, p. 391).

Thus, in order to provide an effective explication of what living beings are, the reference to mechanical and chemical laws needs to be integrated with other kinds of laws which allow for accounting to the specific features of life. In this article, I showed that these laws have a teleological character and that they involve some kinds of paradoxical dynamics, where the living organism is in a relation of exclusion, but at the same time of integration of what is other than itself.

Therefore, from a Hegelian perspective, the only option left to us is to leave the coherent and linear meshes of standard logic – which Hegel calls the logic of understanding [*Verstand*] – in which there is no room for the paradoxical dynamics such as that of backward causality. This rupture with standard thought does not necessarily mean the reference to vitalism or to some kind of principle standing outside the scope of scientific research and rationality. On the contrary, Hegel invites us to extend the sphere of rationality and to articulate a new paradigm of reason [*Vernunft*], which is to be the paradigm of speculative thinking:

> *Speculative thought* consists only in this, in holding firm to contradiction and to itself in the contradiction, but not in the sense that, as it happens in ordinary thought, it would let itself be ruled by it and allow it to dissolve its determinations into just other determinations or into nothing (Hegel 1978, p. 383).

In this sense, speculative thought is nothing but dialectical thinking, that is to say, it is a new paradigm of rationality which pretends to be a type of thinking capable of accounting for complex phenomena such as life. In order to do that, dialectical thinking must be open to the challenge of conceiving contradiction as the constitutive structure of life and of the self-developmental process that characterises it.[34]

Thus, Hegel's dialectic can shed light on the complex phenomena characterising reality in general and, more specifically, life. This can only result from the effort to develop a logic that reflects and attempts to make explicit the circular, paradoxical and contradictory structures characterising the organisation of living beings and which is thus not only an abstract logic of our thinking, but a concrete logic that permeates and structures reality in its most complex and articulated forms, that are the forms of natural and spiritual life.

---

[34] On the ontological and determining function of contradiction in Hegel's logic cf. Bordignon 2015.

# References

Arndt, Andreas/Cruysberghs, Paul/Przylebski, Andrzej/Fischbach Franck (Hrsg.) (2006): *Das Leben denken*. Hegel Jahrbuch, Berlin: Akademie Verlag.
Bach, Thomas (2004): "Leben als Gattungsprozeβ. Historisch-systematische Anmerkungen zur Unterscheidung von Pflanze und Tier bei Hegel". In: Wolfgang Neuser/Vittorio Hösle (Hrsg.): *Logik, Mathematik, Natur im objektiven Idealismus*. Wurzburg: Königshausen & Neumann, pp. 175–190.
Barbagallo, Ettore (2012): "Alterità e autodeterminazione del vivente nel pensiero di Hegel. La dialettica e la finalità interna". In *ACME – Annali della facoltà di Lettere e Filosofia dell'Università degli Studi di Milano* 65(2), pp. 241–260.
Baum, Manfred (1990): "Kants Prinzip der Zweckmäßigkeit und Hegels Realisierung des Begriffs". In: Hans-Friedrich Fulda/Rolf-Peter Horstmann (Hrsg.): *Hegel und die "Kritik der Urteilskraft"*. Stuttgart: Klett-Cotta, pp. 158–173.
Beisbart, Claus (2009): "Kant's Characterization of Natural Ends". In *Kant Yearbook* 1, pp. 1–30.
Bordignon, Michela (2015): *Ai limiti della verità. Il problema della contraddizione nella logica di Hegel*. Pisa: ETS.
Breidbach, Olaf (2004): "Überlegungen zur Typik des Organischen in Hegels Denken". In: Wolfgang Neuser/Vittorio Hösle (Hrsg): *Logik, Mathematik, Natur im objektiven Idealismus*. Wurzburg: Königshausen & Neumann.
Breitenbach, Angela (2009): "Teleology in Biology. A Kantian Approach". In *Kant Yearbook* 1, pp. 31–56.
Breitenbach, Angela (2006): "A Mechanical Explanation of Nature and its Limits in Kant's Critique of Judgment". In *Studies in History and Philosophy of Science* 37(4), pp. 694–711.
Brinkmann, Klaus (1996): "Hegel on the Animal Organism". In *Laval théologique et philosophique* 52(1), pp. 135–153.
Chiereghin, Franco (1990): *Finalità e idea della vita. La recezione hegeliana della teleologia di Kant*. In *Verifiche* 19(1–2), pp. 127–229.
Cohen, Alix A. (2007): "A Kantian Stance on Teleology in Biology". In *South African Journal of Philosophy* 26(2), pp. 109–121.
Dahlstrom, Daniel (1998): "Hegel's Appropriation of Kant's Account of Teleology in Nature". In: Stephen Houlgate (Ed.): *Hegel and the Philosophy of Nature*. Albany, N.Y.: State University of New York, pp. 167–188.
DeVries, Willem (1991): "The Dialectic of Teleology". In *Philosophical Topics* 19(2), pp. 51–70.
D'Hondt, Jacques (1986): "Le concept de la vie, chez Hegel". In: Rolf-Peter Horstmann/Michael John Petry (Hrsg.): *Hegels Philosophie der Natur*. Stuttgart: Klett-Cotta, pp. 138–150.
Düsing, Klaus (1986a): "Die Idee des Lebens in Hegels Logik". In: Rolf-Peter Horstmann/Michael John Petry (Hrsg.): *Hegels Philosophie der Natur*. Stuttgart: Klett-Cotta, pp. 276–289.
Düsing, Klaus (1986b): *Die Teleologie in Kants Weltbegriffe*. Bonn: Grundmann.
Düsing, Klaus (1990): "Naturteleologie und Metaphysik bei Kant und Hegel". In: Hans-Friedrich Fulda/Rolf-Peter Horstmann (Hrsg.): *Hegel und die "Kritik der Urteilskraft"*. Stuttgart: Klett-Cotta, pp. 139–157.

Engelhardt, Dietrich von (1986): "Die biologischen Wissenschaften in Hegels Naturphilosophie". In: Rolf-Peter Horstmann/Michael John Petry (Hrsg.): *Hegels Philosophie der Natur*. Stuttgart: Klett-Cotta, pp. 121–137.

Ferrini, Cinzia (2009): "From Geological to Animal Nature in Hegel's Idea of Life". In *Hegel Studien* 44, pp. 45–93.

Findlay, John N. (1984): "Hegelian Treatment of Biology and Life". In: Robert S. Cohen/Marx W. Wartofsky (Eds.): *Hegel and the Sciences*. Dordrecht – Boston – Lancaster: Reidel Publishing Company, pp. 87–100.

Findlay, John N. (1971): "Hegel's Use of Teleology". In: Warren E. Steinkraus (Ed.): *New Studies in Hegel's Philosophy*. New York: Holt, Rinehart and Winston, pp. 92–107.

Fisher, Mark (2008): *Organisms and Teleology in Kant's Natural Philosophy*. Atlanta: Emory University.

Friedman, Michael (1992): "Regulative and Constitutive". In *The Southern Journal of Philosophy* 30, pp. 73–102.

Garelli, Gianluca (2015): *Dialettica e interpretazione. Studi su Hegel e la metodica del comprendere*. Bologna: Pendragon, pp. 177–193.

Garelli, Gianluca (2008): "Lo spirito dell'inquietudine". In: G.W.F. Hegel, *Fenomenologia dello spirito*, a cura di Gianluca Garelli. Torino: Einaudi, pp. XXXII–XXXIII.

Giacché, Vladimiro (1990): *Finalità e soggettività. Forme del finalismo nella "Scienza della logica" di Hegel*. Genova: Pantograf.

Ginsborg, Hannah (2006): "Kant's Biological Teleology and its Philosophical Significance". In: Graham Bird (Ed.): *A Companion to Kant*. Oxford: Blackwell, pp. 455–469.

Ginsborg, Hannah (2001): "Kant on Understanding Organisms as Natural Purposes". In: Eric Watkins (Ed.): *Kant and the Sciences*. Oxford: Oxford University Press, pp. 231–258.

Goy, Ina (2008): "Die Teleologie in der organischen Natur (§§ 64–68)". In: Otfried Höffe (Ed.): *Immanuel Kant. Kritik der Urteilskraft*. Berlin: Akademie Verlag, pp. 223–239.

Goy, Ina (2014): "Kants Theory of Biology and the Argument from Design". In: Ina Goy, Eric Watkins (Eds.): *Kant's Theory of Biology*. Berlin/New York: Walter de Gruyter, pp. 203–220.

Greene, Murray (1976): "Hegel's Modern Teleology". In: Ute Guzzoni/Bernhard *Rang*/Ludwig Siep (Hrsg.): *Der Idealismus und seine Gegenwart*. Hamburg: Meiner, pp. 176–192.

Hegel, Georg Wilhelm Friedrich (1978): *Wissenschaft der Logik*. Erster Band: *Die objektive Logik*. Zweites Buch. *Die Lehre vom Wesen* (1812–16). In: *Gesammelte Werke*. In Verbindung mit der Deutschen Forschungsgemeinschaft, hrsg. v. der Rheinisch-Westfälischen Akademie der Wissenschaften, Hamburg: Meiner, 1968- (from now on *GW*). Bd. 11, hrsg. von Friedrich Hogemann e Walter Jaeschke. Hamburg: Meiner, pp. 233–409; transl. by Arnold V. Miller. *Science of Logic*. London: Allen & Unwin; New York: Humanities Press 1969, pp. 387–571.

Hegel, Georg Wilhelm Friedrich (1981): *Wissenschaft der Logik*, Zweiter Band: *Die subjektive Logik. Die Lehre vom Begriff* (1816). In: *GW*. Bd. 12, hrsg. von Friedrich Hogemann e Walter Jaeschke. Hamburg: Meiner; transl. by Arnold V. Miller *Science of Logic*. London: Allen & Unwin; New York: Humanities Press 1969, pp. 573–844.

Hegel, Georg Wilhelm Friedrich (1992): *Enzyklopädie der philosophischen Wissenschaften im Grundrisse* (1830). Ester Teil. *Die Wissenschaft der Logik*. In *GW*. Bd. 20, hrsg. von Wolfgang Bonsiepen and Hans-Christian Lucas. Hamburg: Meiner, pp. 61–231; transl. by. Klaus Brinkmann and Daniel O. Dahlstrom *Encyclopedia of the Philosophical*

*Sciences in Basic Outline*. Part I: *Science of Logic*. Cambridge – New York: Cambridge University Press, 2010.
Hegel, Georg Wilhelm Friedrich (1970): *Enzyklopädie der philosophischen Wissenschaften im Grundrisse* (1830). Zweiter Teil. *Die Naturphilosophie*. Mit den mündlichen Zusätzen. Auf der Grundlage der *Werke* von 1832–1845 neu edierte Ausgabe Redaktion Eva Moldenhauer und Karl Markus Michel, Bd. 9. Frankfurt am Main: Suhrkamp; transl. by Michael J. Petry *Hegel's Philosophy of Nature*. London – New York: George Allen and Unwin Ltd. and Humanities Press, 1970.
Hegel, Georg Wilhelm Friedrich (2014): *Frühe Schriften II*. In: *GW*. Bd. 2, hrsg. von Walter Jaeschke. Hamburg: Meiner, pp. 341–344; transl. by Thomas Malcolm Knox. *Early Theological Writings*. Philadelphia: University of Pennsylvania Press 1996.
Hösle, Vittorio (1988): *Hegels System*. Band. 2: *Philosophie der Natur und des Geistes*. Hamburg: Meiner.
Hösle, Vittorio (1987): *Pflanze und Tier*. In: Michael John Petry (Hrsg.): *Hegel und die Naturwissenschaften*. Stuttgart-Bad-Cannstatt: Frommann-Holzboog, pp. 377–416.
Huneman, Philippe (Ed.)(2007): *Understanding Purpose. Kant and the Philosophy of Biology*. Rochester: Rochester University Press.
Illetterati, Luca (1995a): *Natura e ragione*. Trento: Verifiche.
Illetterati, Luca (1995b): "Vita e organismo nella filosofia della natura". In: Franco Chiereghin (a cura di): *Filosofia e scienze filosofiche nell'"Enciclopedia" hegeliana del 1817*. Trento: Verifiche, pp. 337–427.
Illetterati, Luca (2014a): "The Concept of Organism in Hegel's Philosophy of Nature". In *Verifiche* 42(1–4), pp. 155–165.
Illetterati, Luca (2014b): "Teleological Judgment: Between Technique and Nature". In: Ina Goy/Eric Watkins (Eds.): *Kant's Theory of Biology*. Berlin/New York: Walter de Gruyter, pp. 81–98.
Jaeschke, Walter (1980): *Logik, (spekulativ-)dialektische*. In: Karlfried Gründer (Hrsg.): *Historisches Wörterbuch der Philosophie*. Bd. L. Basel – Stuttgart: Schwabe, pp. 389–398.
Kant, Immanuel (1908): *Kritik der Urteilskraft*. In: *Kant's gesammelte Schriften*, hrsg. von Königlich Preußischen Akademie der Wissenschaft (= Akademie-Ausgabe). Bd. 5. Berlin: Walter de Gruyter, pp. 165–485; trans. by Paul Guyer and Eric Matthews *Critique of the Power of Judgment*. Cambridge: Cambridge University Press 2000.
Ng, Karen (2018): "Life and the Space of Reasons: On Hegel's Subjective Logic". In *Hegel Bullettin* 40(1), pp. 121–142.
Klotz, Christian (2010): "O conceito de alma na passage da natureza para o espírito subjetivo". In: Konrad Utz/ Marly Carvalho Soares (Orgs.): *A noiva do espírito. Natureza em Hegel*. Porto Alegre: EDIPCRS, pp. 232–240.
Kreines, James (2005): "The Inexplicability of Kant's Naturzweck. Kant on Teleology, Explanation and Biology". In *Archiv für Geschichte der Philosophie* 87(3), pp. 270–311.
Lugarini, Leo (1992): "Finalità kantiana e teleologia hegeliana". In *Archivio di storia della cultura* 5, pp. 87–103.
Manser, Anthony (1986): "Hegel's Teleology". In: Rolf-Peter Horstmann/Michael John Petry (Hrsg.): *Hegels Philosophie der Natur*. Stuttgart: Klett-Cotta, pp. 264–275.
Marmasse, Gilles (2008): *Penser le réel. Hegel, la nature e l'esprit*. Paris: Kimé.

Marques, Victor (2016): "Positing the Presuppositions – dialectical Biology and the Minimal Structure of Life". In: Agon Hamza, Frank Ruda (Eds.): *Slavoj Zizek and Dialectical Materialism*. New York: Palgrave, pp. 113–132.
McFarland, John D. (1970): *Kant's Concept of Teleology*. Edinburgh: Edinburgh University Press.
McLaughlin, Peter (1990): *Kant's Critique of Teleology in Biological Explanation*. Lewiston, NY: Edwin Mellen Press.
Menegoni, Francesca (2008): *La "Critica del giudizio di Kant". Introduzione alla lettura*. Roma: Carocci.
Menegoni, Francesca (1989): "La ricezione della "Critica del giudizio" nella logica hegeliana: finalità esterna e interna". In *Verifiche* 18(4), pp. 443–458.
Mure, Geoffrey (1950): *A Study of Hegel's Logic*. Oxford: Clarendon Press.
Pierini, Tommaso (2006): *Theorie der Freiheit. Der Begriff des Zwecks in Hegels Wissenschaft der Logik*. München: Fink.
Quarfood, Marcel (2006): "Kant on Biological Teleology. Towards a Two-Level Interpretation". In *Studies in History and Philosophy of Science* 37(4), pp. 735–747.
Sell, Annette (2014): *Der lebendige Begriff*. Freiburg: Alber.
Souche-Dagues, Denise (1990): "Téléologie et Métaphysique Hegeliennes". In: Hans-Friedrich Fulda/Rolf-Peter Horstmann (Hrsg.): *Hegel und die "Kritik der Urteilskraft"*. Stuttgart: Klett-Cotta, pp. 191–204.
Spieker, Michael (2009): *Wahres Leben Denken*. Hegel Beiheft, Hamburg: Meiner.
Stanguennec, André (1990): "La finalité interne de l'organisme, de Kant à Hegel. D'une épistémologie critique à une ontologie spéculative de la vie". In: Hans-Friedrich Fulda/Rolf-Peter Horstmann (Hrsg.): *Hegel und die "Kritik der Urteilskraft"*. Stuttgart: Klett-Cotta, pp. 127–138.
Steigerwald, Joan (2006): "Kant's Concept of Natural Purpose and the Reflecting Power of Judgement". In *Studies in History and Philosophy of Science* 37(4), pp. 712–734.
Verra, Valerio (a cura di) (1981): *Hegel interprete di Kant*. Napoli: Prismi.
Vieillard-Baron, Jean-Louis (Eds.)(2004): *Hegel e la vie*. Paris: Vrin.
Wilkins, Burleigh Taylor (1974): *Hegel's Philosophy of History*. Ithaca – London: Cornwell University Press Ithaca – London, pp. 85–142.
Wohlfahrt, Günter (1981): *Der spekulative Satz. Bemerkungen zum Begriff der Spekulation bei Hegel*. Berlin-New York: De Gruyter.
Zebina, Márcia (2010): "Finalidade sem fim: a centralidade da vida no sistema de Hegel". In: Konrad Utz/Marly Carvalho Soares (Orgs.): *A noiva do espírito. Natureza em Hegel*. Porto Alegre: EDIPUCRS, pp. 152–163.
Zebina, Márcia (2015): "Alcance especulativo da vida em Hegel". In *Dois Pontos* 12(2), pp. 99–110.

Luca Illetterati
# Being Rational: Hegel on the Human Way of Being

**Abstract:** In the second paragraph of the last edition of the *Encyclopaedia*, Hegel writes: "it is through thinking that human beings distinguish themselves from the animals". Therefore, according to Hegel, all that is human is "a result of thinking". In the background of this old prejudice and apparent triviality lies a specific conception of the human subject. Thought is indeed a dimension to which human beings belong and is not at the disposal of human arbitrariness. It is human beings that belong to thought, and not thought that belongs to human beings. The aim of my paper is therefore to analyse the specific Hegelian conception of the "nature" of human subjects, following the idea that the specific way of being of humans is realized only in self-transcendence.

## 1 Introduction

The second section of Hegel's 1830 edition of the *Encyclopedia of Philosophical Sciences*, and its related and very important Remark, represent a key moment in Hegel's philosophical system – one in which he anticipates some of the most fundamental and complex nodes of his philosophy. Indeed, it is within this particularly dense paragraph that one finds the following famous proposition, which I would like to take as the starting point for this essay:

> If, however, it is correct (as it probably is) that it is through thinking that human beings distinguish themselves from the *animals*, then everything human is human as a result of and only as a result of thinking [*dadurch und allein dadurch menschlich, dass es durch das Denken bewirkt wird*].[1]

---

This article is a reworking of a talk given at the 31st Conference of the international Hegel-Gesellaschaft (*Erkenne Dich Selbst. Anthropologiche Perspektiven*) held in Bochum from 17 to 20 May 2016.

[1] Hegel 2010b, § 2. Henceforth cited as Enc. and the paragraph number. The letter R. after the paragraph number indicates that the quoted text belongs to the Remark. The acronym Ad. indicates that the citation origination from the Addition. In the latter case, the page number is also shown.

---

**Luca Illetterati,** University of Padova

https://doi.org/10.1515/9783110604665-005

The human, for Hegel, is thus presented as that which is crossed through by thought. My purpose here is to try to understand concretely what it means to say that thought is what makes the human a human – and to try to clarify what such a thing implies about the human's way of being. In other words, what does it mean, according to Hegel, for the human being to be itself?

Perhaps the most immediate and simple reading of the above proposition takes Hegel to be saying that the human is an animal with an additional property with respect to other animals: thought. That the human is constituted by precisely this specific difference. Yet in reality such a reading expresses very little of what Hegel intended to say. In fact, if not properly understood, this proposition risks being read as something that Hegel does not mean at all. For thought for Hegel – and this is the point that needs to be understood – is not an attribute that simply adds on to a given animal nature, offering the body that possesses it greater possibilities than those without it. Rather, thought, as we will see, profoundly redefines the animal being of the human: it reconfigures and transforms it. As I intend to show here, for Hegel the fact that thought constitutes the proper, irreducible attribute of the human means that thought acts on the entire cognitive apparatus of this particular type of animal. Thought thus emerges as an element that in some way establishes its way of being.

This obviously calls into question the traditional Western definition of the human as a *rational animal*. What is interesting is that, while on the one hand Hegel seems to consider this time-honored definition of the human obvious – "It is an old prejudice, indeed a triviality, that human beings set themselves apart from animals through thinking [...] it may seem trivial to remind ourselves of such a longstanding belief, it must definitely seem strange that there should be a need for such a reminder"[2] – on the other, he systematically submits it to a radical revision, giving it a meaning that is not at all obvious or banal.

## 2 Rational Animals

Hegel could in some ways – but only in some ways – be read as taking up the critique of this traditional definition that was proposed by Heidegger, for example in his famous *Letter on "Humanism"*. As has been noted, within a discussion of the concept of *humanismus*, Heidegger asks what exactly the *humanitas* of *homo humanus* is, and in asking this question tries to bring to light the assumptions that the concept of *humanitas* implies: the metaphysical foundation on

---

[2] Enc. §2, R.

which the classical interpretation of *humanitas* rests. According to Heidegger, this underpinning emerges from the presupposition that there is a universal human "essence," which then gets expressed as consideration of human being as a *rational animal*. What Heidegger finds distinctive about this definition is that it already accounts for the human with reference to *animalitas* – that is, the human is already understood as a living being among others that is distinguished from those others (plants, animals, and also from God) by the faculty of rationality, by the possession of a property that distinguishes it from other living beings. In this way, Heidegger says, "metaphysics thinks of the human being on the basis of *animalitas* and does not think in the direction of his *humanitas*"[3].

This conception of the human way of being, according to Heidegger, is quite misleading with regard to what the human actually is: it is a radical misunderstanding of the original ontological structure of what Heidegger calls ek-sistence (*Ek-sistenz*) and considers constitutive of the specific mode of being of the human. According to Heidegger, we cannot think of the specificity of the human starting from *animalitas*. If anything, we can understand the specific animality of the human being only by starting from its specific way of being, or rather from what he calls "the essence of ek-sistence"[4].

This has absolutely radical consequences for Heidegger: it means the human is not simply an animal that, unlike other animals, has characteristics deriving from its being an *Ek-sistence* rather than a being among others. The human is not a particular type of animal that, unlike other animals, lives in the Lighting of Being [*Lichtung des Seins*], blocked from other animals that do not "have" "language".

If anything – even if reasoning in such terms risks upholding the polarities inherent to the metaphysics from which Heidegger claimed to take leave – the opposite is true. *Ek-sistence*, a negation of the way of being of an entity with a given essence[5], acts on its own body structure, on what could still be understand as its own animality, transforming it, molding it, producing a configuration of being that cannot be thought in terms of the simple *animalitas* of the animal. Heidegger says in the *Letter* that:

> The human body is something essentially other than an animal organism (...) The fact that physiology and physiological chemistry can scientifically investigate the human being as

---

3 Heidegger 1998, p. 246.
4 Heidegger 1998, p. 247.
5 "The ecstatic essence of the human being consists in ek-sistence, which is different from the metaphysically conceived *existential*" (M. Heidegger 1998, p. 248).

> an organism is no proof that in this 'organic' thing, that is, in the body scientifically explained, the essence of the human being consists[6].

According to Heidegger, it is precisely for this reason that we cannot interpret language, understood as that which is proper to the human or as the dimension in which openness to being occurs for the human, to start from the organic nature of the human: "In its essence," Heidegger writes, "language is not the utterance of an organism: nor is it the expression of a living thing"[7].

These pages in many ways constitute the manifesto of Heidegger's anti-naturalism and, to use categories that Heidegger would surely consider inadequate to his thought, the manifesto of what might be called his anti-naturalistic anthropology.[8]

In what sense, then, could Hegel in some ways – but again only in some ways – be read as embracing the Heideggerian critique of the traditional definition of the human as a *rational animal*? As a preliminary answer, we could say that for Hegel, as for Heidegger, calling the human being a *rational animal* does not simply mean it is an animal that also has the property of being *rational*. According to Hegel, if it is true that thought produces all that is human in the human character, it cannot merely be understood as an additional property, an extra tool added to the other instruments with which, as an animal, it would already be endowed. To say that thought is proper to the human being, that everything that is properly human begins with and is thought, for Hegel means positioning thought as the origin point from which the human is configured. It means that only starting from thought can we adequately consider the multiple different faculties with which the human being is endowed. In other words: if thought is what belongs to the human, if it is what makes any act performed by a human a human act, then thought is that from which the entire human way of being must be thought. It is therefore also starting from thought that the animality of the human can be thought – which is not the same thing as conceiving an animality enhanced by thought.

---

6 Heidegger 1998, p. 247.
7 Heidegger 1998, pp. 248–249.
8 For a critical discussion of Heidegger's antinaturalism, cf. Rouse 2005.

# 3 The Humanity of the Human

What, in concrete terms, does it mean to say that thought is what makes the human as such? Obviously, this does not mean that thought and reflective activity in the human substitute that which in the animal is instead governed by sensation, feeling, or instinct. Having thought does not mean that the human lacks sensations, feelings, and instincts. Instead, according to Hegel, a feeling, intuition, desire or any other content of consciousness is human only if it is crossed by thought. A feeling is never a feeling for the human in the same way it would be for an animal. It is always something else, with its own peculiar structure: a feeling accompanied by thought, impregnated with thought.

For Hegel, the human does not experience pure, unintelligent sentiment, devoid of any relationship to thought. Or, even if it does, certainly this is not what makes the human as such. In this sense, Hegel maintains, against any form of romanticism, that it is ridiculous to think of the human in terms of some original uncontaminated naturalness or pre-reflective purity. On the contrary, those who found human activity on the possibility of returning to this kind of pre-noetic state inevitably place human activity within a generic animal sphere, thus disregarding precisely what is specific about the human character.

Yet saying that feelings, intuitions, and representations are human because they are crossed through by thought does not mean that these forms exhaust all possible types of thought. Thought in the human can of course can also appear in the pure form of the concept. But what Hegel is interested in emphasizing in the introductory paragraphs of the *Encyclopedia* is the fact that thought is mostly present in the human in forms other than the concept.

Outlining § 2 of the *Encyclopedia* and its corresponding Remark, Hegel's argument can be summarized in these 4 points:
1) Every human activity is properly human (i.e., specific to the human) insofar as it is characterized by thought. This means that the specifically human nature of what is human is determined by thought. It also means that an act the human shares with another animal is specifically human only insofar as it is "accompanied" or traversed by thought. Or rather, only inasmuch as the act is not in fact identical in the human and the animal.
2) The type of thought that makes an activity human, however, does not for the most part appear in the pure form of the concept. Thought, in the ordinary activity of man, is mostly given as a mixture of feeling, intuition, and representation.
3) These forms of thought (feeling, intuition, representation), which characterize the human's openness to the world, are not other to thought or ways of

accessing reality that are alternative to thought. To the extent that they are traversed and permeated by thought, they in fact constitute something that is properly human.

4) However, although they are not other to thought, feelings, intuitions, and representations should not be understood as thought as such. They should therefore be distinguished from thought as form[9].

According to Hegel, when feelings, intuitions, and representations are considered privileged forms of thought guaranteeing the human authentic access to the world and are therefore nominated as good candidates for revealing an original, pre-noetic form of openness to the world, we find ourselves embracing an inadequate conception of the faculties of the human and the role played by thought. For Hegel, this is a distorted and distorting vision of what the human is.

For Hegel, such emphasis on the role of sentiment and intuition, as opposed to and alternative to thought, is based in reality on a narrow, restricted, and one-sided conception of thought. He saw this emphasis widely embraced in his time, embodied, for example, by the so-called sentiment theologies maintaining that God cannot be thought, only felt, and by the forms of venerating immediacy that claimed to isolate extranoetic intuition as a way of accessing the divine. Such conceptions identified thought solely with reflective activity, or rather, with the abstract procedure of the intellect. When thought is identified with reflective thought that which is extraneous and not reducible to such activity immediately becomes other to thought. Thus, operating within this dichotomy between thought and that which is other to it, these perspectives come to argue that what is not graspable by reflective thought, what is therefore impervious to the reflection of the intellect, can instead be grasped by extranoetic modes of access to the world, such as feeling or intuition. Precisely because they are extranoetic, these activities are understood to exceed the constraints and limitations typical of abstract intellectual thought.

However, according to Hegel, founding religion, art, morality, or any other human and spiritual activity on something foreign to thought is like saying that these activities are not specifically and properly human:

> In this kind of separating it is forgotten that only human beings are capable of religion and that animals no more have religion than they have law and morality[10].

---

9 Cf. Soresi 2007; Illetterati 2016.
10 Enc., §2, R.

What Hegel aims to disavow, evidently, is the possibility of an extra-rational foundation for the spiritual activity of the human, and in particular for knowledge of what is not immediately objectifiable by reflective thought.

This disavowal finds its condition of possibility in the conviction – made explicit at the beginning of Hegel's philosophical system – that thought cannot be reduced to reflective activity and that such a reduction is in fact instrumental to establishing an extra-rational foundation for knowledge. In other words, for Hegel, emphasis on the elements that thought positions as 'other' and identification of thought with the reflective activity of the intellect are two sides of the same coin, mutually constitutive of each other and supported by the same ground. And this is a point worth emphasizing because it has metaphilosophical implications beyond the horizon of meaning to which Hegel referred, for all those theories expressing what has been called the philosophical discourse of otherness.[11]

Hegel thus displaces this idea by making the point that sentiment (or intuition, fantasy, etc.) is not other to thought and that thinking is not reducible to the form of reflective thought. In saying this, Hegel does not intend to eliminate the human specificity of sentiment or to consider it irrelevant, nor does he disregard the force of reflective thought, which he maintains constitutes a decisive and inescapable general mode of knowing. Rather, he intends to show the one-sidedness and self-contradiction of a position that, absolutizing the reflective procedure of the intellect, opens the door to forms of sentimentality and intuitionism that risk evading any rational control.

## 4 Hegel's Philosophy of Mind

On the basis of what has been said, we can begin to grasp, though still in elementary terms, the basic structure of what might be called the Hegelian philosophy of mind. For Hegel, mental contents, the contents of our consciousness, can take the form of feeling, intuition, image, or even thought and concept. Feeling, intuition, image and concept are all forms of thought. They are mostly given in mixed forms. When thought is given in pure form one has the form of the concept.

Describing the function of our mental activities – and therefore also their dysfunction and malfunction – in what we could call phenomenological terms, Hegel notes that even if the content of a mental act is "one and the

---

11 Cf. Labarierre, 1983.

same" [*ein und derselbe*] but it appears in different forms, that same content corresponds in the different forms to different intentioned objects. This is because "the determinacies of these forms [and therefore the specific ways of being of sensation, of intuition, and of the concept] convert themselves into part of the content": the way of being of the forms through which a content is grasped gets reflected in the content itself, thus producing a particular appearance. So, Hegel concludes, "what is in itself the same, can take on the look of a different content"[12].

Put differently, depending on whether an object is felt, intuited, or thought, it will appear as a different object. But what are different are the forms within which that object is caught, not the object as such. In this sense, for example, we can let ourselves be carried away by the feeling that an object of art induces in us, or we can treat the art object in a detached way, as a simple object perceived by the senses: measuring its dimensions, weight, chemical composition, etc. Or we can reflect upon the meaning the art object expresses, upon the thought content that it incorporates. And it is in this sense that we can say that the object intended in these three different activities can be thought of as three different objects (a bit like the evening star and the morning star referred to by Frege).

At the same time, however, all these approaches refer to the same reality and – to the extent that the different approaches through which we can address the same reality express our way of being, allow us to relate to the world – they are all in some way traversed by what characterizes the human as such, that is, by thought. Indeed, thought is what prevents each individual approach, each particular point of view, from giving itself as the autonomous and self-subsisting reality, or at least guarantees the possibility of going beyond the partiality and perspective offered by each of the different ways of accessing the real.

Although each has its own ordering structure and characterizing element that distinguishes it from the others, the organizing structures of experience (i.e., intuition, feeling, representation) are not watertight compartments or, to use a metaphor dominant in discussions on relativism, are not incomparable and incommensurable points of view, precisely because they are also thought. In this sense, thought is what prevents unilateral points of view from becoming fixed as absolute visions of the world and is what is able to put in communication and connect different experiences and approaches of reality[13].

---

**12** Enc., §3.
**13** On this point, it could be interesting to extend this discussion by situating the Hegelian perspective in relation to the Davidsonian one. Davidson, as has been noted, shows within his fa-

The separation of feeling and thought, as faculties meant as extra-noetic and noetic, for Hegel reflects a totally inadequate anthropological model, according to which the human emerges as a composition and combination of a series of faculties that remain independent of, and separate from, one another. To understand better why this compositional model does not work for Hegel, and in what sense he rejects conceiving thought as a property to be simply added to those already possessed by animals, it may be useful to compare Hegel's comments on forms of animal openness to the world in *Philosophy of Nature* with his account of the same forms of openness to the world in the human in *Philosophy of the Subjective Spirit*.

## 5 The Animal's Openness to the World

The concept that profoundly characterizes the sphere of *animalitas*, according to Hegel, is the concept of subjectivity – a notion that finds its first real thematization in his system within a naturalistic context. What does Hegel mean by stating that the animal's way of being is that of a subjectivity? The animal is a subject for Hegel because it contains within itself the fulcrum of its movement. Or rather, its movement does not tend towards some external power that directs and dominates it and upon which it depends, but instead tends toward the direction of itself. In other words, its end is in itself. Even when the animal organism's activity is turned outwards – when, moving from need, and therefore from a lack that it feels in itself – it moves beyond its own singularity and toward what is other to itself, it always tends to realize itself.

Precisely because the center around which its activity revolves is in itself and not in another, animal subjectivity is, in Hegel's language, a concrete unity: this is distinct from a merely formal unity, like that of the plant, whose parts are all autonomous with respect to the whole, in that they are able to survive separation from the whole and give rise, as parts, to new organic totalities. The unity of the animal is a concrete unity, in that it is realized in difference and articulation: the parts are members of a whole, such that, if separated and divided from the whole, they are no longer what they are and lose

---

mous essay how this "dominant metaphor" in the ambit of conceptual relativism conceals within itself a paradox that is difficult to circumvent: "Different points of view make sense, but only if there is a common coordinate system on which to plot them; yet the existence of a common system belies the claim of dramatic incomparability" (Davidson 1973/1974, p. 6). The function that Davidson here attributes to the "common system" is the function that Hegel attributes to the universalizing power of thought and its systematic articulation.

their specific organic meaning. This concrete unity makes the animal an individual in the proper and concrete sense of the term. Or rather, makes it a way of being that cannot be divided without canceling out its own ontological structure – a structure that, in articulating itself (i.e., in its internal self-direction and in its becoming something other than itself), nonetheless remains in unity with itself[14].

As a subject, or rather, having within itself and not outside itself the fulcrum of its unification, the animal is capable of self-movement – it is capable of withdrawing, even if only partially, from external control – and of self-determination on the basis of needs and motives found within itself. It is thus no coincidence that in *Philosophy of Nature*, in the last section on organic physics, the concept of freedom makes its appearance in Hegel's discussion of animal subjectivity.

The concepts of subject and freedom are intimately connected in Hegel; the two words often appear as explanations of each other. In fact, Hegel explains the animal's ability to change dwelling by making reference to the unique relationship, with respect to the rest of the natural world, that the most complete organismal forms have with time. Unlike the plant, which is subjected to the external power of light, above all for its movement, and is likewise dependent on the cyclical time of nature for its growth, nutrition, and reproduction, the animal, as a subjectivity, has a way of being that Hegel significantly calls "liberated time [*freie Zeit*]"[15]. This expression implies some form of independence with respect to the external and to purely natural time. It therefore suggests, for the first time, a capacity for autonomy and self-determination in the realm of nature.

If this "liberated time" manifests itself in the animal's self-movement, this capacity for self-movement should not only be understood as the possibility for spontaneously changing place, but also, to the extent that one can speak of an "ideal form" of self-movement, as the condition capable of accounting for and giving rise to all the characteristics that specifically mark the way of being of this form of natural subjectivity. Self-movement thus constitutes all the particular phenomena which will be further and more specifically articulated at the level of the spirit: the vocal faculty [*Stimme*], animal heat [*animalische*

---

[14] The animal organism, according to Hegel, is the concrete realization of life in nature, insofar as "it is the unit which holds the free parts bound within it; it sunders itself into these parts, communicates its universal life to them, and holds them within itself as their power or negative principle" (Hegel 1970, § 342, Ad., p. 41).

[15] Hegel 1970, § 351.

*Wärme*], interrupted intussusception [*unterbrochene Intussuzeption*], and, above all, feeling [*Gefühl*].

In fact, the voice, according to Hegel, is the organism's expression of "a free vibration *within itself* [ein freies Erzittern *in sich selbst*]"[16] and is in this sense an expression of its subjectivity. Of course, the *Stimme* characterizing animal subjectivity is not yet crossed through with the symbolic production that leads to the spiritual level of spoken language [*Sprache*]. Yet, as a manifestation of animal subjectivity's capacity for expressing itself – exteriorizing its inner self and giving a form of external existence to its intimate conditions of pain, satisfaction, or need – *Stimme* can be read as one natural precondition, necessary, even if not sufficient, of the symbolic capacity of language, which will find its articulation only at the level of the spirit.

The voice is not simply the consequence of some mechanism internal to the organism. It is a form of the animal's self-movement: its self-production, a phenomenon through which the animal succeeds in giving objective form to its own subjectivity, to its own *feeling*, to its own *Gefühl*.[17] The *Gefühl* [feeling] and the *fühlen* [feel] are perhaps the most original expressions of animal subjectivity. Indeed, only in so far as it feels is the animal able to express, through and in the voice, what could be called, without attributing any consciousness to it, its self.

This subjective structure of the animal is further explained by Hegel in relation to how the animal's relationship with exteriority, or the world around it, is articulated – a relationship constituted by the *Assimilation-process*, the process, born with *feeling*, that the animal maintains with the world. Here feeling is always meant to be understood as subjective, in that it is connected to the animal and above to all a feeling of lack that *affects* it. Starting from feeling, animal subjectivity is consequently explained as action emerging in relation to this perception of lack and "the *drive* to overcome it"[18].

The assimilative process is thus born from a need and structural deficiency within the organism itself, but, even before, at a more basic level, it is born from the capacity – unique to the living animal being and which determines its intimately subjective structure – to feel this need and this lack.[19] Therefore, as a subjectivity, the animal organism is not simply 'the lacking being' but 'the being that

---

**16** Hegel 1970, § 351.
**17** Hegel 1970, § 351.
**18** Hegel 1970, § 359.
**19** "Only a living existence is aware of deficiency" (Hegel 1970, § 359 R.). Petry's English translation risks leading us to believe that Hegel attributes consciousness to living beings as such. The German text says: "Nur ein Lebendiges fühlt *Mangel*." A more fitting translation might be "Only a living (being) feels deprivation."

is capable of feeling and of living lack within itself'. And it is precisely because of this capacity to live and feel its own state of destitution, and consequently its own contradiction and wound, as determining elements of its ontological constitution, that it is indeed a subject:

> The *subject* is a term such as this, which is able to contain and *support* its own contradiction; it is this which constitutes its *infinitude*[20].

The infinity that Hegel here attributes to the subject's mode of being should not be understood as possibility on the part of the animal subject to move beyond the concrete forms of lack and need that are constitutive of it, or to extend itself beyond its nature. The infinity instead consists in the animal subject's ability to feel its being finite, to experience its own negativity, to live its limit as a lack and as a drive to overcome that lack.

In this sense, the infinity of the subject reveals itself as the finite subject's ability to transcend itself in the very act in which it first perceives itself as finite. Precisely because it is a subject in this sense, the animal is able to extend itself toward what is presented to it as something else in the form of an assimilative relationship. The first way the animal enters into relationship with its environment, indeed, the very condition that allows the animal to have an environment, to transform what surrounds it into *its* environment, is what Hegel calls 'the theoretical process' [*theoretischer Prozess*]: the form of assimilation of the surrounding world effected by the multiplicity of the senses.

This assimilation of the external world does not imply, as in the case of real assimilation, the nullification and dismemberment of perceived objects, but it is nonetheless an assimilative type of process and, as such, implies a form of transformation. The animal, leading itself into that which is other to itself, externalizing itself, at the same time brings that alterity into itself – and through this assimilative movement thus strips itself [*aufheben*, sublate] of its otherness.

How do the senses transform reality? This theoretical assimilation of the senses, Hegel says in Remark 358 of the *Encyclopedia*, is a "reduction of the separated moments of inorganic nature to the infinite unity of subjectivity." However, the unification produced by the senses is only partial – and this is a point of particular relevance for the argumentative line of this essay – since the animal "is still a natural subjectivity". The moments in which the animal articulates itself, and the plurality of senses through which it is placed in relation to and internalizes the world, "still exist separately": they are separate and divided from

---

**20** Hegel 1970, § 359 R.

each other, or rather, they give rise to what can be thought of as a fractional totality.

This suggests that what we mentioned at the beginning of this essay, Hegel's points in § 2 of the *Encyclopedia*, could also hold true for the animal: different ways of considering the same object could produce the impression that what is being considered is not a single object, but different objects. In fact, for the animal, it seems that experience of the object – even if it is not possible, properly speaking, for the animal to speak of experience as such or of the object – is fractioned into the different senses by which the animal stands in relation to the world. The different senses, despite corresponding to a subjective unity, are nonetheless separate from one another.

# 6 The Human's Openness to the World

On the other hand, the function of unifying different sensible experiences, tracing them back to the unity of the object, takes place in the human through thought. Not because thought unites a content that the senses produce as differentiated, unifying the multifaceted material supplied by the senses, but because the senses in the human are already crossed through by thought. They are already, that is, part of an experience that cannot be isolated into separate sections.

Experience, in the human, is always given as distinct forms – reality can be experienced through a desire, a feeling, a volition, or an intellectual consideration – but is also unitary, in the sense that these forms are never separate from one another but are from the beginning connected to one another. Hegel says in the 1827–28 *Lectures on the Philosophy of Spirit*, that:

> Experience means more than mere sensible grasping or mere perception; it already includes a universality within itself. If something is supposed to count as experience, there must be a law, something universal, and not merely a particular perception. This something must be raised by thinking to universality.[21]

This process of being "raised" does not imply passage from sense considerations to a higher plane where the intellect works (i.e., understanding), as the dialectic of *sinnliche Gewissheit* [sense certainty] in the *Phänomenologie des Gesites* shows. Rather, this raising already takes place in sense consideration: it is already active within it. Or in other words, it already takes place in sense consid-

---

[21] Hegel 2007, p. 62.

eration viewed from the perspective of the concreteness of the spirit, the concrete way of being of the human, whose faculties are not simply juxtaposed to one another but are a connected totality of distinct forms. This is why, according to Hegel, only the human really has experiences: "In order to experience, one must be capable of thought and reason"[22].

The spirit is a concrete unity in a sense more radical than for the animal organism. Indeed, dismemberment of the human into a plurality of autonomous functions connected only externally by some higher function would contradict the concreteness, or truth, of the spirit. For Hegel it is only from an abstract standpoint that we could understand human sensations, feelings, volitions, and desires to be separate modules subsequently put together by some organizing function. Hegel locates in the *I* the dissolution and overcoming of precisely this type of separation. Indeed, the *I* is this concrete unity, and this concrete unity is the human; or rather, the *I* is the spirit. "The human being is spirit," Hegel says in the 1827–1828 *Lectures*, immediately afterwards asking, "What is the innermost, concentrated nature, the root of spirit?" His response is clear: "Freedom, I, thinking" (Hegel 2007, 105, 8).

Here it is evident that freedom, I, and thought for Hegel are different ways of saying the same thing: the ability of the subjective spirit to be with itself. Its ability to cut itself off from any form of dependence on the outside, of denying everything. In this sense, saying that the human being is a *rational animal* for Hegel does not mean saying that the human being is an *animal* to which we add, alongside the properties already proper to the *animal* in general, the further ability of also being *rational*. The link posed by the *also* is that typical of abstraction, of the enumerative procedure of an intellect which, after dismembering a unit into its component parts, reassembles it by simply putting things side by side, each in its place. "The connecting 'also' always allows the independence of every activity and their mutual indifference", so that "the soul appears as an external connection of all these diverse types of powers and activities"[23].

The *also* is the typical form of conjunction used by the intellect: once concrete complexity has been decomposed, the intellect recomposes it by adding the parts back to each other, as if concreteness were a set of separable and distinguishable parts.[24] Hegel expresses it in this way:

---

[22] Hegel 1970, p. 63.
[23] Hegel 1970, p. 63.
[24] On this perspective, see Corti 2015. Corti's text contains a pointed analysis of the secondary literature on this question. See also Corti 2016. I owe some of the ideas presented here to discussions with Luca Corti and Sergio Soresi, whom I thank.

> For the understanding, the difficulty consists in ridding itself of the arbitrary distinction between the faculties of the soul, feeling, and thinking spirit, which it has already fabricated for itself, and in realizing that in the feeling, volition and thought of man there is only one reason.[25]

Thinking of the spirit as a totality of faculties, forces, or functions that can be considered separately and independent of the concrete unity in which they act means considering the spirit in naturalistic terms, so to speak. In fact, it is in nature – dominated by exteriority – that different determinations are separate and exterior to each other. Instead, the spirit is *spirit* precisely inasmuch as it is a process of progressively eliminating this exteriority. Thinking of spirit as an accumulation of faculties would mean reducing it to something else – thinking it in a non-spiritual way.

Hegel thus introduces the notion of philosophy of the subjective spirit in § 380 of the *Encyclopedia* in this way:

> The concrete nature of spirit is peculiarly difficult, in that the particular stages and determinations in the development of its Notion do not remain behind together as particular existences [...] In the case of external nature they do however [...]. The determinations and stages of spirit occur in the higher stages of its development essentially only as moments, conditions, determinations.

By saying that the human is essentially thought, Hegel deconstructs the idea of subjectivity understood as a set of separable faculties. Since this separateness is not given in the concreteness of the spirit, a description of the human that begins from this compositional model is revealed to be not only intellectual artifice but also, and above all, a reduction of spiritual complexity to natural ontology.

# 7 The Nature of Thought

This idea that thought is not an additional faculty but a fluidifying and transformative element is confirmed by the Preliminary Conception of the Science of Logic in the *Encyclopedia*, where Hegel, wanting to introduce an idea of a science of pure thought, moves from ordinary considerations of thought to considering thought scientifically. Here Hegel emphasizes not only the idea of thought as a unifying element of subjectivity but also the radical consequences inherent to recognizing the human as essentially thought.

---

25 Hegel 1987, § 471, R.

According to Hegel, thought is ordinarily meant as
1) one of the spiritual faculties or activities that belongs to the subject "*alongside* others" such as sensing [*Sinnlichkeit*], intuiting [*Anschauung*], imagining [*Phantasie*], desiring [*Begehren*], or will [*Wollen*][26];
2) the *product* [*Produkt*] of a faculty [*Vermögen*];
3) the *I*.

Hegel does not intend to make sense of these conceptions, but rather to show that they are all rooted in a conception of thought that serves as their condition of possibility and which they also transcend.

To this list of ordinary considerations of thought Hegel adds that of thought as a faculty. Without denying that thought can also be thought of as a faculty, he maintains that it is not simply one faculty among others. Rather, thought is a structure that redefines all the other faculties, putting them in connection with each other and that also passes through them, so they are no longer isolated and fragmented.

Moreover, thought for Hegel is an activity [*Tatigkeit*], a productive activity – whose product is nothing other than thought itself. One therefore cannot distinguish thought as faculty from thought as product. It is both activity and product. Thought, in this sense, produces itself; it is a form of self-movement and is therefore essentially subjectivity. This means that thought is never simply a given, it is not something that exists independently of its exercise. Thought is thus always essentially an activity. To say that thought is a subject and essentially activity means for Hegel that thought is thinking: "the simple expression for a concretely existing *[existierenden]* subject that thinks is *I*".[27]

Yet the *I* is not simply the seat of thought, a place or entity characterized by possession of the faculty of thought. Kant had already deconstructed the assumptions upon which such a representation rests, and Hegel did not intend to revive them. Rather than the substratum of its different properties, for Hegel the *I* is thought and is constituted in thought. It is constituted in the relation to itself, as consciousness of itself, as an activity that self-determines itself.

This does not, however, relegate the *I* to the realm of the purely subjective. On the contrary, thought is universal precisely inasmuch as it overcomes subjective particularity – the condition of possibility for the liberation of the finite subject from the constraints in which every particular subjectivity is inevitably immersed. The thinking subject is, despite the paradoxicality of the expression,

---

26 Cf. Enc., § 20.
27 Enc., § 20.

the process through which the finite subject frees itself of its own particular subjectivity.

The product of this activity is the universal, which Hegel says – in a not unproblematic expression – contains "the value of the basic matter [den Wert der Sache], the essential, the inner, the true".[28]

Since it is active, since it captures and produces the universal, thought transforms empirical content; "it is only by means of [*vermittels*] an alteration that the *true* nature of the *object* emerges in consciousness".[29] This true nature, however, is not a product of the subject, at least certainly not in the sense that the subject, as a unilaterally "idealistic" reading would suggest, somehow "creates" the object. If anything, such a reading is typical of what Hegel calls "the sickness of our time";[30] the trait of an era that has reached the despairing point of recognizing only the subjective as true and, in turn, of viewing the subjective as the ultimate term beyond which no subject is able to go. According to Hegel, within such despair there also lies a criticism, namely that the nature of the object is true only because the subject transcendentally gives it the characteristics of truth.

The task of philosophy, according to Hegel, is precisely that of taking back this subjectivist presupposition and showing that the true nature of the object, which the subject reaches through reflection, is true not because the subject makes it true, but because the subject, through thought, is able to go beyond the subjectivist limits of its own experience – because the subject, in thought and through thought, transcends the subjectivist limits of its experience of the thing to grasp the thing's essence. It is within this complexity that we should understand Hegel when he says that, because reflection gives us the true nature of things and reflection is an activity of the subject, "true nature is equally the *product of my* spirit insofar as the latter is a thinking subject (…), i. e., as an I that is entirely *with itself* – *it* is the product of my *freedom*".[31]

This does not simply support, as a naive idealist might believe, the dissolution of the objectivity of things in their being represented. Precisely because thought, through reflection, grasps not a simulacrum of the thing but the essential, it allows us to go beyond both the position that poses objectivity as a dimension separate from thought (as if the essence of the thing were completely independent of the reflective process by which it emerges and therefore impermeable to the thought that thinks it) and the position that dissolves ob-

---

**28** Enc., § 21.
**29** Enc., § 22.
**30** Enc., § 22, Addition, p. 56
**31** Enc., § 23

jectivity in conscious representation (as if essence were a product of the subject without any real anchorage in the thing itself)[32].

It is precisely through this twofold overcoming of both independent objectivity and the conscious reduction of objectivity that "thoughts may be called *objective* thoughts".[33] With the expression "*objective* thought [*objektives Denken*]," Hegel seems to want to indicate that thought is not reducible to the expression of a subjective point of view or to the general ability to forge and make one's own reality a reality still constitutively different from the way in which it is conceived. Rather, for Hegel, the expression "objective thought" indicates the universal *logos* that runs through all reality: the rational plot within which are located both the subject – who actively thinks and, in this act, brings things to meaning – and the object, which is thought and yet is not a production of the subject. This universal *logos* should not be understood as tantamount to the idea that the world is in the mind of God, especially if understood as something simply to be uncovered or that the subject must only somehow find. Universal *logos* is constituted through the comparisons that subjects conduct amongst themselves and with the world; it becomes objective in the rational practices which alone enable us to talk about the subject and the world.

To the extent that "objective" in the language of ordinary representation usually characterizes all that is *mind independent,* applying the term to thought removes the implication of representation, also taken in the ordinary sense in which thought is understood to take on meaning only in relation to a mind from which it originated. The expression "objective thought" seems to be used by Hegel to show the need to move beyond precisely this subjectivist perspective of representation, according to which "objective thought" inevitably sounds like an oxymoron. What is objective is, from the subjectivist point of view, "other" to thought, that which is outside the dimension of thought – just as, vice versa, thought is considered an internal and subjective dimension determined by its opposition to the sphere of objectivity.

---

[32] On this point allow me to refer to Illetterati 2014 and Illetterati 2018.
[33] Enc., § 24. In this sense, §§ 20–24 tend to show how, starting from the notion of thought understood as *activity of the subject*, one arrives, by analyzing the production of this thought as a reflection aiming to capture the universal and therefore the essence of things, at the notion of thought as *objective thought* – and, consequently, at the identification of logic, as a science of thought, with metaphysics.

"Objective thought" is therefore an expression that Hegel simultaneously wants to mean
1. that the dimension of thought is not closed within the subjectivist perspective, and
2. that the sphere of objectivity is not extraneous to thought in itself.

"Objective thought" for Hegel consequently means there is no fracture between thought and the world that needs to remedied through some sort of mutual adaptation to each other. Although not simply the product of the thinking activity of a finite subjectivity, and therefore not independent of the subject (*mind independent*), the world is nonetheless conceivable by the subject: it has within itself the conditions of its own thinkability, which are how the subject can indeed think of it. Both the subject and the world thus belong to a common *logos* that transcends the perspective of any particular subjectivity, emerging instead from the transcendence that subjectivity achieves with respect to its own particularity.

If, on the one hand, the human subject *is as it is thought*, on the other, thought, according to Hegel, does not belong to the subject as his property. The subject is not the "master" of thought. In fact, not only is thought not reducible to a production of the subject, thought so little belongs to the human as its possession – or as sort of instrument it has for grasping the world – that, if anything, "it is they [the determinations of thought] who have us in their possession."[34] Concepts of things and determinations of thought are not intellectual instruments or prostheses through which to subjugate the world, understood as the sphere of the other and of that which is separated from thought. Determinations of thought and concepts of things rather constitute the horizon within which our thinking moves, such that "our thought must accord with them, and our choice or freedom ought not to want to fit them to its purposes".[35]

This "realism" of thought is what makes Hegel's idealism "anti-idealistic", so to speak. But it is an "anti-realistic" realism, because it recognizes in thought (which is never subjectivistically or idealistically understood) the element from which it is possible to begin thinking of the subject's access to the world. In other words, Hegel's realism, in response to constructivist delusions about modernity, does not simply contrast subjectivism with reality posed as a sort of impregnable and impenetrable fortress for subjectivity. Instead, it goes beyond such a subjectivist and instrumentalist conception of thought, in

---

**34** Hegel 2010a, p. 15.
**35** Hegel 2010a, p. 16. G. di Giovanni's translation is perhaps less strong than the original German. Hegel in fact says here that our thought is limited [*beschränkt*] by the determinations of thought in which it moves.

which the reality about which the subject speaks and to which it relates is always and only one of its constructions – the simulacrum of something whose truth remains inaccessible – as well as beyond a conception of reality as that which is other to, foreign, and opposed to thought.

# 8 Conclusions

The human is a being *in itself* rational, Hegel says. But in saying this, he also says that the human being is not *immediately* rational – that being rational is an activity, a process and not a given. Rationality is not a substance *out there* in the world the subject somehow has to make its own, nor is it something the subject has always already simply possessed. Rationality is a process realized in and with thought, in and with reality. Rationality, in general terms, is that which human beings must realize in order to be themselves.

The idea of the human being as a *rational animal* is taken up again in an important way in the *Science of Logic*, in the part on objective logic that is dedicated to quality. This is where Hegel discusses the notions of determination, constitution, and limit [*Bestimmung, Beschaffenheit und Grenze*]. As an example of determination, and to clarify the difference between determination and determinateness [*Bestimmtheit*], Hegel takes up the Fichtian theme of the Bestimmung des Menschen, the vocation or destination of the human being. "The *determination of the human being*, its vocation," Hegel says, "is rational thought [*denkende Vernunft*]"[36]. If thinking [*Denken*] is the determinateness [*Bestimmtheit*] of the human, that which distinguishes it from pure natural being, being rationally thinking is more radically its determination [*Bestimmung*]: not just a property that affects the human being and differentiates it from what is other to it, but that which it must be to be itself. To say that rational thinking is the determination of the human being for Hegel means that this is the human's destination, its vocation; the human is itself when it acts in the direction of the *denkende Vernunft* [thinking Reason] that constitutes it as a human being.

In the *Anthropology*, Kant calls the human being *animal rationabile*, i. e. "an animal endowed with the capacity of reason", and he explains such an expression by saying that the human being is an animal that "can make out of himself a rational animal [*der aus sich selbst ein vernünftiges Thier* [animal rationale] *machen kann*]".[37]

---

[36] Hegel 2010b, p. 96.
[37] Kant 2006, p. 226.

Making a *rational animal* out of itself, according to Hegel, means realizing oneself as a thought. And this realization implies self-liberation from one's own subjective nature – transcending oneself, knowing that there is no given human nature but that human nature is precisely this transcending action, this taking oneself away from oneself.

Such transcendence is not contingent upon something external, something other to subjectivity. Subjectivity transcends one's being simply a finite subject in the very act of realizing one as a subjectivity. Rational nature is therefore not a given towards which the human must strive to reach or match. It is what is realized in precisely the movement of transcendence that subjectivity performs upon itself.

# References

Corti, L. 2015. *Pensare l'esperienza. Una lettura dell'Antropologia di Hegel*. Bologna: Pendragon.
Corti, L. 2016. "Conceptualism, Non-Conceptualism and the Method of Hegel's *Psychology*". In S. Hermann-Sinai, L. Ziglioli (Eds.). *Hegel's Philosophcal Psychology*. New York: Routledge, pp. 228–250.
Davidson, D. 1973/1974. "On the Very Idea of Conceptual Scheme". *Proceedings and Addresses of the American Philosophical Association*, Vol. 47, pp. 5–20.
Kant, I. 2006. *Anthropology from a Pragmatic Point of View*. Translated and edited by R.B. Louden, with an Introduction by M. Kuehn, Cambridge: Cambridge University Press.
Hegel, G. W. F. 1970. *Philosophy of Nature*, vol. III. Edited and translated, with Introduction and Explanatory Notes by M. J. Petry. London-New York: Allen & Unwin.
Hegel, G. W. F. 1987. *Philosophy of Subjective Spirit*, vol. 3 (Phenomenology and Psychology). Edited, translated, and with an Introduction by M. J. Petry. Dordrecht: Reidel.
Hegel, G. W. F. 2007. *Lectures on the Philosophy of Spirit (1827–1828)*. Translated and with an Introduction by Robert R. Williams. New York: Oxford University Press, 2007.
Hegel, G. W. F. 2010a. *Science of Logic*. Edited and translated by G. di Giovanni. Cambridge: Cambridge University Press.
Hegel, G. W. F. 2010b. *Encyclopedia of Philosophical Sciences in Basic Outline. Part I: The Science of Logic*. Translated and edited by D.O. Dahlstrom and K. Brinkmann. Cambridge: Cambridge University Press.
Heidegger, M. 1998. "Letter on 'Humanism'". In M. Heidegger, *Pathmarks*. Edited and translated by W. McNeil. Cambridge: Cambridge University Press, pp. 239–276 (the text of the Letter presented in this edition was originally translated by Frank A. Capuzzi and then revised by W. McNeill and D. Farell Krell).
Illetterati, L. 2014. "The Semantics of Objectivity in Hegel's *Science of Logic*". *Internationales Jahrbuch für Deutschen Idealismus*, 12, pp. 139–164.
Illetterati, L. 2016. "The Thought of Logic". In L. Fonnesu, L. Ziglioli (Eds.), *System und Logik bei Hegel*. Hildesheim: Olms, pp. 105–129.

Illetterati, L. 2018. "Der einzige Inhalt der Philosophie. Ontologie und Epistemologie in Hegels Begriff der Wirklichkeit". In L. Illetterati, F. Menegoni (Eds.). *Wirklichkeit. Beiträge zu einem Schlüsselbegriff der Hegelschen Philosophie*. Frankfurt a.M.: Klostermann, pp. 11–42.

Labarierre, P.-J. 1983. *Les discours de l'altérité. Une logique de l'expérience*. Paris: PUF.

Rouse, J. 2005. "Heidegger on Science and Naturalism". In G. Gutting (Ed.), *Continental Philosophy of Science*. Oxford: Blackwell, 2005.

Soresi, S. 2007. "*Denken, Nachdenken* e *objektiver Gedanke* nella filosofia di Hegel". *Verifiche*, XXXVI, 1–4, pp. 61–92.

Federico Sanguinetti
# Hegel and the Question "What Characterizes Human Beings qua Animal Organisms of a Specific Sort?"

**Abstract:** In this chapter I will sketch Hegel's answer to the question "What characterizes the human individual qua animal organism of a specific sort?". I will also claim that, by looking at the way in which Hegel methodologically develops his answer, one can extract an argument that aims at avoiding confusions between *philosophical* approaches and approaches that are, at best, *scientific*. In particular, I will claim that Hegel would not consider evolutionary-shaped, anthropogenetic accounts as good *philosophical* answers to the above mentioned question. In the final part of the chapter, I will show that Hegel's take on these issues is similar to McDowell's.

A philosophical consideration of the realm of what is human can be seen as the overarching theme of the third part of Hegel's *Encyclopaedia*, the Philosophy of Spirit. In the second section of the Philosophy of Spirit, called Objective Spirit, Hegel discusses what he takes to be the inner rationality of the juridical, moral and political features of modern, European society, conceived as a result of the historical development of humanity. In the third section, called Absolute Spirit, he discusses what he takes to be the nature of the distinctive cultural productions of the human beings (art, religion and philosophy). Before dealing with such topics, in the first section, called Subjective Spirit, Hegel approaches the more fundamental question: "What characterizes the human individual qua animal organism of a specific sort?"[1] (from now on HI, for brevity's sake), in the sense of what characterizes the rational existence that is realized by the biological organism pertaining to the species *Homo sapiens*.

In this chapter I will try to sketch Hegel's answer to HI. I will also claim that, by looking at the way in which Hegel methodologically develops his answer to

---

[1] The term 'specific' does not have here a biological meaning. In other words, the relevant question, for Hegel, is not what are the biological features that make a determinate organism a member of the species *Homo sapiens*. As I will claim, being human is considered by Hegel as a metaphysical (not biological) determination.

**Federico Sanguinetti,** Federal University of Rio Grande do Norte

https://doi.org/10.1515/9783110604665-006

this question, one can extract an argument that aims at avoiding confusions between *philosophical* approaches and approaches that are, at best, *scientific*. As a token of this confusion I will consider accounts that put forward anthropogenetic claims concerning how the transition from merely animal life to spiritual animal life[2] took place as part of philosophical answers to HI. From the standpoint of a contemporary interpreter, reading Hegel's methodology in a similar way could be tempting, since such a reading would reflect a currently popular way of bringing results from the natural sciences into philosophical accounts of the relationship between human individuals and their environment at both a natural and a social level.[3] However, I will claim that this is not what Hegel does in answering HI and that Hegel would argue that these approaches are not good philosophical programs. In the final part of the chapter, I will show that Hegel's take on these issues is similar to McDowell's, highlighting both that McDowell's philosophy can be used to interpret Hegel's position on HI and that Hegel's argument is still present in the contemporary debate.

# 1 What is Hegel's answer to the question about HI?

## 1.1 Restricting the Focus: Spirit as a Whole and Spirit as HI

In § 377 of his *Encyclopedia*, Hegel says that what is "substantial" in human beings is "spirit." (ES, § 377)[4] If we are to translate Hegel's talk of spirit in a non-Hegelian vocabulary, we could say that spirit, in an extended sense (a sense that encompasses subjective, objective and absolute spirit), can be translated as "rationality + its realizations in the world". However, Hegel's way of dealing with the question "what is spirit?" is far more complex than that. In this section, I will try to give an overlook of Hegel's take on the question and to restrict the focus to how he conceives of spirit qua HI.

---

[2] I use here the terms "anthropogenetic/anthropogenesis" in a (broadly) evolutionary sense, a sense which merely indicates that the human being, qua spiritual animal, developed from non-spiritual animals over time.
[3] This seems to be the core of Moss's (2016, 2017) reference to Hegel as an inspiring source for his philosophical anthropology. Traces of this reading seem to be present in Testa (2010), Pinkard (2012, 2017a), even though it is far from my intention to present them as full-blown advocates of it.
[4] See also NE 1827/28, p. 60. Throughout this chapter, I will modify Inwood's translation by substituting "mind" with "spirit".

When Hegel asks himself "what is spirit?", he says that this question can be interpreted as having either one or different things as an answer.[5]

In one sense, spirit is said to have a distinctive determination [*Bestimmung*], which Hegel further characterizes as its "truth". "Truth" is conceived here in a material/normative fashion: as indicating a relation of correspondence between something and its own concept, conceived of as its *essence*.[6] And if we ask what the "truth" of spirit is, the answer, for Hegel, is simple: *to be free*.[7] What does "to be free" mean in this context? Hegel has a broad, 'logical' concept of freedom, which can be glossed as "self-reference", "not being constrained from the outside".[8] Freedom defines activities not limited by something external to themselves.[9] In particular, the kind of freedom that characterizes spirit is a specific instantiation of the concept that involves self-consciousness.[10] Freedom involving self-consciousness (= "self-consciousness-involving self-reference", "not being constrained from outside in a way that involves self-consciousness"), then, is the standard against which every manifestation of what is human is measured.

In a second sense, the different things [*verschiedenes*] that constitute the second answer to the question "what is spirit" are what Hegel calls *figure[s]* [*Gestalt[en]*] of spirit – where "figures" means ways in which a concept (in our case: freedom qua concept of spirit) manifests itself in existence, or in reality [*Realität*].[11] In the Philosophy of Spirit as a whole Hegel organizes these figures according to an increasing level of adequacy to freedom qua concept of spirit. Thus, the ways in which spirit is instantiated can be more or less adequate to

---

**5** "What is spirit? The sense of this question is: what does the truth [*das wahrhafte*] of spirit amount to, or the determination [*Bestimmung*] of spirit; to this last question there is only one answer; to the question what is spirit the answer is: different things [*verschiedenes*]." (NS 1827/28, p. 559, *my translation*)
**6** See Ikaheimo (2017, pp. 427–428). On Hegel's concept of truth, cf. Halbig (2002), Miolli (2016).
**7** See NE 1827/28, pp. 65–66: "The essence of spirit is freedom[.]" See also NS 1827/28, p. 556, NE 1827/28, p. 66 and p. 68.
**8** As we shall see in Section 2, this concept does not merely apply to spirit.
**9** As examples of instantiation of this logical structure, we can take the fall of bodies and universal gravitation (EN, §§ 267 and 269 A) – the system of celestial bodies can be analysed in terms of an "activity" which is not limited by something external – as well as living beings in general and non-spiritual animals in particular (EN, § 351) – in cognitively (e.g., sensing) and practically (e.g., digesting) assimilating their environment, non-spiritual animals prove themselves as "free", insofar as in these activities the environment of the animal is not something absolutely other to them. The environment is a constitutive element of – not a merely external constraint for – the activities that make out their animality.
**10** I will come back to what it does mean in section 1.2.
**11** See ES, § 553.

the standard of "self-consciousness-involving self-reference", "not being constrained from outside in a way that involves self-consciousness".

In the Philosophy of Subjective Spirit, Hegel seems to restrict his analysis to the consideration of spirit qua embodied rationality of the individual human being – the embodied rationality of the individual human being as a *figure* (or better yet, a "set" of figures) of spirit. And spirit considered qua embodied rationality of the individual human being seems to be conceived of (at least to a certain extent) in an Aristotelian fashion, qua the peculiar aspect that qualifies our way of being animals.[12] *I take spirit, in this restricted meaning, to be what characterizes human beings qua animal organisms of a special sort, and in this chapter I will focus on the way Hegel conceives of spirit according to this specification.* When necessary, in what follows I will use the notation 'spirit$_{HI}$' in order to circumscribe my talk to this aspect of spirit.[13]

Now, how are we to conceive of spirit$_{HI}$ against the background of these general remarks about spirit?

Hegel specifies his views in the beginning of his *Lectures on the Philosophy of Subjective Spirit*, in the course of which he tells us what spirit$_{HI}$ is. In order to unpack Hegel's specifications, I will start from two quotations and try to isolate the claims that are relevant to my reconstruction.

First:

> *What is spirit?* This we have to assume as a premise, and to appeal to the representation. [iii] What spirit is, the concept of spirit, can at first only be something entirely formal, because [i] what spirit is is the concern of our entire <discipline> [= *the Philosophy of Spirit*, FS]. [iv] This part [= *the Philosophy of Subjective Spirit*, FS] can only contain the formal universal essence, the substance of spirit, and this is freedom. How this determination is connected to nature, we shall see afterwards [*I will discuss this in Section 2*, FS]. [ii] The essence of spirit is freedom; [...] [v] the entire discussion that follows will constitute the explanation and demonstration of this proposal. (NE 1827/28, pp. 65–66)

---

**12** Hegel's high appreciation for Aristotle, in particular for his treatment of the human being, is well known – see, among others, Ferrarin (2001), in particular Ch. 8, Dangel (2013). "Hegel and Aristotle on the Metaphysics of Mind" is also the title of a call for paper for a special issue of the *Hegel-Bulletin* 2020. Recently, the extent of Hegel's Aristotelianism has been put at the center of a lively debate, which focuses on the way Hegel conceives of the relation between norms, freedom and nature – see, among others Pippin (2008, ch. 2), Pinkard (2012, 2017b); Stern (2016, 2017); Peters (2016), Alznauer (2016), McDowell (2017), Stein, (2018), Kern (*forthcoming*). In this paper I tend to follow McDowell's and Stern's perspective.

**13** I am not claiming here that individual human beings can be the animals of the specific sort they are in absolute abstraction from the social dimension – see McDowell (2017, 16–17 and 23).

Second:

> [i] What is spirit? The concept about this can be only incomplete; we *will* know it only in philosophy itself [= *in the whole Philosophy of Spirit*, FS]; [iii] here [*in the Philosophy of Subjective Spirit*, FS] what is spirit can be said only formally and universally; if we could define it already here, all the rest would be unnecessary. [ii] The essence of spirit is what is simply substantial, freedom, this lies in our consciousness [...]. [iv] But this is only formal freedom; the concrete is the further self-determining within itself. – Freedom is the foundation from which all other activities come from [*ausgehen*]; [the foundation] to which all other [activities] are subordinate. (NS 1827/28, p. 567–568, *my translation*)

The relevant claims here are:
i. 'What is spirit?' is the central question of the entire Philosophy of Spirit.
ii. The "essence of spirit is freedom."
iii. In the Philosophy of Subjective Spirit, we will not know what spirit is in its entirety (= we will not see to what extent all the *figures* of spirit correspond to freedom qua its essence), but we will merely know what spirit is "only formally and universally." As specified above, I take this to mean that in the Philosophy of Subjective Spirit we will know what $spirit_{HI}$ is – not what the realizations of $spirit_{HI}$ in the world are.[14]
iv. In the Philosophy of Subjective Spirit we will know freedom only as "formal freedom."[15] This seems to mean that freedom, qua formal freedom, must be intended in a different sense with respect to other specific forms of freedom and from spirit's freedom in general qua 'self-consciousness-involving self-reference', 'not being constrained from outside in a way that involves self-consciousness' – I will specify what does it mean in Section 1.2.
v. The claim that "the essence of spirit is freedom" is, at first, a proposal the "explanation and demonstration" of which is contained in "the entire discussion that follows".

From i-v), then:
vi. Hegel conceives of formal freedom qua essence of $spirit_{HI}$.
vii. The Philosophy of Subjective Spirit contains the "explanation and demonstration" that formal freedom is the essence of $spirit_{HI}$.

In the next sections I will try to explain what I take the claims vi) and vii) to mean.

---

14 This is the topic of Objective and Absolute Spirit.
15 I take the assertions "the formal essence of spirit is freedom" and "the essence of spirit is formal freedom" as being equivalent.

I will try to do that by applying to spirit$_{HI}$ Hegel's general claim that the question 'what is spirit?' can be interpreted as having either one or different answers.

## 1.2 Spirit$_{HI}$ Is, Essentially, to Be Formally Free

In the previous section, I analysed Hegel's claim according to which, when we ask what spirit is, we can give a unique answer or different things as an answer. In this section and in the next one, I will claim that, if we restrict our focus and ask what spirit$_{HI}$ is, we can *also* give a unique answer or different things as an answer. I will now discuss the unique answer.

If we give a unique answer, we are specifying what the essence of spirit$_{HI}$ is, and this is formal freedom. What does it mean, then, that human beings, qua animal organisms of a special sort, are, essentially, formally free? I believe that Hegel's talk of spirit's formal freedom can be interestingly translated in contemporary terms as *'responsiveness to norms as reasons'*.[16] That is: human beings are essentially formally free to the extent that they are in the position to give reasons for what they think and do. But what does it mean that human beings are essentially formally free insofar as they are responsive to norms qua reasons? As a way to illustrate what does it mean to be responsive to norms qua reasons, we can start by considering what Hegel says about the relation of correctness [*Richtigkeit*]. When discussing this concept, Hegel says:

> We have in general two levels [*when we talk about correction*, FS], an objectual and a subjective one, i.e. thinking. In theoretical matters, the object provides the norm, while what is spiritual [*das Geistige*] lays down the norm in practical matters. (LL 1831, p. 17, *translation modified*)

This means that, for Hegel, our theoretical performances are normatively constrained by how things are, and our practical performances are governed by norms we set for how things in the world ought to be.[17] Moreover, Hegel considers human beings as beings for which norms are *self-consciously known*. This dis-

---

[16] See, for instance, McDowell (2010), Pinkard (2012), Pippin (2008, Ch. 2). See also Knappik (2013, Ch. 6.2) for textual evidences that could support this claim.
[17] According to the theoretical dimension of correctness, "[t]he object lays down the law to which thinking adapts. But in practical life we have an aim, an obligation [*Sollen*], a plan. When we build a house, it is we who establish the norm [i.e., the blueprint by which matters are measured]. Things are not themselves the norm, but rather must conform to our chosen representations. Matters are correct when we make them adapt to our chosen norm" – see LL 1831, p. 16, *translation modified*.

tinguishes us from non-spiritual animals. Non-spiritual animals are too, according to Hegel, responsive to normativity. However, they do not know norms as reasons.[18] By self-consciously knowing the norms that govern our thoughts and actions, these norms can become *reasons* we can give for what we think or do. And this is what makes out the freedom that characterizes us qua animal organisms of a special sorts, since norms, insofar as they are known, are not conceived of as external powers which limit us.[19]

## 1.3 What Does It Mean that the Philosophy of Subjective Spirit, Contains the "Explanation and Demonstration" that Formal Freedom is the Essence of spirit$_{HI}$?

If we now consider that the question "what is spirit$_{HI}$?" has different things as an answer, we have to specify what these "different things" are.

The claim that formal freedom is the essence of spirit$_{HI}$, is programmatic.[20] Hegel seems to tell us that, in order to understand what concretely means that human being are, essentially, formally free, we require a more refined understanding that refers to a plurality of determinations. These determinations are meant to transform the programmatic claim that the essence of spirit$_{HI}$ is formal freedom into the explained and demonstrated meaning of what spirit$_{HI}$ is. I believe that a very promising way to make sense of this movement is to take seriously what Corti (2016a, 2016b, 2018) calls a "reconstructive" stance.[21] What are

---

[18] See EN, § 360: "Since the impulse can only be fulfilled through wholly determinate actions, this appears as instinct, since it seems to be a choice in accordance with a determination of an end. However, because the impulse is not a known purpose, the animal does not yet know its purpose as a purpose." See also EL, § 19, Ä, 80 (apud Pinkard 2012) and NH 1822, p. 5. See on this Pinkard (2012).

[19] This is the kind of self-consciousness involving freedom (which is in turn a kind of "logical" freedom qua "self-reference", "not being constrained from the outside") that characterizes us qua animal organisms of a specific sort.

[20] With respect to spirit in general, Hegel says that "if we could define it [= *what spirit is*, FS] already here, all the rest would be unnecessary." NS 1827/28, p. 567–568, *my translation*.

[21] Corti opposes a "reconstructive" reading of the *Philosophy of Subjective Spirit*, which "denies a picture of the mind as constituted by multiple layers that would correspond to the different sections of the *Philosophy of Subjective Spirit* (Corti 2016b, pp. 241–242), to a "descriptive" reading, characterized by the "1) *[s]eparability,* 2) *instantatiability,* and 3) *additivity*" (Corti 2016b, p. 236) of the layers. Corti is mainly interested in reading Hegel's treatment of the determinations of the Philosophy of Subjective Spirit as "parts of a global *retrospective reconstruction* of the con-

the relevant determinations that transform the programmatic claim that formal freedom is the essence of spirit$_{HI}$ into the "concrete" meaning of what spirit$_{HI}$ is? These determinations are – pre-eminently[22] – our practical and theoretical capacities such as sensing (§§ 399–402), habituating oneself (§§ 409–410), being conscious of an external world (§§ 413–423), being self-conscious (§§ 424–437), having intuitions of objects in space and time (§§ 446–450), having representations that are not bound to the present empirical situation (§§ 451–457), speaking a language (§§ 458–464), comprehending concepts, judging, making inferences (§§ 465–468), having practical feelings (§§ 471–472), practical inclinations, and the capacity of making choices (§§ 473–477). And how do these "different things" transform the programmatic claim that the essence of spirit$_{HI}$ is formal freedom into the "concrete" meaning of what spirit$_{HI}$ is? In order to understand this transformation we have to analyse in further detail how these "different things" relate to formal freedom qua essence of spirit$_{HI}$. Hegel writes:

> [f]reedom constitutes the essential determination of spirit, and we can say that freedom is the concept of spirit. By 'concept' is understood first of all the simple determination that constitutes the distinctive characteristic of something. (NE 1827/28, p. 67)

Hegel however goes on saying that this distinctive characteristic:

> [...] is a simple determinacy that includes difference as suspended, so that the subject is rendered determinate, and incorporates the difference in such a way that in this difference it has returned to itself. The concept includes the difference, but the difference is at the same time transparent [...]. In this science we have to grasp the finite firmly in both its difference and its unity. (NE 1827/28, p. 67)

Formal freedom qua concept of spirit$_{HI}$ is a determination that involves in itself features that concretely articulate it. Thus, in the first place, we can say that between the concept of spirit$_{HI}$ and its different determinations there is a relation of inclusion. In order to concretely spell out what spirit$_{HI}$ concretely is – not to give a mere, abstract definition – it is necessary to refer to internal articulations of the concept. Hegel conceives of these internal articulations as a "self-determining within itself" of formal freedom qua concept of spirit$_{HI}$. On the one hand these specifications represent internal differences, internal *partial* aspects of the concept in question which makes it determinate. On the other

---

ditions for a cognitively contentful experience" (Corti 2016b, p. 239). However, I believe that his take can also be fruitfully adopted in considering the broader topic of formal freedom qua HI.
**22** I do not take into account here the determinations discussed by Hegel at the beginning of the Anthropology, such as the Natural Qualities and Natural Changes. On this, see Corti (2016a).

hand, each of them, taken individually, does not exhaust the meaning of the concept in question and we cannot understand them in abstraction from the concept in question – we cannot understand the spiritual capacities *qua spiritual capacities* independent from the consideration that spirit$_{HI}$ is essentially formally free.

Secondly, Hegel claims that the exposition of the self-determining of the concept of spirit$_{HI}$ through its internal differences is not arbitrary, as it follows particular structure. We can unpack this structure by analysing its a) beginning, b) end, and c) middle.

a) Hegel says that:

> There is a distinction between the determination of something and the original being of something; the determination represent an aim, a goal [...]. (NS 1827/28, p. 559, *my translation*)

So, with what must the beginning of the exposition of the self-determining of the concept of spirit$_{HI}$ be made? The beginning of the exposition coincides with the internal articulation that is less adequate to account for the full meaning of what it means to be formally free. This is spirit$_{HI}$ in "its original being", namely the soul.[23] Human being is considered here according to a determination (the concept of soul) that thematizes the human being in its corporeality and bodily capacities such as sensation, feeling, habit.

b) The end, the result of the exposition of the Philosophy of Subjective Spirit – where Hegel says what spirit$_{HI}$, in its essence, *concretely* is – is free spirit. Free spirit is "freedom itself" (ES, § 481), "will as free intelligence." (ES, § 481)

c) The beginning and the end of the exposition are linked by a teleological process striving to reach the concrete meaning that spirit$_{HI}$ is, essentially, to be formally free.

> Only when we consider spirit in this process of the self-actualization of its concept, do we know in its truth (for truth just means agreement of the concept with its actuality). In its immediacy, spirit is not yet true [...], has not yet converted its actuality into the actuality appropriate to its concept. The entire development of spirit is nothing but its self-elevation to its truth, and the so-called soul-forces have no other meaning than to be the stages of this elevation. (ES, § 379 A, 7)

---

23 See ES, § 387 A: "We must begin our treatment with immediate spirit; but this is *natural spirit, soul*. To suppose that we should begin with the mere concept of spirit would be a mistake; for as we have already said, spirit is [...] actualized concept."

Formal freedom is conceived of as the goal of an expositive process according to which various determinations necessarily follow one another until an adequate determination is reached – "will as free intelligence". This determination includes in itself all the preceding determinations qua its internal difference. The preceding determinations are still to be conceived as manifestations of formal freedom, modes according to which our formal freedom (our capacity to be responsive to reasons) conceptually articulates itself: sensation, feeling, habit, and so on, in their human specificity, articulate the answer to the question "what does it mean that human beings are free insofar as they are responsive to norms as reasons?". And even though any of them, taken separately, accounts for the full meaning of our formal freedom (= any of them, taken separately, accounts for our capacity to be responsive to norms as reasons), these determinations are answers to the question "what is spirit?"[24] – they are *figures* [*Gestalten*] of spirit.

In Section 3.1 we shall further analyse how this teleological movement towards the full understanding of what it means that the essence of spirit$_{HI}$ is formal freedom must be conceived of from a methodological point of view. In particular, I will explore in further detail what is meant by saying that "the [...] discussion that follows" the initial assertion that the essence of spirit$_{HI}$ is formal freedom represents, on the one hand, the "explanation" and, on the other hand, the "demonstration" that the essence of spirit$_{HI}$ is formal freedom. Before doing that, I shall discuss in Section 2 a further element of Hegel's treatment that is key to understand his argument, namely his understanding of the (self-)exposition of formal freedom qua concept of spirit$_{HI}$ as a "liberation from nature" [*Befreiung von der Natur*] through different "stages". This will help us to shed light on the way Hegel conceives of the relation between spirit$_{HI}$, on the one hand, and both external nature and our own biological nature, on the other hand.

---

[24] They are part of the "different things" (the "*verschiedenes*") that – as Hegel writes in the passage cited in 1.1 – provide the answer to the global question "what is spirit?"

## 2 The Process of Liberation from Nature Through Stages

Hegel characterizes spirit by distinguishing it from nature[25]:

> Determined is only that which determines itself against an other, therefore we must here talk, at the same time, of that from which spirit distinguishes itself – this is nature. So, if one wish to determine spirit one must indicate its difference from nature. (NG 1825, p. 165)

In order further to understand how Hegel conceives of the relation between formal freedom qua the concept of spirit$_{HI}$ and spirit$_{HI}$'s internal articulations, it is useful to see how Hegel embeds them into the spirit's relation to nature. First (2.1), I will characterize spirit's relation to nature in an oppositional, but non-dualistic way; secondly (2.2) I will show how Hegel articulates this relation as a process of liberation from nature.

### 2.1 Spirit's Formal Freedom Stands in Opposition to Nature

If we consider spirit$_{HI}$ starting from his essence, the relation between spirit$_{HI}$ and nature is embedded into an opposition between spirit's freedom and nature's necessity.[26] However, this opposition is not rigidly dualistic. As a matter of fact, the global relation between nature and spirit is framed into a progressive reconstruction of reality as a whole that displays increasing degrees of freedom – this latter concept being intended now in the extended, logical meaning, qualifiable by the gloss "self-reference", "not being constrained from outside". According to this broad meaning, non-human nature displays degrees of freedom too[27] – the natural levels of freedom are lower, less adequate to the conceptual standard of "self-reference", "not being constrained from outside", than the specifically human ones. Still, there is a metaphysical shift between the non-spiritual natural world and human beings – a shift determined by the fact that human beings self-consciously know the norms that govern their thoughts and actions as reasons.

---

**25** Nature is defined by Hegel as "the Idea in the form of *otherness*" (EN, § 247). I cannot dwell here on what this formulation means. In general terms, we can say that Hegel holds that nature coincides with the subject matters of Mechanics, Physics, and Organic Physics – sciences which, *very roughly*, cover the fields of inquiry of physics, chemistry and biology.
**26** See NH 1822, p. 10, 11 and NG 1825, p. 176; 197.
**27** See above fn. XX.

This shift justifies Hegel's talk of necessity with respect to non-spiritual nature. "Necessity" seems to be here a shortcut for "not free in the same way as spirit is". Nature's necessity is therefore the term in contrast to which the talk of spirit's freedom obtains its determinate meaning. In particular, Hegel claims that spirit's freedom is, precisely, freedom from nature.[28] However, it must be noted that, even though Hegel conceives of "freedom as fundamental essence of spirit" in contraposition to nature, he does not conceive of the human being as something "super-natural". Metaphysical difference does not mean ontological otherness: in order to claim that human beings are animal organisms which instantiates a form of freedom which is not instantiated by other animals does not necessarily imply that we should cut off human beings from the realm of what is natural.[29] In Hegel's words, "freedom from [the realm of] what is natural" remains a freedom "within [the realm of] what is natural." (NS 1827/28, p. 575, *my translation*)[30]

## 2.2 The Internal Development of Spirit's Formal Freedom Coincides with the Liberation from Nature Through Stages

The opposition between the spirit's freedom and nature's necessity, if merely asserted, is still an abstract way of considering the spirit's relation to nature – a way that limits itself to the *not yet articulated* (= the not yet philosophically "expla[ined]" and "demonstrate[ed]") consideration of "freedom" qua essence of spirit. Hegel therefore introduces a second relation of the spirit to nature, which coincides with the process of development of spiritual freedom. If we restrict our talk to formal freedom, this process coincides with the development of the concept of spirit$_{HI}$ we have seen above in Section 1.3. Hegel calls this process a process of "liberation from nature." (EN, § 376R)[31]

> We have a twofold relation of spirit [to nature] before us: the first relation is that we compare it with nature [*this is the relation of opposition I have just discussed in 2.1, FS*], the second relation is that we observe it while it detach itself from nature, this is its coming out

---

**28** "We consider freedom as fundamental essence of spirit, freedom from [the realm of] what is natural [...]." (NS 1827/28, p. 575, *my translation*)
**29** See McDowell (2017). McDowell's remarks in Lectures IV and VI of McDowell (1996) are helpful to understand this point.
**30** See also See NH, 1822, p. 8, 151, and NS 1827/28, p. 556.
**31** See also ES, § 381R, § 386, FSS 1822/25, 92.

> from nature, the self-production of spirit, this is what the dissolution of what is natural into spirit is, it is [nature's] striving to be spirit, to reach its own truth. (NG 1825, p. 165, *my translation*)[32]

Spirit, in the beginning of the exposition, in its "original being", is not yet formally free and becomes formally free only at the end of a development of *Befreiung von der Natur*. This "liberation", qua "detach[ment] [...] from nature", takes place through "stages" (ES, § 386).[33] As I have already stated, the beginning of the exposition of the Philosophy of Subjective Spirit is spirit$_{HI}$ in its "original being", namely the soul. The concept of the soul or natural spirit is for Hegel an internal specification of formal freedom, the specification under which he conceives of the spiritual (= substantially human) determinations that are in closest connection with (= are more conditioned by) the natural dimension – more specifically, the dimension of our body.[34] And this is why Hegel equates the concept of soul and the concept of natural spirit [*Naturgeist*].[35] Starting from the spirit in its "original being" (namely the concept of soul), Hegel exposes the further internal articulations of formal freedom that account for the detachment from the natural dimension:

> Our discipline shows the way of liberation. Spirit is free, but first it is merely implicitly free in itself. It has to bring forth what it is implicitly in itself. This process is the content of our discipline: to liberate oneself, i.e., to liberate oneself from nature. (See NE 1827/28, p. 71) We begin from natural spirit and we see the succession of stages of its liberation until the point in which its freedom becomes its object, it grasps its freedom, [it grasps that] man as such, according to its own nature, is free. (NG 1825, p. 205, *my translation*)

At the end of the exposition of the Philosophy of Subjective Spirit we grasp the human being qua a natural organism of a special sort – namely, a formally free one, in the sense of a being that is responsive to norms as reasons. But how are we to read the self-determining of freedom qua liberation from nature from a methodological point of view?

---

[32] See also NS 1827/28, p. 576, *my translation*, and NG 1825, p. 165, the passages that precede the one I have quoted.

[33] Again, these stages are the internal articulations of formal freedom discussed in Section 1.3.

[34] See ES, § 387 A: "We must begin, therefore, with spirit still in the grip of nature, related to its bodiliness, spirit that is not as yet together with itself, not yet free." See also NG 1825, p. 153, *my translation:* "[I]t is spirit in its natural being, in its corporeality, there nature is what is overwhelming [*das Uebermächtige*], spirit is not yet what is free in it [= *in nature*, FS]."

[35] See NH 1822, p. 18, ES, § 387 and § 392 A. See also NH 1822, p. 42, NG 1825, pp. 152–153, and p. 202.

# 3 How to Read (and How Not to Read) Hegel's Approach to the Claim the Development of Spirit's Formal Freedom Is a Process of Liberation from Nature

## 3.1 How to Read Hegel's Approach to the Claim that the Development of Spirit's Formal Freedom Is a Process of Liberation from Nature

At this point, it is necessary to explain in further detail how Hegel conceives of his treatment of spirit$_{HI}$ from a methodological point of view.[36]

As I have argued in Section 2.2, the spirit's liberation from nature through different stages coincides with the development exposed by Hegel in the Philosophy of Subjective Spirit. This process constitutes, on Hegel's view, the *authentically philosophical manner* to answer HI.

On the one hand, Hegel takes this movement as being the "explanation"[37] of formal freedom qua essence of spirit$_{HI}$. Explanation here means an answer to the question "what does it mean (to be) X?". Questions of the form "what does it mean (to be) X?" are intended here to be *constitutive questions*. Constitutive questions can be also spelled out in Why-terms – in our case "why are human beings animal organisms of a specific sort – namely, rational ones?". However, the "because" of a constitutive answer is different from the "because" of a causal answer.[38] Causal answers account for the causal history that led something to be there.[39] On the contrary, for constitutive answers I here mean answers that provide for "the very understanding of the key *terms* of the question" and define the "conditions on something's being what it is. [...] Constitutive conditions ground explanations of something's nature, the aspect of what it is that could not possibly be different if it is to be and remain what it is."[40] In our case, formal freedom *explains* what it means when we say that we are animal organisms of a specific sort, namely spiritual ones. The "discussion that follows" the programmatic

---

[36] Once more, this take on Hegel's thought is directly based on the reading of Corti (2016a, 2016b, 2018).
[37] NE 1827/28, pp. 65–66
[38] See Dasgupta (2017).
[39] I will come back on this in Sections 3.2 and 4.
[40] See Burge (2010, p. 5). See also Burge (2010, pp. 533–534) for further clarifications.

statement that formal freedom is the essence of spirit$_{HI}$ *explains* what does it mean that we are formally free – and it does so by showing which are the conditions of possibility for us to be formally free. In order for animal organisms like us to be formally free, in the sense of our being responsive to norms as reasons, we must be endowed with the features exposed by Hegel in the course of the Philosophy of Subjective Spirit.

On the other hand, and at the same time, this movement is the "demonstration" (NE 1827/28, pp. 65–66) that formal freedom is the concept of spirit$_{HI}$. Demonstration here means that we cannot describe the series of the conceptual conditions for something to fall under the concept in question in an arbitrary way. Rather, we must introduce the conceptual conditions for something to fall under the concept in question *according to a specific method*, that Hegel takes to be the self-unfolding of the concept:

> In the philosophical consideration, we have to follow the development of the concept; the concept as such, only as concept is at first the universal immediate; but we must now also consider the different figures of spirit, these are as determinate concepts as well, degrees in the development of the concept. However, we must not expose [*aufführen*] merely a gallery of shapes, because in that case we would proceed empirically, the series of our shapes must be a series of concepts in their necessity. When we go on in the development, we bump into the further determinations of the concept, and this progress of the concept is what is actually philosophical. (NH 1822, p. 22, *my translation*)

This method, as I have claimed in Section 1, orders the constitutive conditions from the condition (the figure [*Gestalt*] of spirit$_{HI}$) that is more distant and less adequate to account for the concept of spirit$_{HI}$ through determinations that correspond in an increasingly higher way to this concept.[41] In what follows, I will try to show how Hegel's philosophical treatment of spirit$_{HI}$ distinguishes itself from claims – conducted from a (broadly conceived) evolutionary, anthropogenetic perspective – about the *factual steps* in the transition from mere animal life to spiritual animal life.

---

[41] The increasing adequacy is merely conceptual. Otherwise, it would be hard to make sense of the fact that "practical feeling" follows "thought", and that "intuition" follows "reason" and "self-consciousness".

## 3.2 How Not to Read Hegel's Approach to the Claim that the Development of Spirit's Formal Freedom Is a Process of Liberation from Nature

What I have proposed up to now seems to rule out that Hegel's *philosophical* treatment of HI – a treatment which coincides with the transition from nature to free spirit through a process of liberation from nature – can be interpreted as involving anthropogenetic claims that should answer the question of "how has it come to be" that responsiveness to norms as reasons emerged from nature.[42] The answer to such a question would require reconstructing the conditions that fill the gap between the realm of (non-responsive-to-norms-qua-reasons) nature and the emergence of responsiveness to norms qua reasons. These conditions would be interpreted as some kinds of 'properties', 'events' or 'processes' and would be organized into a narrative which conceives of them as following one another in time.

There are many Hegelian formulations that seem to characterize the process of the spirit's liberation from nature as a process that describes a sort of *factual* emergence of spirit from nature, an emergence that happened in time.[43] However, this way to answer the question HI would not be a *constitutive explanation*. It would rather be a *causal explanation*, in the sense of a reconstruction of a *causal history* that led human beings to appear – and, for Hegel, this is not the *philosophical way* to answer the question about HI.[44] As a matter of fact, the reading I have reconstructed attributes to Hegel a *synchronic* treatment both of spirit's detachment from nature and of the stages of spirit in its liberation. Hegel's *philosophical* treatment of these stages cannot be interpreted as indicating that the lower stages of spirit *chronologically* precede the higher ones.[45] As Hegel writes:

---

42 In addition, I claim that Hegel does not formulate elsewhere a separate philosophical theory to the extent of explaining how human beings qua animal organisms of a specific sort could emerge from nature.
43 Hegel often talks of an "emergence" [*herauskommen*] of the spirit from nature (see, for instance, NG 1825, pp. 152–153; 165). This "emergence" is further characterized as a "detach[ment]" (NS 1827/28, p. 576), a "motion" (FSS 1822/25, p. 93), a "process of coming out from nature and to liberate from it" (NS 1827/28, pp. 151–152 and 165) and the topic of the philosophy of Subjective Spirit is described as "the *history* [*italics mine*, FS] of its liberation" (NE 1827/28, pp. 60–61).
44 This would probably be for Hegel, at best, the "external appearance" of the passage from nature to spirit's freedom – see NG 1825, p. 180, *my translation*. See also PhR, § 258, p. 156.
45 I do not claim here that *every* philosophical explanation, in Hegel's view, rejects the diachronic dimension as being irrelevant (take, for instance, Hegel's philosophical treatment of history). Moreover, it is important to underline that diachronic philosophical explanations do not coincide with merely causal-historical explanations. Here I merely claim that Hegel does not

"All of this which belongs to the sphere of history must be accepted as fact; it does not pertain to philosophy. If, now, this is to be explained, we must understand the way in which this must be treated and considered." (EN, § 339 A) Thus, an explanation that involves claims about the causal history that led something to appear is not properly a *philosophical* account of something.

To better understand Hegel's point, it could be useful to look at a remark on the origin of the state:

> As far as the Idea of the state itself is concerned, it makes no difference what is or was the *historical* origin of the state in general (or rather of any particular state with its rights and determinations) – whether it first arose out of patriarchial conditions, out of fear or trust, or out of corporations etc., or how the basis of its rights has been understood and fixed in the consciousness as divine and positive right or contract, habit, etc. In relation to scientific cognition [= *philosophy*, FS], which is our sole concern here, these are questions of appearance, and consequently a matter [*Sache*] for history [...]. The philosophical approach deals only with the internal aspect of all this, with the *concept as thought* [*mit dem gedachen Begriffe*]. (PhR, § 258, p. 156)

I take what Hegel says here about the Idea of the State to be methodologically valid for Hegel's answer to HI as well. Accordingly, anthropogenetic claims about the process that made us from merely natural organisms into organisms responsive to norms as reasons are not even part of Hegel's *philosophical* answer to HI.[46] His account – as shown in the previous sections – seems rather to be a reconstruction of the conceptual requirements of our self-understanding qua for-

---

offer a diachronic philosophical explanation in his treatment of the metaphysical nature of the individual human being. In addition, we might notice that Hegel distinguishes between time and history, attributing history only to spiritual beings. Non-spiritual, natural beings do not properly have an history, they merely are in time – on Hegel's opposition between history and nature see NG 1825, p. 180; 195; 228; NH 1822, p. 14. With this distinction in mind, we can see that Hegel rules out that philosophy's task is to make claims about: i) How nature produced spirit$_{HI}$ in *time* – see ES, § 381 A, pp. 14–15 and EN, § 376 A. This seems to me to speak against Moss's (2016, 2017) claim according to which his own project of philosophical anthropology has a Hegelian echo. ii) How spirit$_{HI}$ produced itself in *history*. I align myself here with McDowell's criticism of Pippin's idea that, since spirit is "product of itself", it is "a kind of socio-historical achievement." (Pippin 2008, p. 42). See McDowell (2017), p. 18 and pp. 23–24. On this debate see also Pippin (2018), McDowell (2018), in particular p. 253, and Schuringa's (2018) defence of McDowell's interpretive take.

**46** We will assess in Section 4 whether Hegel would have considered that full-blown anthropogenetic accounts could legitimately constitute a *scientific* way to answer the question HI.

mally free beings, the constitutive conceptual requirements necessary for us to be responsive to norms qua reasons.[47]

## 4 Conclusion: What Would Hegel Think About Anthropogenetic Answers to the Question HI?

I would like to conclude by trying to shed further light on how I believe Hegel would think of anthropogenetic approaches to the question about HI. I will do this by making reference to what John McDowell says about the distinction between evolutionary and philosophical ways of conceiving this question and its answer.

As Carl Sachs (2011) has persuasively highlighted, McDowell would consider evolutionary interpretations of the question of HI (such as "[h]ow has it come about that there are animals [responsive to norms as reasons]?"[48]) as perfectly sound and legitimate. However, according to him, these are scientific questions that are not to be confused with philosophical questions.[49] According to Sachs, McDowell conceives of philosophical questions as transcendental questions that basically ask "what it is" to be something – in our case, "what it is to be a being which is responsive to norms as reasons?". Sachs contrasts this question with the question "[h]ow has it come about that there are animals [responsive to norms as reasons]?".

> Of course something can be said about the emergence, over the prehistoric timescale, of rational animals. If the question is put in a scientific vocabulary, it becomes one of how human culture evolved from the proto-cultures of our hominid and hominoid ancestors. Consider the evolutionary process as we now understand it: a process that led from extinct Miocene apes through the australopithecines and early species of *Homo*, to later species of *Homo* and the emergence of *Homo sapiens*. This certainly seems to be a process which begins with mere animals and passes through apparently 'proto-rational animals' and results in fully fledged rational animals. Thus the question about how rational animals emerged

---

47 In this vein, we can extend to the issue at stake the methodological attitude of Hegel's claim that historical/causal accounts of concept acquisition by an individual cannot be confused with a philosophical account – see SL, p. 519.
48 This is a paraphrase of McDowell (1996), p. 123, apud Sachs (2011), p. 72: "[h]ow has it come about that there are animals that possess the spontaneity of understanding?" On this, see McDowell (1996), pp. 123–124.
49 See McDowell 1996, p. 123. See also McDowell (1996, p. 124). For a discussion of this topic in McDowell, see, among others, Bernstein (2002), McDowell (2002), Lovibond (2008), McDowell (2008), Perini-Santos (2018) and McDowell (2018, p. 247 ff).

from "mere" animals is a perfectly fine question for the empirical sciences to raise and answer, as best as possible, given the multiple and fragmentary lines of evidence drawn from paleontology, comparative genetics, morphology, and psychology. But this question is not only a perfectly acceptable question by McDowell's lights; it is also "as close a good question can come to the bad questions" that he urges us to exorcise. (Sachs 2011, p. 74)

What does this last claim mean? What are the "bad questions" McDowell wants to exorcise? McDowell's bad questions seem to be questions that present themselves as philosophical, start from a dichotomy, and need to fill a gap between the relevant terms through pieces of constructive theory. Historical/causal explanations are pieces of constructive theory – in the sense that they aim to *solve* a puzzle through the *construction* of a theory. And to fill a gap in historical/causal terms between a level in which responsiveness to norms qua reasons is absent and one in which it is present is precisely what evolutionary shaped accounts of HI try to do. Given this framework, McDowell's remark that the evolutionary question about how responsiveness to norms as reasons emerged from nature is "as close a good question can come to the bad questions"[50] seems to mean that the question at stake, which is perfectly legitimate within a scientific framework, is in some sense dangerous because it is liable to become a "bad question" insofar as it is taken to be a *philosophical* one.[51]

I believe that McDowell's distinction between the two approaches to the question about HI can be useful for shedding light on Hegel's standpoint. As noted in Section 3.2, Hegel would not consider anthropogenetic questions as philosophical questions – and, correspondingly, anthropogenetic claims as philosophical claims. To put it in McDowellian terms, Hegel would consider an anthropogenetic interpretation of the question about HI (a question such as "how the transition from mere responsiveness to norms and responsiveness to norms as reasons took place in time/history?") as a "philosophically bad" question. Why so?

Let us consider again what Hegel says about the origin of the State. Hegel claims that the question about the origin of the State (a question which plainly is to be responded to through a reconstruction of a series of causal conditions that occurred in time – be they "patriarchial conditions", "fear", "trust", "corporations", or other things) cannot be confused with the philosophical, constitutive question of *what is* the State (= what characterizes the State – as it was at the

---

50 McDowell (1996, p. 124, fn. 12).
51 See Sachs (2011, p. 76). This would probably be, by paraphrasing an expression McDowell uses in another context, to try "to make [responsiveness to norms as reasons] out of items that are in themselves less than that." (McDowell 2002, p. 273).

time Hegel wrote – as a community of a specific sort). Extending these considerations to the issue at stake in this chapter, I believe that is possible to approximate Hegel to McDowell, and attribute to Hegel the claim that the question about the emergence from nature of an animal which is responsive to norms qua reasons (a question that, analogously to what happens in the case of the State, should be responded to through an historical/causal account) *cannot be confused* with the philosophical, constitutive question about HI. The answer to the latter question is not the exposition of a series of events chronologically ordered according to the "external appearance". (NG 1825, p. 180, *my translation*) Rather, it is the internal development of a concept, qua its (self-)explanation, which coincides with the articulation of the conditions of possibility on something's falling under it and, at the same time, the necessary justification of that concept.[52]

Would Hegel follow McDowell in considering the question about the emergence from nature of an animal organism which is responsive to norms qua reasons as being a scientifically "good question"? The main obstacle to approximating Hegel and McDowell's views on this issue is not in principle methodological. As a matter of fact, Hegel, just like McDowell, conceives of questions that have historical/causal explanations as an answer as legitimate *scientific* questions. More precisely, Hegel considers both chronologically shaped explanations of natural phenomena and historical explanations of human events as being accounts that can pertain to *positive sciences*.[53] The answer whether Hegel would follow McDowell in considering the question about the emergence from nature of an animal which is responsive to norms qua reasons as being a scientifically

---

[52] I do not aim to investigate whether Hegel shares McDowell's argument that philosophy should not be a constructive enterprise. For my purposes, it is sufficient to argue convincingly that they share the claim that the historical/causal question of the origin of responsiveness to norms as reasons is not philosophical.

[53] See EL, § 16, where Hegel treats history and natural history as being *positive sciences*. The § 16 of the *Encyclopaedia* is particularly relevant for understanding how Hegel conceives of the differences between philosophy and the practices of positive sciences – there is no room here for a detailed reconstruction, Inwood (2003, Ch. 4) offers an extended discussion. Given what Hegel says there, it seems to me that anthropogenetic accounts have two features that characterize them (at best) as "positive-scientific" accounts, in contrast with philosophical accounts. As a matter of fact, like positive sciences, anthropogenetic accounts deal with "empirical singularit[ies] and actuality" (EL, § 16); and just like positive sciences, anthropogenetic accounts do not follow the development of the concept as epistemological ground of cognition but a "ground of cognition" which can be "partly argumentation; partly feeling, belief, the authority of others; and, in general, the authority of inner or outer intuition" (EL, § 16). The distinction between philosophy and positive sciences does not absolutely mean that Hegel underrates the importance of the latter.

"good question" depends specifically on how we interpret a series of Hegelian assertions that *prima facie* seem to rule out the plausibility of the idea of an anthropogenetic reconstruction of "[h]ow has it come about that there are animals [responsive to norms as reasons]."[54] The claims which are relevant to discussing this issue are the following:
1. Hegel believes that the different levels into which nature can be categorized, levels which can be considered as a system of increasing degrees of complexity, cannot be seen as naturally producing one from the other.[55]
2. In particular, living organisms (including human beings) must not be considered as the result of a natural, chronological process of formation and mutation and successively developing in time.[56]

These claims, however, can be interpreted in two ways. On the one hand, one can interpret these claims as showing that Hegel denies that there has been something like an anthropogenetic process of emergence from nature – that meaning that the evolutionary-shaped question about "[h]ow has it come about that there are animals [responsive to norms as reasons]" does not make any sense at all.[57] The question about "[h]ow has it come about that there are animals [responsive to norms as reasons]" would be, in this case, a bad question tout court. According to this interpretation, Hegel's position would not be exactly like McDowell's, since it would not recognize the scientific legitimacy of the question about "[h]ow has it come about that there are animals [responsive to norms as reasons]".[58] On the other hand, one can interpret these claims as showing that Hegel is prescribing a sort of methodological/explanatory prohibition to philosophy[59], namely the prohibition to deal with evolutionary-shaped questions such as "[h]ow has it come about that there are animals [responsive to norms as reasons]". In other words, Hegel would simply be trying to preserve philosophy

---

54 We may say that at the *type*-level (the general level of questions which have historical/causal accounts as an answer) there is a convergence between Hegel and McDowell's views. As we will see, doubts can be cast whether there is convergence also at the relevant *token*-level (the specific case of the question "[h]ow has it come about that there are animals [responsive to norms qua reasons].")
55 See EN, § 249.
56 See EN, § 339 A. See also EN, § 249 A, pp. 20–21.
57 This seems to be the position – among others – of Pinkard (2012, Ch. 1, fn. 25), Fritzman and Gibson (2012). I have myself held this thought, even though I am not inclined to do so now.
58 Drees (1992, p. 54) holds that "Hegel agrees that evolution factually occurs in nature, but simultaneously refuses to accept any significance of this kind of evolution for nature's scientific and philosophic comprehension".
59 For such a position, see Houlgate (2005, pp. 173–174).

from determinate (bad, in his view) ways of practicing it – namely ways that confuse historical/causal explanations with constitutive explanations. If we refer what Sachs says about McDowell to Hegel, we could say that these claims are just meant to remind us that we "should maintain a clear distinction between the intellectual vocation of scientific explanations [= *historical/causal explanations*, FS] and the intellectual vocation of transcendental description [= *philosophical, constitutive explanations*, FS]." (Sachs 2011, p. 76) By this interpretation, Hegel's position could be, in principle, very close to McDowell's, since it makes room (from the points of view both of method and content) for a scientific (= historical/causal) understanding of "[h]ow has it come about that there are animals [responsive to norms as reasons]". One must however keep in mind that such a claim never explicitly appears (at least to my knowledge) in Hegel's texts.

For this reason, we can say that Hegel seems to share with McDowell the claim of a separation between the question HI and the question "[h]ow has it come about that there are animals [responsive to norms as reasons]?". They seem to agree that the first is a distinctively constitutive (= philosophical) one – even though it is not straightforward whether Hegel would consider the second a legit scientific one, as McDowell does. Their common point seems to be that philosophy has a specific contribution to offer to our comprehension of what we are, a contribution which cannot be reduced to the explanatory practices of positive sciences.

# Abbreviations

| | |
|---|---|
| Ä | Hegel, Georg Wilhelm Friedrich (1988): *Aesthetics: Lectures on Fine Art*. Trans. T. M. Knox, 2 vols. Oxford: Clarendon Press. |
| EL | Hegel, Georg Wilhelm Friedrich (1991): *The Encyclopaedia Logic (with the Zusätze)*. Trans. T.F. Geraets, W.A. Suchting, and H.S. Harris. Indianapolis/Cambridge: Hackett. |
| EN | Hegel, Georg Wilhelm Friedrich (1970): *Hegel's Philosophy of Nature*. Trans. A.V. Miller. Oxford: Clarendon Press. |
| ES | Hegel, Georg Wilhelm Friedrich (2007): *Hegel's Philosophy of Mind*. Trans. A.V. Miller, with revisions and commentary by M.J. Inwood. Oxford: Clarendon Press. |
| FSS 1822/25 | Hegel, Georg Wilhelm Friedrich (2011): "A Fragment on the Philosophy of Spirit (1822/5)". In: *Hegel's Philosophy of Subjective Spirit*. Vol 1. Ed. M. J. Petry. Dordrecht/Boston: Reidel, pp. 91–139. |
| LL | Hegel, Georg Wilhelm Friedrich (2008): *Lectures on Logic*. Trans. C. Butler. Bloomington and Indianapolis: Indiana University Press. |
| NE 1827/28 | Hegel, Georg Wilhelm Friedrich (2007): *Lectures on the Philosophy of Spirit*. Trans. R. R. Williams. Oxford: Oxford University Press. |

| | |
|---|---|
| NG, 1825 | Hegel, Georg Wilhelm Friedrich (2008): "Philosophie des Geistes vorgetragen vom Professor Hegel. Sommer 1825. Nachgeschieben durch Griesheim." In: *Vorlesungen über die Philosophie des Subjektiven Geistes*. In: *GW*, vol. 25,1. Ed. C. Bauer. Hamburg: Meiner, pp. 147–544. |
| NH, 1822 | Hegel, Georg Wilhelm Friedrich (2008): "Philosophie des Geistes nach dem Vortrage des Herrn Professor Hegel. Im Sommer 1822. Berlin. H Hotho". In: *Vorlesungen über die Philosophie des Subjektiven Geistes*. In: *GW*, vol. 25,1. Ed. C. Bauer. Hamburg: Meiner, pp. 3–144. |
| NS 1827/28 | Hegel, Georg Wilhelm Friedrich (2011): "Psychologie oder Philosophie des Geistes nach Hegel". In: *Vorlesungen über die Philosophie des Subjektiven Geistes*. In: *GW*, vol. 25,2. Ed. C. Bauer. Hamburg: Meiner, pp. 553–918. |
| PhR | Hegel, Georg Wilhelm Friedrich (1991). *Elements of the Philosophy of Right*. Ed. A. Wood, trans. H.B. Nisbet. Cambridge: Cambridge University Press. |
| SL | Hegel, Georg Wilhelm Friedrich (2010): *Science of Logic*. Trans. G. di Giovanni. Cambridge: Cambridge University Press. |

# References

Alznauer, Mark (2016): "Hegel's Theory of Normativity". In: *Journal of the American Philosophical Association*, pp. 196–211.

Bernstein, Jay (2002): "Re-enchanting Nature". In: Nicholas Smith (Ed.): *Reading McDowell. On Mind and World*. London: Routledge, pp. 9–24.

Burge, Tyler (2010): *Origins of Objectivity*. Oxford: Oxford University Press.

Corti, Luca (2016a): *Pensare l'esperienza. Una lettura dell' "Antropologia" di Hegel*. Bologna: Pendragon.

Corti, Luca (2016b): "Conceptualism, Non-Conceptualism, and the Method of Hegel's Psychology". In: Susanne Hermann-Sinai and Lucia Ziglioli: *Hegel's Philosophical Psychology*. London: Routledge, pp. 228–250.

Corti, Luca (2018): "Senses and Sensations. On Hegel's Later Account of Perceptual Experience". In: Federico Sanguinetti and André Abath: *McDowell and Hegel. Perceptual Experience, Thought and Action*. Switzerland: Springer, pp. 97–116.

Dangel, Tobias (2013): *Hegel und die Geistmetaphysik des Aristoteles*. Berlin: DeGruyter.

Dasgupta, Shamik (2017): "Constitutive Explanation". In: *Philosophical Issues* 27, pp. 74–97.

Drees, Martin (1992): "Evolution and Emanation of Spirit in Hegel's Philosophy of Nature." In: Hegel-Bulletin 13. No 2, pp. 52–61.

Ferrarin, Alfredo (2001): *Hegel and Aristotle*. Cambridge Mass.: Cambridge University Press.

Fritzman, J.M and Gibson, Molly (2012): "Schelling, Hegel, and Evolutionary Progress." In: *Perspectives on Science* 20. No. 1, pp. 105–128.

Halbig, Christoph (2002). *Objektives Denken. Erkenntnistheorie und Philosophy of Mind in Hegels System*. Stuttgart-Bad Cannstatt: Frommann-Holzboog.

Houlgate, Stephen (2005): *An Introduction to Hegel, Freedom, Truth and History*. Malden MA: Blackwell.

Ikaheimo, Heikki (2017): "Hegel's Psychology". In: Dean Moyar (Ed.): *The Oxford Handbook to Hegel*. Oxford: Oxford University Press, pp. 424–449.

Inwood, Michael (2003): *Hegel*. London: Routledge.

Kern, A (*forthcoming*): "Life and Mind: Varieties of Neo-Aristotelianism: Naive, Sophisticated, Hegelian". *Hegel Bulletin*.
Knappik, Franz (2013): *Im Reich der Freiheit*. Berlin/Boston: DeGruyter.
Lovibond, Sabina (2008): "Practical Reasons and Its Animal Precursors". In: Jakob Lindgaard (Ed.): *John McDowell: Experience, Norm, and Nature*. Malden: Blackwell, pp. 112–123.
McDowell, John (1996): *Mind and World*. Cambridge, Mass: Harvard University Press.
McDowell, John (2002): "Responses". In: Nicholas Smith (Ed.): *Reading McDowell. On Mind and World*. London: Routledge, pp. 269–305.
McDowell, John (2008): "Responses". In: Jakob Lindgaard (Ed.): *John McDowell: Experience, Norm, and Nature*. Malden: Blackwell, pp. 200–267.
McDowell, John (2010): "Autonomy and Its Burdens". In: *The Harvard Review of Philosophy* 17. No 1, pp. 4–15.
McDowell, John (2017): "Why Does It Matter to Hegel that Geist Has a History?" In: Rachel Zuckert and James Kreines (Eds.): *Hegel on Philosophy in History*. Cambridge: Cambridge University Press, pp. 15–32.
McDowell, John (2018): "Responses". In: Federico Sanguinetti and André Abath: *McDowell and Hegel. Perceptual Experience, Thought and Action*. Switzerland: Springer, pp. 231–258.
Miolli, Giovanna (2016). *Il pensiero della cosa*. Trento: Verifiche.
Moss, Lenny (2016): "The Hybrid Hominin: A Renewed Point of Departure for Philosophical Anthropology". In: Phillip Honenberger (Ed.): *Naturalism and Philosophical Anthropology*. Basingstoke and New York: Palgrave Macmillan, pp. 171–182.
Moss, Lenny (2017): "Detachment Theory: Agency, Nature and the Normative Nihilism of New Materialism". In: Sarah Ellenzweig and John Zammito: *Materialism and the New Politics of Ontology: History, Philosophy, Science*. Abingdon and New York: Routledge, pp. 227–249.
Perini-Santos, Ernesto (2018): *A Second Naturalization for a Second Nature*. In: Federico Sanguinetti and André Abath: *McDowell and Hegel. Perceptual Experience, Thought and Action*. Switzerland: Springer, pp. 177–192.
Peters, Julia (2016): "On Naturalism in Hegel's Philosophy of Spirit". In: *British Journal for the History of Philosophy* 25. No 1, pp. 111–131.
Pinkard, Terry (2012): *Hegel's Naturalism. Mind, Nature, and the Final Ends of Life*. New York: Oxford University Press.
Pinkard, Terry (2017a): Does History Make Sense? Cambridge Mass./London: Harvard University Press.
Pinkard, Terry (2017b): "The Form of Self-Consciousness". In: Rachel Zuckert and James Kreines (Eds.): *Hegel on Philosophy in History*. Cambridge: Cambridge University Press, pp. 106–120.
Pippin, Robert (2008): *Hegel's Practical Philosophy*. Cambridge et al.: Cambridge University Press.
Pippin, Robert (2018): "Reason in Action. A Response to McDowell on Hegel." In: Federico Sanguinetti and André Abath: McDowell and Hegel. Perceptual Experience, Thought and Action. Switzerland: Springer, pp. 211–227.
Sachs, Carl (2011): "The Shape of a Good Question: McDowell, Evolution, and Transcendental Philosophy". In: *The Philosophical Forum* 42, No 1, pp. 61–78.

Schuringa, Christoph (2018): "Review of Terry Pinkard (2017): *Does History Make Sense? Hegel on the Historical Shapes of Justice*." In: *European Journal of Philosophy* 26, No 1, pp. 679–682.

Stein, Sebastian (2018): "The Metaphysics of Rational Action: Kantian and Aristotelian Themes in Hegel's Absolute Idealism". In: Michael Thompson (Ed.): *Hegel's Metaphysics and the Philosophy of Politics*. London and New York: Routledge, pp. 142–175.

Stern, Robert (2016): "Why Hegel Now (Again) – and in What Form?" In: *Royal Institute of Philosophy Supplement* 78, pp. 187–210.

Stern, Robert (2017): "Freedom, Norms, and Nature in Hegel: Self-Legislation or Self-Realization?" In: Rachel Zuckert and James Kreines (Eds.): *Hegel on Philosophy in History*. Cambridge: Cambridge University Press, pp. 88–105.

Testa, Italo (2010): *La natura del riconoscimento. In Riconoscimento naturale e ontologia sociale in Hegel (1801–1806)*. Milano-Udine: Mimesis.

Yusuke Akimoto
# Marx's Philosophy on Natural History

**Abstract:** The purpose of this article is to clarify how Marx criticized the idealist concept of nature. By analyzing Marx's texts, I will focus on their relation to Hegel's teleology, in which nature is taken as an intermediate stage of process wherein the idea as a logical category gradually begins to actualize itself in the objective world. By distancing himself from Hegelian idealism, Marx reached his own concept of natural history wherein nature is always considered to remain non-identical to human beings. As I will argue, Marx's critical concept of nature is motivated by his critique of the notion that humans are somehow dominant over nature. Marx's claim that nature cannot be completely dominated by human individuals led him to a reflection on the existence of human individuals and how they cannot be properly understood outside of the dialectical relationship that they entertain with nature. As I will claim in the conclusion, Marx's understanding of the human individual is made up of a dialectical relationship between individual, nature, and society. Furthermore, I will also claim that this dialectical relationship can prevent us from accepting a reductive and organicist understanding of human beings.

After the environmentalist movement emerged, many began to argue that the philosophy of Karl Marx approved of a view in which mankind dominates nature. This is due to the fact that he appears to argue that the expansion of human productive power is the path to the "sphere of freedom". According to traditional Marxists, Marx understood that human beings were able to bring out their productive potential only if they were able to free themselves from their exploitation at the hands of the capitalists. If we follow this view, freedom in human life would be found in the liberated power of production, which could be controlled under the "rational" and restrained plan of the socialist community. This portrait of Marx's work admittedly contains an affirmation of man's dominance over nature. Moreover, we have seen examples in history in which real socialist states have readily destroyed the natural environment.

However, we should ask whether or not it is really the case that Marx made such a serious error. Today, especially after John Bellamy Foster's work made it clear that Marx's concept of nature cannot be understood without reference to his ecological intentions, there are no longer any serious scholars who accept

**Yusuke Akimoto,** Kyoto University

the image of Marx as fetishizing productive power. As Foster has shown, Marx grasped human labor as a metabolism [*Stoffwechsel*] between nature and man. From this ecological point of view, capitalism must be abolished, because capitalist production "disturbs the circulation between man and the soil" (Marx 1990, p. 441). Thanks to Foster's interpretation, Marx's ecology became understood as a criticism of capitalism, which opens up a rift in this metabolism. With that said, it is worth noting that Foster presupposes an ontological possibility of a unity between human beings and nature.

Foster conceives of the older Marx's theory of metabolism as "a more solid and scientific expression" of the "metabolic relation between human beings and nature" for which the young Marx argued (Foster 2000, p. 158). Actually, the problem of the relationship between human beings and nature is a theme that was at the core of Marx's early work. However, we should not understand this to mean that Marx simply insisted on the possibility of unifying both beings. While Hegel, from his standpoint of absolute idealism, intended to show us the way to a path that could bring about an identity between nature and spirit, Marx could not accept this way of thinking. Foster does not pay attention to this point because he regards Marx's critique of Hegelian philosophy as nothing more than a protest motivated by his adherence to a standpoint of anti-idealism or materialism (Foster 2000, p. 15 f.). Foster also stresses the importance of dialectics for confronting a mechanical materialism, but he seems to treat the term dialectics as if it were a theoretical panacea when he does so. This brings us to the crucial question of how the content of the very term "dialectics" should be understood. It was not scientific positivism, but rather a philosophical reflection on idealism that led Marx to think about the relation between human beings and nature.

The purpose of this article is to clarify how Marx criticized the idealist concept of nature. In analyzing Marx's texts, I will focus on their relation to the problem of Hegel's teleology, which constitutes the essence of Hegel's own philosophical system. Through my interpretation, we will see that Marx reached his own concept of natural history by distancing himself from Hegelian teleology, which reduces nature into one step in the realization of the absolute. Therefore, my interpretation differs from the account given by Michael Quante, who conceives of Marx's concept of nature as a reversed version of the Hegelian "spirit".

The first section describes the framework of Marx's ideas on natural history by analyzing the text of the *Paris Manuscripts of 1844*. If looked at superficially, the thesis of the unity between human beings and nature could lead one to believe that Marx espouses the same teleological concept of nature as Hegel. However, by focusing on *Enzyklopädie* as a philosophical system, Marx was able to criticize Hegel's concept of nature as being too abstract, insofar as it is was locked into a teleological framework. The second section interprets Marx's criticism of Hegelian teleology in order to highlight the fact that Marx considered nature to always remain non-identical to human beings. The third section claims that Marx's concept of nature, which is liberated from this teleology, contains within itself an attempt to provide a critique of the notion of human dominance over nature.

# 1 Human Beings and Nature: Society as the Resurrection of Nature

Marx's early theory of alienation can be described by the following thesis: "fully developed naturalism, equals humanism" (Marx 1975, p. 296). According to Quante, by reversing what Hegel grasped as the alienation [*Entäußerung*] of the absolute idea, Marx gave nature its place as first substance, which can then develop into the "Spirit" through the process of alienation. Thus, Marx comprehended nature as a self-relation which sublates itself over the alienation and through a series of developments (Quante 2009, p. 303). The naturalistic materialism of Ludwig Feuerbach was hugely impactful on Marx. Yet Marx saw that Feuerbach lacked the kind of dialectical method that Hegel offered. Hence, Marx created a complex of materialism and dialectics. In this complex, the overcoming of self-alienation by means of abolishing the capitalist notion of "private property" leads to the recovery of the human being, who can thus renew his or her life as a *natural* life. If the systematic position of nature is indeed contrary to Hegel's philosophy, we ought then to ask whether Marx followed Hegel's teleological understanding of history. To this question, Quante answers that Marx also took a goal-oriented view of history, in which human beings are seen as the highest stage of nature (Quante 2009, p. 303).

As is well known, for Marx, labor does not only go in one direction, i.e., from the human being to nature. According to him, the human being is one part of the whole of nature, and for human beings, nature is an "inorganic body [*unorganischer Leib*]" (Marx 1975, p. 276). Because of this, the labor done by the human being is itself a self-production by nature, insofar as the subject of labor is one part of nature and the subjectivity of labor is the mediation of nature's relationship to itself. In this way nature develops into man through human labor and the human being develops itself as a natural being. On the basis of this subject-object relationship, Marx determines labor to be a praxis of "objectification [*Vergegenständlichung*]", in which he finds the means of self-reflection of nature itself. It is said that in objectification, the subject objectifies his or her own internal essence and knows him or herself by reflecting on the result of the act of producing, which is expressed by the product. Naturally, the result of labor in each instance does not always coincide with the form that the subject originally imagined. However, both the result of producing, which opposes the intention of the subject, and the result, which was not imagined in advance, reflect the state of subjects of labor. Therefore, reflecting on the product of labor is an opportunity for the producing subject to know him or herself, and the next stage of production is also stimulated by this reflection. According to Marx, "the product of labor is labor which has been embodied in an object, which has become material: it is the objectification of labor" (Marx 1975, p. 272)[1]. When the human being "sees himself in a world that he has created" (Marx 1975, p. 277), the self-reflection of the human being through his or her own objectification is regarded as a self-reflection of nature itself, because man is at the same time a natural being.

Here, Marx shows the special position of the human being within the whole of nature by expanding Feuerbach's theory of consciousness to the sphere of praxis. According to Feuerbach, the reason why the human being can be distinguished from other living things consists in the nature of human consciousness. In reaction to Feuerbach's anthropology, Marx shows the "character of species [*Gattungscharakter*]" of humanity, by observing not only human consciousness, but also by observing the connection between human needs and praxis, because for Marx "the productive life is the life of the species" (Marx 1975, p. 276). For Marx, human beings are able to behave without any purpose of self-preservation, while "the animal is immediately one with its life activity" (Marx 1975, p. 276). Or rather, because human beings can behave without reference to their physiological needs, and moreover are able to form objects "in accordance with the laws of beauty" (Marx 1975, p. 277), they can be distinguished from other living things. Since human praxis does not connect directly with the immediacy of life activity, it can be referred to as a "conscious life activity" (Marx 1975, p. 276) and a conscious self-producing of nature which is mediated by the "human natural being" (Marx 1975, p. 337).

On the other hand, the capitalist notion of "private property" is criticized as a source of estrangement, because it corrupts human labor by reducing it to an unspontaneous action which is undertaken for the purpose of individual self-preservation. This effectively causes labor to become subordinated to one's phys-

---

[1] Please note that I have adjusted the spelling in all translations to match with American spelling conventions.

ical needs as an end itself. For Marx, it is "private property" that hinders a constant circulation between nature and human beings.

(1) Namely, the product on which labor objectifies its own essence is necessarily exploited as a commodity and the laborer is thus only obliged to work without any enjoyment of their own products, upon which they were originally able to reflect themselves (in other words, this is the alienation of laborers from their own products).

(2) In this way of working, laborers are forced to make an involuntary exchange, and in so doing, the laborers experience an external character of their own activities, because he or she "only feels himself outside his work, and in his work feels outside himself' (Marx 1975, p. 274) (alienation from the productive action).

(3) However, so long as "free, conscious activity is man's species-character" (Marx 1975, p. 276), and this conscious activity is external to the subjects, they, i.e., the subjects of labor, are already alienated from their "species-being" as human beings (alienation from the species-being).

(4) And one who has been alienated from the species-being must also be alienated from any coexistence between human beings (in other words, this is the alienation of the human being from others). Generally speaking, the description in the part of *Estranged Labor* from the first manuscript intends to show not only the failure of objectification, but also to provide a framework that can clarify in what sense private property is reproduced by this alienated labor. In this framework, the unity of nature and man is forced to destroy itself.

As a critique of alienation, Marx acclaims Communism: "as the *positive* transcendence of *private property* as *human estrangement*, and therefore as the real *appropriation* of the *human* essence by and for man" (Marx 1975: p. 296). Marx believed that, through communism as a movement, human beings can once again obtain [*aneignen*] their essence as a natural being. Needless to say, the relationship between human beings and nature is not solitary, but rather social, thus "the human aspect of nature exists only for social man" (Marx 1975, p. 298). By overcoming this alienation from nature, human society can be renewed as a unity between nature and man, according to Marx's "programmatical formulation" (Quante 2009, p. 306). Thus *society* is the complete unity of man with nature – the true resurrection of nature – the accomplished naturalism of man and the accomplished humanism of nature. (Marx 1975, p. 298)

In this way, in the *Manuscripts of 1844*, natural history is portrayed as having four stages of development: (1) nature itself, (2) the unity of nature and human beings, (3) the alienation of human beings from nature, (4) the overcoming of alienation and reunification of both human beings and nature. As Quante pointed out, this portrayal of natural history appears to inherit a kind of teleological thinking from Hegel, in which history is constructed as a process aiming toward realizing the highest stage of itself. In addition, according to Quante, Marx's "realism" can be seen in his praise for industry and the natural sciences. On this point, Quante gives a thumbs-up to the interpretation that Marx regarded natural science as a relevant measure for the technical domination of nature (Quante 2009, p. 309).

## 2 The Problem of Teleology in Marx

Marx writes, "history itself is a *real* part of *natural history* – of nature developing into man" (Marx 1975, p. 303 f.). If we only look at this formula, then it may seem as if Marx views natural history in the same teleological light that Hegel did. According to Quante, although the teleological aspect of his concept of nature contradicts his realist manner of thinking, in which nature is conceived of as independent from human beings, Marx did not resolve this problem (Quante 2009, p. 310). On the one hand, the traditional Marxist orthodoxy regarded this aspect of his concept of nature as immature and thus cut it off from his mature work, i.e., his so-called "scientific socialism". On the other hand, being confronted with a series of abortive developments in real socialist states, philosophers such as Ernst Bloch , who was otherwise taken by Habermas as a "Marxist Schelling", intended to restore the teleological concept of nature because of its affinity with his philosophy of nature. However, both interpretations overlook the critical fact that Marx does not accept any meaning of Hegel's teleology. It was not his "realism" that led Marx to his own concept of natural history. It was rather his own philosophical reflection.

In his *third manuscript*, Marx discusses "the *positive* aspects of the Hegelian dialectic in the realm of estrangement" (Marx 1975, p. 341). Marx, on the one hand, appreciated that "Hegel conceives of labor as man's act of self-genesis" by grasping the structure of "*Aufheben* as an objective movement of *reacting* the alienation *into self*", but on the other hand he criticized that Hegel's grasp remains "within the sphere of abstraction" (Marx 1975, p. 342). According to Marx, nature, within the abstract framework of speculative idealism, was destined to lose any sense of reality. "As a result therefore one gets general, abstract *forms of abstraction* and, consequently, valid for, all content – the thought-forms or logical categories torn from *real* mind and from *real* nature (Marx 1975, p. 343). Marx explains this point while following along Hegel's texts.

As is well known, Hegel built up his philosophical "System" as a trinity of Logic, Nature and Spirit. It was his intention to describe the totality, in which the absolute idea develops itself in conformity with *logos* and comes back into itself. In the Hegelian system, nature is positioned as an intermediate stage wherein the idea as a logical category gradually begins to actualize itself in the objective world. As long as it remains at the level of material, nature is regarded by Hegel as imperfect. He conceives of nature as a step which must be overcome before human history, in which freedom could become actual, can come about. Hence it can be seen that Hegel accounts for nature with his teleological framework.

Marx did admire Hegel's "dialectic", which was defined by the movement of the idea;

> The outstanding achievement of Hegel's *Phänomenologie* [*des Geistes*] and its final outcome, the dialectic of negativity as the moving and generating principle, is thus first that Hegel conceives the self-creation of man as a process, conceives objectification as *Entgegenständlichung*, as alienation and as transcendence of this alienation; that the thus grasps the essence of *labor* and comprehends objective man – true, because real man – as the outcome of man's *own labor* (Marx 1975, p. 333)[2]

However, the "divine" dialectic in Hegel remains a circulation of the identical idea. Nature, as it is discussed in Hegel's philosophical system, is "the idea in a form of otherness" (Hegel 1970, p. 205). Even if it would succeed in overcoming otherness, this overcoming is only a realization of what is going to be realized from the beginning [*Anfang*]. According to Marx, it would not be until Hegel also experienced nature that he could abstract his concept of nature from that of reality. If Hegel believed it possible to create entities from pure abstraction, then his concept of nature could not be equivalent to reality. As Marx explains, "[t]o him [i.e. Hegel], therefore, the whole of nature merely repeats the logical abstractions in a sensuous, external form" (Marx 1975, p. 345). All of the entities within Hegel's system are already contained in the beginning of their development, hence the all is oriented toward the end of development, i.e., to the *telos*.

However, Marx's materialistic standpoint does not integrate nature into the closed system of idealism. The reason why he stresses – as Feuerbach had – the character of human beings as sensual, could be understood on this point: "as a natural, corporeal, sensuous, objective being he is a suffering [*leidend*], condi-

---

[2] Martin Milligan and Dirk J. Struik translated the German word, "Entgegenständlichung" as "loss of the object". Contrary to their translation, Raya Dunayevskaya – who was secretary to Leon Trotsky in Mexico— adopted "contra position" as the translation for this word (Dunayevskaya 1958, p. 309). Her translation of choice was motivated by her understanding the concept of objectification as the act of "*entgegen-stehen*" of subjectivity toward the objects, not as the act of "*ent-gegenständlich (zu machen)*". When viewed in this context, it seems more rational to follow to Dunayevskaya's suggestion here.

tioned and limited creature, like animals and plants" (Marx 1975, p. 336). That the human being is "suffering", means that they come into conflict with nature, i.e., they come into conflict with an object that is independent of themselves. Within Hegel's system, there is no heterogeneity, because a large number of its contents are the result of splitting from the one beginning of the system. That is why Hegel takes pride in his own philosophy as an "identity of identity with non-identity" (Hegel 1991, p. 74). But for Marx, "neither nature objectively nor nature subjectively is directly given in a form adequate to the *human* being (Marx 1975, p. 337). In light of this idea, it does not seem that Marx only reproached Hegel for the abstract structure of his philosophical system. He criticized him insofar as all the contents of his system are identical. In this sense, he did not only criticize Hegel because his dialectic "is standing on its head". Hence the task of his critique cannot be reduced to making sure that the dialectic "be turned right side up again", i.e., it cannot be reduced to an attempt to turn the idealistic dialectic into a concrete dialectic, as Marx suggested in *Das Kapital* (Marx 1990, p. 27). In short, the Marx's early thoughts should be not characterized by a teleology which hopes to achieve the recovery of identity, but instead ought to be characterized as a dialectic which contains permanent conflict between non-identical entities.

Marx accounts for his own standpoint in the following way: "a non-objective being is a *non-being* [*Unwesen*]" (Marx 1975, p. 337). Namely, human beings cannot confirm their existence without *real, sensuous objects*, with which they can express their lives. For Marx, human beings come into being by the act of engaging with the real objective world, and the place wherein they are able to come into being is none other than history. This conception of objective nature can be found in Hegel's philosophical system, because he integrated objectivity into the totality – of which the first principle is ideal. Hegel was able to clarify the dialectic of history by grasping that "the true is the whole" (Hegel 1977, p. 11), and hence establishing the end of history. Within his idea of totality, all of non-identity functions as a means of completing the steps in the development of identity, wherein the last point of the system is determined as the identity between thinking and being. History teleologically understood in this way is, for Marx, none other than the self-circulation of one idea, which has essentially nothing outside of itself. Hence, in Hegel's system, the subject as the substance of totality has no object of primary meaning. For Marx, history is found in the activity in which human beings are ceaselessly relating to nature outside of themselves. In other words, the movement in which they express their life "in real, sensuous objects" (Marx 1975, p. 336). He describes it as follows:

> And as everything natural has to *come into being,* *man* too has his act of origin – *history* – which, however, is for him a known history, and hence as an act of origin it is a conscious self-transcending act of origin. History is the true natural history of man (Marx 1975, p. 337)

So, if Marx wishes to preserve the possibility of admitting any difference between human beings and nature, then it would be quite inappropriate to look for a teleological claim within the thesis of "fully developed naturalism, equals humanism". It only appears to be teleological, because Marx operates outgoing from a viewpoint in which the relationship between human beings and nature is conceived of as a totality of natural history, without making the conflict of both indistinct. Certainly, as has already been quoted, the older Marx said that Hegel's dialectic is "standing on its head" and should be "turned right side up again". However, this problem is not as simple as Marx's description would make it out to be. Even if the idealistic dialectic of Hegel were to be turned materialistic, its teleological structure could still continue to exist within materialism. The reason why Marx's concept of natural history should be strictly distinguished from Hegelian teleology can be explained by his account of the ontological position of human beings. For Marx, the factum that "man is directly a natural being" (Marx 1975, p. 336) and that only humans experience conscious life activity does not mean that they are able to represent nature as a whole. That would be impossible, because nature is not given for the purpose of human existence.[3] As natural history, Marx submits an intricate relationship between the natural and the artificial. From this realistic way of viewing this relationship, a teleology which regards human history as the highest step of natural history would be, for Marx, nothing more than the reflection of a desire to dominate nature.

## 3 The Idea of Natural History: Nature with a Capital 'N'

Up to this point, I have argued that Marx's concept of natural history, even in his early period, contains a criticism against man's attempt to dominate nature, which is based on his insights concerning the non-identity between human beings and nature. Marx obtained his own position by confronting the teleology of Hegel. The following considerations will further investigate how Marx's later

---

[3] As opposed to the interpretation of Alfred Schmidt, the early work of Marx can be taken to claim that, "there is no reconciliation" of human beings and nature, because there is no nature which is given specifically for the sake of human existence (Schmidt 1971, p. 158).

theory works out the problem of man's domination of nature. Although Foster's contribution focused on the "metabolism" between human beings and nature, we should instead pay attention to the concept of natural history, which is at the core of a central methodological aspect of *Das Kapital*. By conceiving of the framework of capitalistic social economy as the process of natural history, Marx was able to explain the movement of capital as a natural law, by which human subjectivities are affected.

First, it is worth noting that the concept of natural history is a methodological category, by which capitalist society can be analyzed in its totality. For Marx, the persons who appear and act in this capitalist society are no more than a function of the logic of capitalism, therefore his criticism against capitalism is not based on any moral principles. Marx thus refers to capitalists as "capital personified", and claims that their "soul is the soul of capital" (Marx 1990, p. 201), because they are not the subject of profiting, but the object of the movement of capital as it pursues surplus values. There is no person who could play the role of subject in this schema. Hence, Marx conceives of the movement of capital as natural law.

> I paint the capitalist and the landlord in no sense *couleur de rose*. But here individuals are dealt with only in so far as they are the personifications of economic categories, embodiments of particular class-interests. My standpoint, from which the evolution of the economic formation of society is viewed as a process of natural history, can less than any other make the individual responsible for relations whose creature he socially remains, however much he may subjectively raise himself above them (Marx 1990, p. 17 f.)

What Marx wishes to achieve is the abolishment of this natural law, which carries on through itself by making subjective intentions into a mere medium for its own realization. This is to say, he hopes to liberate human history from the form of natural history. Along with the law of value, which first appears in capitalistic society, the socially determined labor-time is *post factum* distributed without the mediation of subjectivities. This is because the quantum of necessary labor is determined by the *fulfilled* exchange of commodities, while the *failed* exchange reduces the labor that makes these commodities into non-value-making labor. Within this social framework, the commodities present themselves as a "fetish," as if they are dressed in their own values. Thus, in a commodity, "the social character of men's labor appears to them as an objective character stamped upon the product of that labor" (Marx 1990, p. 62). Capitalist society, whose unit of being is a single commodity, must present itself as "the fantastic form of relation between things" (Marx 1990, p. 63). Although it is true that in the social relationships between human beings, it appears as if it were a natural phenomenon which could be changed by any subjectivity. By explaining this natural appearance as the unconscious process of social relations, Marx intends to find a way to abolish the "natural" character of society. To be sure, even though the natural law of society could be clarified as a social relationship between human beings, according to Marx, "it [society] can neither clear by bold leaps, nor remove by legal enactments, the obstacles offered by the successive phases of its normal development". However he says here, "it can shorten and lessen the birth-pangs" (Marx 1990, p. 17). So, what does Marx mean by this idea of shortening "the birth-pangs"?

Capital works by making the capitalist's subjective desire for profit into its own medium. Capitalists are able to maximize surplus value, when they can utilize "the capabilities, of his [man's] species" (Marx 1990, p. 285), i.e., the capabilities of co-operation [*Zusammenwirken*], as one of the "natural forces of social labor", which "cost nothing" (Marx 1990, p. 337). Capitalists desire to maximize surplus value, and the competition for their existence moves as the power that sets capitalist society in motion. In this society, unplanned and limitless production opens up a rift in the metabolism between human beings and nature. As Foster explained in detail, Marx condemns this capitalist destruction of nature. However, it does not mean that the recovery of the metabolism between human beings and nature is a core point of his criticism of capitalism. Because of the dual aspects of his concept of nature, the critical theory of Marx contains a paradoxical thesis: harm is wrought upon nature within natural history. While the former is, in a narrow sense, the object of human being, the latter, i.e., Nature with a capital "N", is conceived of by Marx as the force of history. Hence, to shorten "the birth-pangs" means leaving from the realm of necessity, to which human society is firmly tied.

What Marx concretely refers to is "the forces of production". In the third volume of *Das Kapital*, Marx calls the sphere in which human beings make their own metabolism with nature "a realm of necessity" (Marx 1998, p. 807). For him, capitalist society is tied up with natural necessity, insofar as it dominates the entirety of human life. Even if capitalists buy labor-power on an hourly basis, time spent out of labor is, for the workers, little more than the time needed to reproduce the power of labor. Marx says:

> In fact, the realm of freedom actually begins only where labor which is determined by necessity and mundane considerations ceases; thus in the very nature of things it lies beyond the sphere of actual material production (Marx 1998, p. 807)

In any form of society, it is necessarily the case that human beings must have a relationship with nature for the sake of their own existences. However, according to Marx, the "wrestle with nature", or the effort to sustain human life, can be minimized by maximizing "the forces of production", namely, by increasing the production of use-values within determinate terms. Instead of remaining in a state of being dominated by the force of natural necessities, Marx, as we all know, proposes that we regulate the production accomplished by human beings:

> Freedom in this field can only consist in socialized man, the associated producers, rationally regulating their interchange with Nature, bringing it under their common control, instead of being ruled by it as by the blind forces of Nature; and achieving this with the least expenditure of energy and under conditions most favorable to, worthy of, their human nature (Marx 1998, p. 807)

In this context, his interest in our "wrestle with Nature" can be, for many readers, understood as an affirmation of mankind's technical domination of nature. However, as Foster explained, scholars who found a sort of "Prometheanism" within Marx's theory had only "transposed" Proudhon's analysis of machinery and modernity into a critique of Marx himself (Foster 2000, p. 135). The term "force of production" implies rationalism, which prioritizes the sustainable metabolism between human beings and nature. In spite of this, it is worth examining a series of critical readings by Theodor W. Adorno. His criticism against Marx is, for Foster, no more than a critique which was "almost entirely culturalist in form, lacking any knowledge of ecological science" (Foster 2000, p. 245). Due to his narrow interpretation, Foster lost sight of the philosophical aspect of Marx's concept of nature.

## 4 In Place of a Conclusion: Natural History as a Critical Concept

So far, we have seen that Marx's early theory intended to provide an insight into the non-identity between human beings and nature, while Hegel integrated nature into the teleological system of his absolute idealism. Marx developed his own concept of nature by making a framework of natural history, within which the problem of human dominance over nature is interpreted as the self-destruction of Nature with a capital "N". Moreover, it opens up a new problem, because Marx seems to demand that we dominate nature in order to be liberated from the form of natural history. About this paradoxical idea, Adorno offered a detailed examination. According to him:

> In Marx the principle of the domination of nature is actually accepted quite naïvely. According to the Marxian way of seeing, there is something of a change in the relation of domination between people [...] but the unconditional domination of nature by human beings

is not affected by this, so that [...] the image of a classless society in Marx has something of the quality of a gigantic joint-stock company for the exploitation of nature (Adorno 2008, p. 58).

From Foster's standpoint, Adorno's critique can only be looked at as a result of identifying the Marxian theory with that of Proudhon. However, the main objective of Adorno's interpretation is not to identify what does not work within Marx's theory, but to find the manner of ideology-critique from the concept of nature in Marx. In accordance with this aim, Adorno focuses on the concept of natural history, by contrasting it with the Hegelian concept of the "trick of reason". As is already indicated, within the Marxian concept of natural history, the persons are functioning as the personifications of economic categories. According to Adorno, Marx's conception of natural history can be distinguished from Hegel's concept, because the former regards the natural character of history as an ideological appearance, while the latter fails to dissolve the "trick of reason" thoroughly. In Hegelian philosophy, there is not any kind of route that could help us escape from the structure of this "trick", so that "in this way, the unconscious history of nature is continued" (Adorno 2006, p. 118). Adorno stresses that "the 'natural laws of society' are ideology inasmuch as they are claimed to be immutable" (Adorno 2006, p. 118). Ideology here means neither a perceived notion nor false consciousness in the sense that Karl Manheim used the phrase. Instead, the illusion of immutable natural laws appears as if it is real, because real society holds a form of natural history. Adorno aimed to conceive of the Marxian concept of natural history as a critical term pertaining to the semblance of historical reality, which is both ideological and real. Adorno explains Marx's concept of "realm of freedom" in the following way:

> In contrast to the prevailing belief that Marx had a positive view of the natural laws of society and that one needs only to obey them to obtain the possibility of the right kind of society – in contrast to this belief, Marx wishes to get beyond them into the kingdom of freedom, i.e., to escape from the notion of history as natural history (Adorno 2006, p. 116).

As is well known, the notion that the complex between the domination of nature and social domination is a strategy of enlightenment [*Aufklärung*] is a core aspect of Adorno's philosophy. According to him, nature that is suppressed by rational institutions turns toward barbarism, in which enlightenment combines with destructive impulses and brings upon us a social catastrophe. This is "[w]hy mankind, instead of entering into a truly human condition, is sinking into a new kind of barbarism" (Adorno / Horkheimer 1979, p. xi). From this viewpoint of general history, we can see that the truth of the matter is that Marx stressed the possibility of abolishing "natural law", not the "metabolism" between human beings and nature. Marx does call for us to dominate nature, but only in a fully conscious and regulated form of domination. In doing so, it is possible for us to be liberated from our current unconscious form of dominating nature. For Adorno, this paradoxical way of thinking has to be seen as the true expression for the difficulty of criticizing our domination of nature.

Given the hint that Adorno's work can provide to us, we can now see how crucial it is that we are able to read the way that Marx intended to show how we can escape from the unconscious form of dominating nature. Thus, arguments about whether or not Marx managed to achieve some form of an ecology miss the point. Marx conceived of capitalistic society within the category of "natural history", and in the background of this idea we can still hear the reverberations of his early thought echoing that: "man is directly a natural being" (Marx 1975, p. 336). No matter how we look at it, there is no possibility to unify human beings with nature. As long as nature is not "directly given in a form adequate to the *human* being" (Marx 1975, p. 337), our relationship with nature is deeply embedded in human society. "Nature is a social category" (Lukács 1972, p. 239), and "if nature is a social category, the inverted statement that society is a category of nature is equally valid" (Schmidt 1971, p. 70).[4] At any rate, Marx's philosophy on natural history offers us an abundance of insights concerning the dialectics of society and nature. It shows us that there is a dialectical relationship between human being and nature, which can never be reduced to just one of the two poles.

Such relationship is basically constituted by labor, i.e. a social category and phenomenon, which is essential to the very realization of the individual human beings as human being. In this thought, Marx's theory allows us to realize that the understanding of human individual cannot be assimilated to the one of a purely natural being or organism, such as an animal, but it rather has to be characterized via a triadic relationship between the human individual, her society, and nature. If this is correct, then, not only the understanding of human society should not follow an organistic model, but also the human individual should not be reduced to a natural organism. With this in mind, one can better appreciate Marx's contribution to the issue of a metaphysics of human individuals, as it were, without falling pray of a kind of anthropocentric bias or delusion.

---

[4] As Schmidt indicates here, the philosophy of the Frankfurt scholars can be seen as an inversion of Georg Lukács's formula.

# Acknowledgement

I would like to express my sincere gratitude to Richard Stone for his continuous support. And this work was supported by JSPS KAKENHI Grant Number 19 J00446.

# Bibliography

Adorno, Theodor W. / Horkheimer, Max (1979): *Dialectic of enlightenment*, translated by John Cumming, London: Verso.

Adorno, Theodor W. (2006): *History and freedom: lectures 1964–1965*, translated by Rodney Livingstone, Cambridge: Polity.

Adorno, Theodor W. (2008): *Lectures on negative dialectics: fragments of a lecture course 1965–1966*, translated by Rodney Livingstone, Cambridge: Polity.

Dunayevskaya, Raya (1958): *Marxism and freedom... from 1776 until today*, New York: Bookman Associates.

Foster, John Bellamy (2000): *Marx's ecology: materialism and nature*, New York: Monthly Review Press.

Hegel, Georg W. F. (1970): *Hegel's Philosophy of Nature*, edited and translated with an introduction and explanatory notes by M.J. Petry, London : Allen & Unwin.

Hegel, Georg W. F. (1977): *Phenomenology of spirit*, translated by Arnold V. Miller, Oxford: Clarendon Press.

Hegel, Georg W. F. (1991): *Hegel's Science of Logic*, translated by Arnold V. Miller, London: Allen & Unwin.

Lukács, Georg (1972): *History and class consciousness: studies in Marxist dialectics*, translated by Rodney Livingstone, London: MIT Press.

Marx, Karl (1975): "Economic-philosophical manuscripts", In: Marx, Karl / Engels, Friedlich (1975): *Collected works*, vol. 3, Moscow: Progress publishers, pp229–346.

Marx, Karl (1990): *Marx-Engels Gesamtausgabe* (MEGA²), Abt. 2, Bd. 9: Capital. a critical analysis of capitalist production, London 1887, Berlin: Diez.

Marx, Karl (1998): *Collected works*, vol. 37: Capital, vol. III, New York: International publishers.

Quante, Michael (2009): "Kommentar", In: Marx, Karl (2009): *Ökonomisch-philosophische Manuskripten*, Frankfurt am Main: Suhrkamp, S. 209 ff.

Schmidt, Alfred (1971): *The concept of nature in Marx*, translated by Ben Fowkes, London: NLB.

Caroline Angleraux
# From Monads to Monera

**Abstract:** Leibniz developed his concept of monad clearly stating that monads are metaphysical entities. More precisely, monads should be understood as simple entities with an intensive content that provide an ontological substratum to bodies. Moreover, in order to offer an exhaustive account of the world, Leibniz also connected monads to considerations about living beings. During the following centuries, a series of interpretations was developed that naturalised the monads and turned them into simple biological entities. *Via* Wolff, Baumgarten, Kant, and Schelling, the monads progressively became intensive units at the basis of the living world; *via* Bonnet, Bourguet, and O. F. Müller, the monads were transformed into the primitive living blocks of the universe. The two interpretative traditions intersect each other and find one of their peaks in Haeckel's concept of monera. This paper will outline this process of naturalisation, from monads to monera, focusing on the works of Leibniz, Schelling and Haeckel.

## Introduction

Between the seventeenth and nineteenth centuries, a paradigmatic change in the way living beings were seen occurred: at first considered as entities of the world characterised by their individual unity, living beings were later perceived as self-governed processes that should be entirely explained by physical laws. In the very beginning of the eighteenth century, which was the period during which Leibniz developed his "mature philosophy", living beings were included into a more general explanation of the world, and even though Leibniz paid particular attention to them, they should be seen as part of a wider philosophical endeavour. By the second half of the nineteenth century, biology was quite well established and living beings were studied for themselves, within specific methodological frameworks such as the cell theory or the theory of evolution by natural selection. In short, in one century and a half, there was a main theoretical shift which changed the scientific goals, the methodology and the way of understanding living beings.

In this context, the conceptualisation of the smallest living entity may appear to translate a modern need to classify living beings and to place them in an evolutionary context. On the other hand, the Leibnizian monads, considered

---
**Caroline Angleraux,** Université Paris 1 Panthéon-Sorbonne and University of Padova

https://doi.org/10.1515/9783110604665-008

as spiritual soul-like entities that exclusively belong to metaphysics, seem to belong to an old paradigm that has no relation with the nineteenth century biological one. And yet, the idea of a fundamental living being that distinguishes the living from the non-living and that concentrates on vital activity in itself, inherits from older paradigms. In this sense, my goal is to show how this idea of a vital nucleus partly results from the naturalisation of a speculative concept which also draws up on the Leibnizian monad.

Indeed, as a big admirer of Darwin, the biologist Haeckel based his concept of the monera on an empirical approach that took the cell theory for granted, within a Darwinian paradigm. In this framework, monera referred to plasmatic substances with no nucleus, entities of primitive living matter, or protoplasm with no nucleus. Therefore, monera were the simplest biological entities and they included all the characteristics of life; they were both the smallest entities in a general taxonomy of life and the epicentres of vital activity. Here, the concept of monera clearly belonged to the nineteenth-century scientific paradigm. And yet, as monera constituted the centrepoints of the vital dynamism, they echoed Schelling's use of monads in his *Naturphilosophie* or Leibniz's use of monads in his mature philosophy. Hence, I will focus on the naturalisation of an entity that went from being metaphysical and then speculative, and that played a major part in the establishment of the idea of a very basic living being.

In order to do so, I will mainly focus on the Leibnizian monad, the Schellingian natural monad and the Haeckelian moneron, given that these three concepts can be seen as the main steps of the progressive naturalisation with which I am dealing. Consequently, I will focus on neither the Kantian turning point (although Kant paid attention to the ways in which living beings were understood, he did not particularly focus on the idea of a basic living being[1]) nor on the empirical and taxonomic meaning of "monad" that Otto Friedrich Müller establishes[2] (considering the format of this article, we will only focus on the speculative side of this conceptual history). I will give a general sketch of this naturalisation that intertwines metaphysical, speculative, and scientific considerations and I will articulate the conceptual, historical and logical threads that constitute this conceptual and historical genesis from monads to monera.

In short, my aim is twofold. First, I will attempt to show from a Leibnizian perspective, how the search for a vital unit, which started with the monad, was

---

[1] "Relative to his contemporaries, Kant was not biophilic; he occasionally mentions his approval of gardens and leafy decorations, and he disapproved of cruelty to animals, but one cannot say he felt, intuitively, the unity of the living world as his romantic successors professed to, and as Leibniz had before them." (Wilson 2010, p. 300).

[2] Cf. Section 3 below.

physicalized. Second, from a Haeckelian perspective, I will aim to see how, in order to become the most simple living beings, the primordial living unit, the moneron drew upon both (a) theories that were mainly based on observation and experiments – such as the cell theory or the Darwinian Theory – and also (b) metaphysical and speculative conceptions such as the naturalisation of the monad.

## The Leibnizian Monads, the Principles of Living Beings

In his mature philosophy, Leibniz characterised living beings as natural machines: contrary to human machines, living beings are self-moving machines in which each part is also a machine, which is also infinitely mechanically structured: "natural machines, that is, living bodies, are still machines in their least parts, to infinity" (Leibniz 1989, p. 221)[3]. Contrary to mere aggregates, each part of the living being is ontologically well founded with a principle of unity, i.e. a subordinate monad, and the whole living being is ontologically unified by a dominant monad. Living beings are mass bodies organised and ontologically unified by a hierarchy of monads. Consequently, monads are the living beings' principle of unity, of activity and of life all at once; this first section deals with this point.

\* \* \*

Based on the standard definition presented in *Monadology* § 1, monads are usually defined as simple substances, i.e. metaphysical entities with no parts. But in

---

[3] "Les Machines de la Nature, c'est à dire les corps vivans, sont encor des machines dans leur moindres parties jusqu'à l'infini" (Leibniz, GP VII, p. 618). As O. Nachtomy notes, the infinite composition of the corporeal body does not refer to the infinite number of its parts, but rather refers to the infinite structuration of its parts. It is in that sense that "une machine naturelle demeure encor machine dans ses moindres parties" (Leibniz, GP IV, p. 482). Thus, highlighting the infinity of the structure and not the infinity of the organs' number also allows us better to distinguish between the organic body and the aggregate: considering that an aggregate is divisible *ad infinitum*, it also has an infinite number of parts. So the real difference between organic bodies and aggregates consists in the infinite structuration. To quote O. Nachtomy: "According to the reading I suggest, what extends to infinity is not the *number* of organs or machines but rather the very *structure* of a natural machine which involves machines within machines. [...] While an artificial machine might also have an infinity of parts, a natural machine has an internal structure that extends to infinity" (Nachtomy 2011, p. 73).

a letter to Sophie, Leibniz also characterised monads as simple entities that have perception: "Your Electoral Highness asks me what a simple substance is. I reply that its nature is to have perception, and consequently to represent composite things"[4] (Leibniz 2006, p. 81). Here, having perception means being able to grasp a multitude into a unity (Leibniz GP III, p. 622; GP III, p. 574; GP VI, p. 608), to turn a diversity of inputs into an understandable series. As A. M. Nunziante explained in a previous chapter in this volume, having perception implies structuring the data from the environment into a homogenous and unified behavioural response; consequently having perception is what gives unity to the animal body. In addition, having perception makes the living being's cognition one. Indeed, for Leibniz, perception is linked to appetition, which he sees as a kind of tendency by which each perception tends towards a new perception. In this sense, having perception implies structuring the different perceptions over time into a temporal web; this perceptive web is specific to each monad and constitutes its specific point of view from which it expresses the whole world; hence, having perception makes the living being's cognition one. Thus considering that monads have perception, monads make the body and the cognition of living beings one: monads are principles of unity, they ontologically found living beings and are what make them real substances.

Moreover, each perception tends towards a new perception via appetition, thus perceptions and appetition highlight the profuse internal activity of monads. With their perceptive/appetitive content, monads contain and initiate change in the world. With a kind of spontaneity, each monad pulls everything from its own content and expresses from its own point of view the entirety of divine Creation. Given that they have perception, monads are both highly individuated (through their own point of view) and endless (unless God annihilates them). As Leibniz wrote in the *De Ipsa Natura:* "that which does not act, which lacks active force, which is robbed of discriminability, robbed finally of all reason and basis for existing, can in no way be a substance"[5] (Leibniz 1989, pp. 165–166). Consequently, if monads turn living beings into real substances, it is because having perception guarantees the monads to be both principles of unity and principles of activity. Monads are active entities at the basis of the living world.

---

[4] "V.A.E. me demande ce que c'est qu'une Substance simple. Je réponds que sa nature est d'avoir de la perception et par consequent de representer les choses composées" (Leibniz, GP VII, p. 566).
[5] "Id quod non agit, quod vi activa caret, quod discriminabilitate, quod denique omni subsistendi ratione ac fundamento spoliatur, substantia esse nullo modo possit" (Leibniz, GP IV, p. 515).

More precisely, in a letter to De Volder, Leibniz explained that monads are made of a primitive active power (the entelechy) and a primitive passive power: "therefore I distinguish: (1) the primitive entelechy or soul; (2) the matter, namely, the primary matter or primitive passive power; (3) the monad made up of these two things"[6] (Leibniz 1989, p. 177). Here, the primitive passive matter is not physical matter but a potential power, and it plays a limitative role inside the active monad. As P. Phemister argued, if monads were to be pure active entities with no limitative force, their perceptive power would have no limit, no specific point of view and, all in all, would be identical to God[7]. Consequently, monads have to be made of a purely active entelechy as well as primitive matter. But it is the active entelechy that makes the nucleus of monads active: in a kind of elision, in his correspondence with Lady Masham, Leibniz made monads and the active entelechy synonymous: "the positive idea of this simple substance or primitive Force perfectly suit"[8] (Leibniz GP III, p. 356); in so doing, Leibniz highlighted the central part of monads as active entities[9]. Once again, in having perception, monads are both principles of unity and principles of activity in living beings.

\* \* \*

It is because monads are principles of activity that they are also the principles of the life of living beings. As Leibniz declared to Des Maizeaux:

> But if there are in Nature other living organic bodies than those of animals, as it seems and of which plants seem to constitute an example, these bodies will also have their simple substances, or Monads, that will give them life, i.e. perception and appetition.[10] (Leibniz GP VII, p. 535)

---

6 "Distinguo ergo (1) Entelechiam primitivam seu Animam, (2) Materiam nempe primam seu potentiam passivam, (3) Monada his duabus completam", (Leibniz, GP II, p. 252).
7 "Creatures must possess this limiting passive force if they are to be distinguished both from their Creator and from each other. A creature whose perceptual states were identical to God's would, since beings are distinguished according to their perceptions and appetitions, be indistinguishable from God and, by the Principle of the Identity of Indiscernibles, would be one and the same with God [...] then, without primitive passive force or primary matter to restrain the individual creature's active force, there would be no variation in the world" (Phemister 2005, p. 242).
8 "L'idée positive de cette substance simple ou Force primitive est toute trouvée" (Leibniz, GP III, p. 356). My translation.
9 "Forces primitives, qui ne contiennent pas seulement l'acte ou le complement de la possibilité, mais encor une *activité originale*" (Leibniz, GP IV, p. 479), my italics.
10 "Mais s'il y a dans la Nature d'autres corps organiques vivans que ceux des animaux, comme il y a bien de l'apparence et comme les plantes nous en semblent fournir un exemple, ces corps

Here, having perception characterises what is alive; hence, monads are the purely living entities. However, living beings are not mere monads but embodied creatures. Consequently, it is necessary to specify that the derivative forces that animate the physical world stem from the primitive forces, i.e. from the genuine activity of the primitive power which characterises the monads. As Leibniz develops in the *Théodicée* §396: "the qualities or derivative forces, or what are called accidental forms, I take to be modifications of the primitive Entelechy"[11] (Leibniz 1952, pp. 360–361). In other words, the concept of force connects the metaphysical living activity with the activity of physiological processes. Therefore as A.-L. Rey argues: "the force appears to be what guarantees the mediation between a corporeal substance and an immaterial substance while guaranteeing the unity of the notion of substance"[12] (Rey 2009a, p. 60); monadic dynamism permeates its corporeal representatives in order to make them living beings.

This indirect impact of metaphysical monads on physical bodies is not an *ad hoc* proposition made to fit the theory with the world as it appears, but a necessity of Leibniz's theory and it is also implied in the definition of the term "monad". Indeed, as we have already quoted, Leibniz wrote to Sophie: "I reply that its nature is to have perception, and consequently to represent composite things." Having perception implies representing the whole universe via a specific point of view, but in order to be *de facto* situated from a specific point of view, the monadic perceptive activity has to have an expressive support that concretely realises this perceptive activity, i.e. an organic body: "every created monad is endowed with some organic body, in accordance with which it perceives and has appetitions"[13] (Leibniz GP VII, p. 502); as an unextended metaphysical entity that is not situated in space, monads have to be linked to a body in order to efficiently realise themselves. Given that the active monad has perception and consequently represents composite things, it must[14] be linked to an organic body and therefore form a living being. As Leibniz wrote to Rémond: "An

---

aussi auront leur substances simples ou Monades qui leur donneront de la vie, c'est à dire de la perception et de l'appétit" (Leibniz, GP VII, p. 535). My translation.

11 "Je conçois les qualités ou les forces derivatives, ou ce qu'on appelle formes accidentelles, comme des modifications de l'Entelechie primitive" (Leibniz, GP VI, p. 352). See also Leibniz, GP II, p. 251.

12 "La force apparaît comme ce qui permet d'assurer la médiation entre substance corporelle et substance immatérielle tout en garantissant l'unité de la notion de substance" (Rey 2009a, p. 60). My translation.

13 "Omnis Monas creata est corpore aliquo organico praedita, secundum quod percipit appetitque" (Leibniz, GP VII, p. 502). Translation by E. Pasini in Pasini 2011, p. 88.

14 "His [Leibniz's] doctrine of simple substances *requires* the existence of corporeal substances" (Levey 2007, p. 63), my italics.

authentic substance (such as an animal) is composed of an immaterial soul and an organic body, and it is the Composite as well as these two elements that we call *Unum per se*"[15] (Leibniz GP III, p. 657). It seems highly meaningful that Leibniz specifies that he is talking about an authentic substance: although substantiality comes from monads – because they are principles of ontological unity, of activity and of life – the idea of an authentic substance implies organic bodies. Ultimately, authentic substances are, for Leibniz, living beings.

Finally, in Leibniz's mature thoughts, monads are metaphysical substances that are the principles of the ontological unity, activity and life of living beings. In other words, monads belong to metaphysics whereas the study of living beings is part of the natural sciences. In Leibniz's argument, metaphysics and epistemology are clearly distinct – with pre-established harmony above everything else. But, as we have just shown, monads contribute to the ontological characterisation of living beings and living beings are authentic substances. Therefore, metaphysics and natural sciences remain highly connected for Leibniz, especially *via* the dynamism that emanates from monads into the living physical world (cf. Rey 2009a).

# The Schellingian Monads, Intensive Basis of Living Beings

In order to realise his dynamic conception of nature, Schelling reinterprets the relationship between metaphysics and epistemology that could be seen as characterising Leibniz's mature philosophy and, through a Spinozist lens, turns the Leibnizian monads into natural entities. Here, inheriting from Herder's reading of Leibniz (Zammito 2017, p. 300–305), Schelling reemploys the main characteristics of the Leibnizian monads but gives them a new meaning; natural monads are still principles of activity and of life in living beings, but they have a new relationship to substantiality and play a new part in the overall approach. This second section shows how Schelling formulated his theory of natural monads based on Leibniz's ideas.

\* \* \*

---

[15] "Une veritable substance (telle qu'un animal) est composée d'une ame immaterielle et d'un corps organique, et c'est le Composé et ces deux qu'on appelle Unum per se" (Leibniz, GP III, p. 657). My translation.

Schelling highlights the central role played by monads in Leibniz's view by proposing that everything emanates from monads. In *Zur Geschichte der neueren Philosophie*, he characterises the Leibnizian monads as the last elements, understood as spiritual, of all material bodies[16] (Schelling 1861, p. 49). In this sense, Schelling perfectly understands the substantial role played by monads in Leibniz's view. He establishes the hierarchical relationship between monads based on their perceptual clarity and explains how pre-established harmony works, but also emphasises the regulative role played by monads in bodily regulation. And yet, in *Zur Geschichte der neueren Philosophie*, Schelling aims to show how both the Spinozian and the Leibnizian approaches regarding substantiality are quite similar for answering the Cartesian mind-body problem. Indeed, first, Schelling explains how the Spinozian monist substance lacks spirit and activity. For him, Spinozian nature is entirely reduced to matter; consequently, all of nature is already developed through a material expression and is immobilised in the inertia of existence, with no possible development of a living becoming: "For him, extension lacks mind, like for Descartes, and for this reason Spinoza's approach to nature, his Physics, is as mechanistic and as unalive than that of his predecessors"[17] (Schelling 1861, p. 38). Similarly, Schelling considers that Leibniz ultimately reduces everything to monads. Monads are substantiality, i.e. an inertial expression of things into existence, with no living becoming. Hence, for Schelling, in the same way that Spinoza's philosophy does not account for the activity inherent to the world, Leibniz's philosophy also fails to account for this activity and dynamism; "the created monads stem from God or from divine Nature passively and with no proper activity, as it is for Spinoza"[18] (Schelling 1861, p. 51). In short, in *Zur Geschichte der neueren Philosophie*, both Spinoza and Leibniz limit substantiality into existence – whichever entity is doing the limiting (matter or monads) – and this is why, for Schelling, both the Spinozian and the Leibnizian philosophies lack the notion of activity.

Consequently, Schelling interprets Leibniz *via* Spinoza and adapts Leibniz's mature philosophy into monist and dynamic considerations of his own. In the *Ideas for a Philosophy of Nature*, he writes:

---

[16] "Leibniz nimmt jene Einheiten, die er Monaden nennt, auch als die letzten, obwohl geistigen Elemente alles Materiellen an" (Schelling 1861, p. 49).

[17] "Das Ausgedehnte ist ihm völlig so geistlos wie dem Cartesius, und Spinozas Ansicht der Natur, seine Physik, ist aus diesem Grunde nicht weniger mechanisch und unlebendig als die seines Vorgängers" (Schelling 1861, p. 38). My translation.

[18] "Die geschaffenen Monaden folgen aus Gott oder der göttlichen Natur ebenso still und ohne eigne That, wie sie nach Spinoza aus ihr folgen" (Schelling 1861, p. 51). My translation.

> What alone Leibniz held to be originally real and actual *in themselves* were *perceptual beings;* for in these alone was that *unification* original, out of which everything else that is called actual *develops* and *goes forth*. For everything which is actual outside us is finite, and so not conceivable without a positive, which gives it reality, and a negative, which sets its limit. This unification of positive and negative activity, however, is nowhere *original* except in the nature of an individual. External things *were* not actual *in themselves*, but have only *become* actual through the mode of presentation of spiritual natures; but that from whose nature all existence first *emerges*, that is, the ideating being alone, would have had to be something which bears the source and origin of its existence in itself.[19] (Schelling 1988, p. 29)

Here Schelling uses the perceptive nature of monads as driving forces, and claims that activity, reality, and existence in the world come from them. In this interpretation, monads remain genuine principles of substantiality. But, with no limitation, all the substantiality would be expressed at once and would lose itself in existence, in such a way that substantiality would be completely inert and all the activity governing the world would be eliminated. Therefore, Schelling needs to consider a limitative element, i.e. external things. He thus conceives monads as spiritual entities out of which reality, and the activity inherent to reality, flow. External things, understood as concrete structuring forms, limit the existential flow of the monads, and, in this sense, prevent substantiality from completely turning into existence all at once. Reciprocally, it is with this monadic support that external things become actual. Without monads, external things would remain inert and even non-existent; monads, as ontological principles, give existence to external things. In this sense, Schelling deals with becoming by joining an active and a limitative pole, a positive and a negative one. It is worth noting that this idea of a convergence between a positive and a negative pole to explain situated becoming is reminiscent of what Leibniz declares in the previously quoted letter to De Volder, when he establishes that monads are made of an entelechy and primitive matter; Leibniz also wished to distinguish between a metaphysical, full positive ontological source, and a limitative metaphysical matter that creates a situated and defined point of view.

---

[19] "Was Leibnitz allein für ursprünglich – real und an sich wirklich hielt, waren vorstellende Wesen, denn in diesen allein war jene Vereinigung ursprünglich, aus welcher erst alles andere, was wirklich heisst, sich entwikkelt und hervorgeht. Denn Alles, was ausser uns wirklich ist, ist ein Endliches, also nicht denkbar ohne ein Positives, das ihm Realität, und ein Negatives, das ihm Gränze giebt. Diese Vereinigung positiver und negativer Thätigkeit aber ist nirgends als in der Natur eines Individuums ursprünglich. Aeussere Dinge waren nicht wirklich an sich selbt, sondern nur wirklich – geworden durch die Vorstellungsweise geistiger Naturen, dasjenige aber, aus dessen Natur erst alles Daseyn hervorgeht, d. h. das vorstellende Wesem allein, musste Etwas seyn, das in sich selbst Quell'und Ursprung seines Daseyns trägt." (Schelling 1803, p. 39).

> In any case, Schelling understands Leibniz through a Spinozist lens. As F. C. Beiser argues: From Leibniz, Schelling took the idea that the essence of matter consists in living force; from Spinoza, he borrowed the concept of nature as a *natura naturans*, a single power and productive force. Hence his *Naturphilosophie* was the monism of Spinoza spelled in vitalistic terms, the vitalism of Leibniz stretched to the universe as a whole. (Beiser 2008, p. 550)

It is important to understand Schelling's reading of Leibniz to understand the *Naturphilosophie* of his youth. Indeed, Schelling considers nature as an unlimited productive force whose expression is canalised into the embodiment of specific living beings. He does not regard nature as an established result of *produced* living beings, but as a *productive* force that is still perpetually active[20]. In this sense, the unconditional at the heart of nature is never totally embodied in a specific quantity of living beings, but is only visible intermittently through them: natural productivity is never confined to the "staticness" of being but remains in perpetual becoming[21], in a never-ending active movement.

\* \* \*

Obviously, this monist Schellingian framework (which is open to possibility) is very far from the Leibnizian approach (in which pre-established harmony and divine preordination are central). However, when Schelling considers nature as a productive force that is constantly torn between embodied stops (the living beings) and a perpetual forward movement (the unconditional natural activity), Schelling's reasoning is similar to Leibniz's : he has to suppose that natural monads are the authentic active nuclei within living beings. Indeed, in *Erster Entwurf eines Systems der Naturphilosophie*, Schelling writes: "Every external being is a being in space. Therefore, something has to come to the fore in experience which, although itself not in space, is yet principle of all occupation of

---

[20] "Iede Thätigkeit aber erstirbt in ihrem Produkte, denn sie ging nur auf dieses Produkt. Die Natur als Produkt kennen wir also nicht. Wir kennen die Natur nur als thätig" (Schelling 1858, p. 13).
[21] "Das Unbedingte kann überhaupt nicht in irgend einem einzelnen Ding, noch in irgend etwas gesucht werden, von dem man sagen kann, daß es ist. Denn was ist, nimmt nur an dem Seyn Theil, und ist nur eine einzelne Form oder Art des Seyns. – Umgekehrt kann man vom Unbedingten niemals sagen, daß es ift. Denn es ift das Seyn selbst, das in keinem endlichen Produkte sich ganz darstellt, und wovon alles Einzelne nur gleichsam ein besonderer Ausdruck ist. [...] "es kann in keinem einzelnen Naturding, als olchem, das Unbedingte der Natur gesucht werden"; vielmehr offenbart sich in jedem Naturding ein Princip des Seyns, das nicht selbst ist." (Schelling 1858, pp. 11–12).

space"[22] (Schelling 2004, pp. 19–20). If, in the world, living beings are material and thus infinitely divisible, the aim is to find a principle outside of spatiality, i.e. an immaterial and intensive principle at the heart of living beings. For Schelling, atomists are right when they establish the atom as the smallest possible and indivisible top-down element, but they are wrong to try to find it within matter, which is infinitely divisible. The aim is rather to find the atomic entity within activity which connects matter and gives it its power and its dynamism.

In this sense, Schelling declares: "Accordingly, our claim can be called the principle of *dynamic atomism*"[23] (Schelling 2004, p. 21). Given that a genuine entity needs to be found and that matter is divisible *ad infinitum*, it is therefore an intensive building block that must be found, in other words, a fundamental active entity that plays the part of an atom (i.e. the primitive entity we cannot go beyond) without being material. Consequently, Schelling concludes: "For us, every original actant is just like the atom for the corpuscular philosopher; truly *singular*, each is in itself whole and sealed-off, and represents, as it were, a *natural monad*"[24] (Schelling 2004, p. 21). The original actants are the intensive building blocks of living nature and their authentic activity, as the productive force at the basis of living beings, makes them play the part of natural monads.

Here, Schelling assigns to his monads the main qualities of the Leibnizian monads: they are indivisible, immaterial, non-spatial, entirely singular, complete, and accomplished. The Schellingian natural monads are also characterised, like the Leibnizian monads, by their genuine active intensity[25]. But, unlike Leibniz, Schelling completely sets aside the Leibnizian vocabulary of perception and appetition: the perceptive approach as Leibniz develops it does not fit in Schelling's framework, because, for him, natural monads are not metaphysical

---

[22] "Alles äußere Seyn nun ist ein Seyn im Raume. Es müßte also in der Erfahrung etwas vorkommen, das, obgleich selbst nicht im Raume, doch Princip aller Raumerfüllung wäre" (Schelling 1858, p. 21).
[23] "Unsere Behauptung kann sonach Princip der dynamischen Atomistik heißen." (Schelling 1858, p. 22).
[24] "Denn jede ursprüngliche Aktion ist für uns ebenso, wie der Atom für den Corpuscularphilosophen, wahrhaft individuell, jede ist in sich selbst ganz und beschlossen, und stellt gleichsam eine Naturmonade vor." (Schelling 1858, pp. 22–23).
[25] "Was untheilbar ist, kann nicht eine Materie seyn, sowie umgekehrt, es muß also jenseits der Materie liegen: aber jenseits der Materie ist die reine Intensität – und dieser Begriff der reinen Intensität ist ausgedrückt durch den Begriff der Aktion. – Nicht das Produkt dieser Aktion ist einfach – wohl aber die Aktion selbst abstrahirt vom Produkt gedacht, und diese muß es seyn, damit das Produkt theilbar sey'." (Schelling 1858, p. 23), quotation from *Einleitung zum Entwurf*.

entities but parts of the natural world that make living beings alive. Hence, the nature of the *impetus* that characterises the natural monads is not the same as in the Leibnizian framework. For Leibniz, the main metaphysical driving forces are the appetition present in the primitive Entelechies, and the derivative forces that animate the physical world stem from these primitive forces. For Schelling, the *impetus* is fundamentally to be found in the unconditional at the heart of nature; natural monads have to be postulated in order to explain the positive/negative two-sided creation of becoming, but they are, for Schelling, just an explanatory tool, not the main topic. For Leibniz, monads are the central concept: they link metaphysics, the physical world, the natural sciences and physics. For Schelling, natural monads are a necessary but secondary concept which helps him complete his general explanation, but nature and living beings remain the central focus. In this sense, we should note that, for Schelling, the existence of natural monads is not demonstrated (contrary to Leibniz who established monads on a kind of *modus ponens*[26]) but only postulated (*if* nature were to achieve its full development – which is concretely impossible – *then* we would analytically reach monadic entities): "These simple actants do not really allow of demonstration— they do not *exist*; they are what one must posit in Nature, what one must think in Nature, in order to explain the originary qualities"[27] (Schelling 2004, p. 21).

Consequently, if natural monads cannot be demonstrated, it is because they are ideal entities for Schelling:

> Our opinion is thus not that *there are* such simple actants in Nature, but only that they are the *ideal* grounds of the explanation of quality. These simple actants do not really allow of demonstration— they do not *exist*; they are what one must posit in Nature, what one must think in Nature, in order to explain the originary qualities. Then we need only prove as much as we assert, namely, that such simple actions must be *thought* as ideal grounds of explanation of all quality, and we have provided this proof.[28] (Schelling 2004, p. 21)

---

26 In the first paragraph of the *Principes de la Nature et de la Grace, fondés en raison* (Leibniz, GP VI, p. 598) and in the second paragraph of the *Monadology* (Leibniz, GP VI, p. 606) in particular, the existence of monads appears like a rational requirement, and almost logical imperative: there is, empirically speaking, composites. But the composite implies indivisible substantial unities. So there *has to be* – "il faut nécessairement" (Leibniz, GP IV, p. 473) – indivisible substantial unites, i.e. monads. This argument may be seen as a *modus ponens* that M. Fichant calls the monadologic thesis (Fichant 2005, p. 32).

27 "Diese einfachen Aktionen lassen sich nicht wirklich aufzeigen – sie existiren nicht, sie sind das, was man in der Natur setzen, in der Natur denken muß, um die urfprünglichen Qualitäten zu erklären" (Schelling 1858, p. 23).

28 "Unsere Behauptung ist also nicht: es gebe in der Natur solche einfache Aktionen, sondern nur, sie seyen die ideellen Erklärungsgründe der Qualität. Diese einfachen Aktionen lassen sich

Natural monads are an explanatory tool, a theoretical notion in order to complete the *raison d'être* of living activity in nature.

And yet, maybe paradoxically, it is by transforming monads into ideal entities that Schelling naturalises them. Indeed, concretely, in nature, primitive, original actants mutually limit each other; each productive force is limited by another productive force and this mutual limitation in a limited area forms the whole organic body:

> Each organism is itself nothing other than the collective *expression* for a multiplicity of actants, which mutually limit themselves to a determinate sphere. This sphere is something perennially enduring—not something merely fading into the background as appearance—for it is that which *originates* in the conflict of actants, the monument, as it were, of those activities prehending one another; it is the *concept of that change itself*, which is the only enduring thing in the change.[29] (Schelling 2004, p. 51)

In this sense, monadic entities with unlimited force are ideal, but their combination and mutual limitation in organic bodies embody and naturalise them. While the existence of original actants considered separately is ideal, their conflict, on the other hand, is real and not simply an appearance. Therefore, Schelling takes a significant step forward: even though monads are singular, complete, sealed-off, they do possess windows[30] and mutually limit one another. Considered in themselves, they are ideal, but their conflict creates becoming and is expressed *via* living bodies. In other words, ideal entities acquire their meaning in the organic products and monads do not exist outside of bodies: it is in their conflict that ideal actants become *natural* monads[31]. And here we should remember that, for Schelling, Leibniz himself did not consider monads as external to bodies. In

---

nicht wirklich aufzeigen – sie existiren nicht, sie sind das, was man in der Natur setzen, in der Natur denken muß, um die urfprünglichen Qualitäten zu erklären. Wir brauchen also auch nur so viel zu beweifen, als wir behaupten, nämlich, daß solche einfache Aktionen gedacht werden müssen als ideelle Erklärungsqründe aller Qualität, und diesen Beweis haben wir gegeben." (Schelling 1858, p. 23).

**29** "Iede Organisation ist selbst nichts anderes als der gemeinschaftliche Ausdruck für eine Mannichfaltigkeit von Aktionen, die sich wechselseitig auf eine bestimmte Sphäre beschränken. Diese Sphäre ist etwas Perennirendes – nicht bloß etwas als Erscheinung Vorüberschwindendes —; denn sie ist das im Conflict der Aktionen Entstaudene, gleichsam das Monument jener ineinander greifenden Thätigkeiten, also der Begriff jenes Wechsels selbst, der also im Wechsel das einzige Beharrende ist." (Schelling 1858, p. 65).
**30** "Jetzt gilt es für Philosophie, zu glauben, dass die Monaden Fenster haben, durch welche die Dinge hinein und heraus steigen." (Schelling 1803, p. 16).
**31** Let's say it again: "und *stellt* gleichsam eine Naturmonade vor" (Schelling 1858, p. 23), my italics.

the *Ideen zu einer Philosophie der Natur*, he writes: "I cannot think otherwise than that Leibniz understood by substantial form a mind *inhering in* and regulating the organized being."[32] (Schelling 1988, p. 35)

Finally, with Schelling, we witness a major requalification of the meaning of the monad: from metaphysical substances that are the principles of ontological unity, of activity and of life in living beings, monads become, with Schelling, the ideal dynamic basis of becoming in nature[33]. And yet, even if monads are ideal, they have windows and therefore interfere with each other; the conflict that results from their interaction constitutes the inherent activity of living beings. In a sense, idealising each monad allows for their naturalisation. Hence, even if they are non-spatial, natural monads are not metaphysical but belong to the natural living world. Simultaneously, as the theoretical dynamic basis of becoming in nature, natural monads do not supply the same ontological weight as the Leibnizian monads do. For Schelling, these simple entities do not exist in themselves, but only concretely through their mutual conflict which gives them the physical concreteness which suits them. Thus, similar to Leibniz's monads, Schellingian monads are indivisible, immaterial, non-spatial, fully singular, complete, and accomplished. But, contrary to his predecessor, Schelling naturalises his monads by making them the concrete driving forces of becoming in nature that are expressed through living beings.

## The Haeckelian Monera, the Mere Living Beings

In combining Spinoza and Leibniz's approaches, Schelling forges a speculative meaning of monads as the active and vital ideal basis of living beings. Such an intensive and dynamic understanding of monads plays a major part in the *Naturphilosophie* approach. Simultaneously, naturalists such as Müller or Ehrenberg later also use the term "monads" in a different context. In his *Vermium Terrestrium et Fluviatilium* for instance, the point for Müller is to create a typology of small living beings; in doing so, he puts a total emphasis on observation only

---

[32] "Ich kann nicht anders denken, als dass Leibniz unter der substantiellen Form sich einen den organisirten Wesen inwohnenden regierenden Geist dachte." (Schelling 1803, p. 51).
[33] "Schelling, meanwhile, was attracted by the idea of a 'natural monad' as an 'original actant' [...] and, despite his tendencies towards Spinozistic absolute monism, he championed a multiplicity of such monads as dynamic atoms or immaterial vital forces, that, while not parts of matter, must nevertheless be posited as the ideal original grounds of the seeds or 'qualities' in Nature and which, abstracted from their "products' (namely the organic bodies themselves) do not exist." (Phemister 2016, p. 55).

and does not attempt to produce an explanatory overview of the world as a whole, as *Naturphilosophie* tries to do. In this context of scientific theories based on observation and experiments, "monads" refer to the simplest living beings, composed of only a drop of mucus – for instance, Müller distinguishes between three kinds of monads (Müller 1773, pp. 25–27). This last part aims to show how, because of this complicated legacy, Haeckel rejects the monads in favour of his monera and how, despite this rejection and his support to the theory of natural selection and the cell theory, his approach is still aligned with *Naturphilosophie* that mainly stems from Schelling.

\* \* \*

The way Schelling naturalises monads and makes them concrete driving forces in nature has a strong impact on the *Naturphilosophie* of which Schelling was a contributory founder. In particular, the *Naturphilosoph* Oken, who partly draws his inspiration from Schelling, considers nature as a deeply active world made of antagonistic forces that complement each other. For Oken, these dynamic exchanges that govern the world are first symbolised in mathematics and, in this mathematical symbolism area, the entity that comes first is the ideal zero, also named the "monas". This homogeneous, numberless, indivisible, eternal, absolute unity is a pure identity that gives birth to the multiplicity of the world. With the monas, Oken reinterprets the main qualities of the pure productivity at the basis of the Schellingian natural monads: contrary to Schelling, the monas is mathematical and it is a single general un-singular entity and not a multiplicity of different forces. Similar to Schelling, this monas is external to matter, and determines individuals via a play of forces which it establishes. From this mathematical basis, Oken also put forth physical and biological considerations. In particular, in the living world, he considers that monads are vesicles of mucus and a certain kind of infusoria, and the basic living building blocks in living beings. More precisely, in the *Elements of Physiophilosophy* (Oken 1844, p. 570), Oken explains that monads are at the same time the units of living matter ("the primary mass of the animal kingdom"), the first compounds of living beings ("the animal body is nothing else than a compound fabric of Monads") and the living semen ("semen of the animal kingdom").

In this way of thinking about active forces in the living world, of joining mathematics, physics and natural sciences, and of assuming living monads at the basis of the living beings, Oken clearly draws upon Schelling. And yet, his way of conceiving natural monads as primitive active living beings also echoes the naturalistic meaning of monad. In particular, in *Die Zeugung*, Oken alludes to Otto von Müller's works (Oken 1805, pp. 7–8), in which Müller characterises monads as very simple, tiny worms with a pointy shape. For Müller, the "monad" refers to a category of living beings that are characterised by their highly simple constitution: composed of a single or a few mucus vesicles, monads are the simplest beings of the living world (Müller 1773, pp. 25–27).

Consequently, in the middle of the nineteenth century in Germany when Haeckel was publishing his works, the term "monad" presents a terminological difficulty. On the one hand, an old tradition that dates back to Leibniz understands the monad as a metaphysical – or a speculative – entity that has something to do with the natural sciences. On the other hand, in observational experiments and taxonomic approaches, the monad refers to the simplest forms of living beings. Moreover, it is quite hard to estimate to what extent the latter has been influenced by the former, especially when the latter, in virtue of a strict experimental and a strict observational approach, rejects the former. In his talk "Öffentliche Sitzung der Akademie zur Feier des Leibnizischen Jahrestages", E. Du Bois-Reymond claimed that many people want to see in the corporeal world what Leibniz only conceptualised about the spiritual world. But for E. Du Bois-Reymond, Müller's use of the term monad was not conceptually linked to Leibniz and the term "monad" is merely a witticism referring to Leibniz. Hence, only readers equipped with a very imaginative mind are fooled into thinking that there is more to read into this mention to Leibniz than a simple quip.[34] Whether Müller's mention of the monad was actually a pleasantry or not, this declaration by E. Du Bois-Reymond shows that the reinterpreted Leibnizian meaning of the monad remained relevant throughout the nineteenth century and that this monadic meaning still carried a complicated conceptual liability, even if scientists like him ostensibly rejected it. More generally, Du Bois-Reymond's declaration emphasises the fact that, from an empirical and reductionist approach to a more speculative and holistic one, there is no hermetic division between the different ways of conceptualising life and living beings, even if it is hard to evaluate how much one way of conceptualising influenced the other.

---

34 "Was für das geistige Auge gemeint war, wollte das leibliche Auge sehen; und wenn man nicht geradezu versuchte, die Monaden mit dem Mikroskope zu entdecken, so glaubte man doch, sie oder etwas ihnen ähnliches beobachtet zu haben, als das Mikroskop wirklich jeden Tropfen einer Infusion von kleinen, scheinbar einfachen Wesen wimmelnd zeigte. Dass Otto Friedrich Müller, unter Hrn. Ehrenberg's Vorläufern einer der bedeutendsten, für dergleichen Formen den Namen *Monas* in die zoologische Nomenclatur einführte, war nur einer jener terminologischen Scherze, wie sie auch bei Linné die Trockenheit des Systemes anmuthig beleben; allein diese Anspielung deutet auf eine damals vorhandene Richtung der Geister, die bei phantasiereichen Persönlichkeiten zu schweren Irrthümern führte." (Du Bois-Reymond 1870).

In any case, Haeckel inherits the monad's terminological ambiguity and rejects it. In the *Protistenreich*, speaking about different Protista such as *Protogenes primodialis*, Haeckel writes:

> One year later, two different organisms very similar to *Protogenes*, and still of an extreme simplicity, have been described by an eminent microscopist, M. Cienkowski. In the first volume of 'Archives of microscopic anatomy', he published the very interesting 'Contributions to the knowledge of Monads'. Among the different Protista which were summarized under the ambiguous, old-fashioned and uncertain notion of 'Monads' by M. Cienkowski, there are two kinds of freshwater microscopic inhabitants, which look like *Protogenes* thanks to their totally simple, unstructured and shiny plasmatic body.[35] (Haeckel 1878, p. 70)

Here, the word "monad" clearly refers to a biological entity, and yet, in characterising it as an uncertain, ambiguous and old-fashioned notion, he rejects it in favour of the "moneron". In addition to wanting to establish his own taxonomy, his rejection of the "monad" aims to prevent the use of a word which has a heavy metaphysical and speculative background.

\* \* \*

The use of "monad" conveys a metaphysical and a speculative background that cannot easily be stamped out, even by a strict empirical approach. Indeed, even if the Haeckelian approach is based first and foremost on the Darwinian theory of evolution by natural selection and on Schwann and Virchow's cell theory (Haeckel 1876, pp. 19–20), Haeckel's general approach to introducing monera may also recall Schelling's dynamic atomism. In the *Perigenesis der Plastidule*, Haeckel writes:

> Thus, in a mechanistic view from monism, we consider every matter as animated, every atomic mass with an eternal and constant atomic soul, we do not fear being accused of materialism. Because this monist point of view is as far from unilateral materialism as it is

---

[35] "Schon im folgenden Jahre wurden zwei verschiedene, dem Protogenes sehr ähnliche, höchst einfache Organismen von dem ausgezeichneten Mikroskopiker Cienkowski beschrieben. Im ersten Bande des Archivs für mikroskopische Anatomie veröffentlichte derselbe sehr interessante "Beiträge zur Kenntniss der Monaden." Unter den verschiedenen Protisten, die Cienkowski hier unter dem alten, vieldeutigen und daher sehr unsicheren Begriffe der "Monaden" zusammenfasst, befinden sich zwei mikroskopische Bewohner des süssen Wassers, welche in der vollkommen einfachen und structurlosen Beschaffenheit ihres kernlosen, strahlenden Protoplasma-Körpers dem Protogenes gleichen" (Haeckel 1878, p. 70). My translation.

from empty spiritualism. We may find in it the conciliation between a crude atomistic view and a dynamic view of the world.[36] (Haeckel 1876, p. 39)

Haeckel inherits the Schellingian approach that unifies materialism and spiritualism, materialist atomism and energetic dynamism, in order to join atomism and dynamism and, consequently, turn toward a kind of dynamic atomism. In this dynamic atomism, as we will see, the monera play the part of basic active living entities.

Correlative to his inheritance of Schellingian dynamic atomism, we may note that Haeckel also inherits with Goethe the Schellingian monistic and naturalistic reading of Leibnizian monads. And, in a sense, Haeckel takes this naturalisation of Leibnizian monads a step further since he considers them to be quite similar to atoms and intrinsically united with organic bodies *via* a substantial link, i.e. God (Haeckel 1904, pp. 125–126). In other words, considering that monads are equated with atoms, they have to have the atom's features, i.e. having a soul and a body. In this sense, Haeckel embodies monads in linking them to an intrinsic natural dynamism; conversely, he gives a soul to extensive atomic bodies[37]. Here, there is no doubt that monads are attached to bodies in Haeckel's view, the Leibnizian monads directly enter the natural world.

Consequently, even if Haeckel rejects the word "monad", the metaphysical and speculative background it conveys is quite hard to remove completely. In this sense, the theory of monera he establishes results both from an empirical and rigorous observation-based approach and from a speculative and holistic background (Richards 2008). In the *Perigenesis der Plastidule*, he relates for instance the discovery of Monera as follows:

> The discovery of Monera allowed us to advance our knowledge of elementary organisms. In 1864, I observed for the first time in the Mediterranean Sea, in Nice, an extremely simple organism, whose whole body, not only during its growth but also at an advanced stage

---

[36] "Indem wir so von dem mechanischen Standpunkte des Monismus aus alle Materie als beseelt, jedes Massen-Atom mit einer constanten und ewigen Atom-Seele ausgerüstet uns vorstellen, fürchten wir nicht den Vorwurf des Materialismus auf uns zu laden. Denn dieser unser monistischer Standpunkt ist ebenso weit von einseitigem Materialismus, wie von leerem Spiritualismus entfernt. Ja wir können in ihm allein die Versöhnung der rohen atomistischen und der inhaltsleeren dynamischen Weltanschauung finden" (Haeckel 1876, p. 39). My translation.

[37] "Kein Geringerer als der hochangesehene Philosoph Leibniz schloß sich diesen Ausführungen an und verwertete sie für seine Monadenlehre; und da dieser zufolge sich Seele und Leib in ewig unzertrennlicher Gemeinschaft befinden, übertrug er sie auch auf die Seele" (Haeckel 1921, p. 34).

of development and of free movement, was made of a structureless, homogeneous, piece of protoplasm with no nucleus and no form. This *Protogenes primordialis* proved for the first time that there are simpler organisms than unicellular organisms: living beings with a body that still has not attained the formal quality of the simplest cell, and that presents a crystalline homogeneity.[38] (Haeckel 1876, p. 25)

Haeckel recounts his discovery of monera, emphasising the place, the date, adding external proofs, precisely detailing the monera's structure and highlighting that this discovery resulted from an empirical approach, i.e. microscopic observations. In addition, considering that monera are plasmatic substances with no nucleus, the elementary composition of monera makes them simpler living beings than cells. In this sense, for Haeckel, the theory of monera completes the cell theory: since monera are the simplest building blocks of life, they are simpler than the cells and they are henceforth conceived as the morphological structure that distinguishes between life and non-life – as underlined by the reference to crystals. In addition, since cells were considered as the first chronological living beings that emerged and since monera, for Haeckel, chronologically emerged before cells, monera now have taken on the role of the cell in the theory of evolution; the theory of monera also connects with the theory of evolution (Haeckel 1904, pp. 220–221). In short, the theory of monera completely connects with the current theories of its time.

However, as we have already seen, for Haeckel, extended matter is soul-joined. Consequently, considering that monera are the primitive building blocks of life and the primitive living beings, they must contain an intrinsic spiritual and living pole which confers them this status of the basis of life. And given that monera are pieces of protoplasm with no nucleus and no form, the protoplasm must contain what makes living beings alive; as Haeckel declares, "the protoplasm is the true 'life-giving substance'"[39] (Haeckel 1876, p. 23). And yet, the protoplasm is amorphous so the criteria for characterising "aliveness" can-

---

[38] "Ein weiterer Fortschritt in unserer Erkenntniss der Elementar-Organe wurde durch die Entdeckung der Moneren herbeigeführt. Im Jahre 1864 beobachtete ich im Mittelmeer bei Nizza zum ersten Male einen einfachsten Organismus, dessen ganzer Körper nicht blos während seiner Entwickelung, sondern auch in vollkommen entwickeltem und frei beweglichem Zustande aus einem homogenen und structurlosen Stückchen Protoplasma ohne Kern und ohne alle differenten Formtheile bestand. Dieser "Protogenes primordialis" führte also zum ersten Male den Beweis, dass es noch einfachere Organismen, als die einzelligen giebt: Lebewesen, deren Körper noch nicht einmal den Formwerth einer einfachsten Zelle erreicht, sondern in sich so gleichartig und homogen erscheinen, wie ein Krystall." (Haeckel 1876, p. 25). My translation.

[39] "Das Protoplasma ist die eigentliche "Lebens-Substanz"" (Haeckel 1876, p. 23). My translation.

not be linked to a morphological structure, but to a chemical composition. Thus, Haeckel distinguishes in the protoplasm, before the atomic level, the plastidules: defined as "the true active factors of the vital process"[40] (Haeckel 1876, p. 77), plastidules truly make a living being alive; they are the ultimate "living particles"[41] (Haeckel 1876, p. 65). By referring to "active factors" (i.e. dynamic agents constituting the very foundation of living beings), this explanation of what makes the living world alive echoes the aims of the more speculative monadic approach: in monera, which are the first building blocks of life, plastidules establish the activity and the inherent "livingness" of living beings, in the same way that natural monads establish the activity and the inherent "livingness" of the living world.

Haeckel, through his theory of monera, embraces different frameworks: he adopts a kind of dynamic atomism as the main theoretical approach and he inherits from *Naturphilosophie* and the naturalistic reading of Leibniz. Within the nineteenth-century methodological framework in which the main biological discoveries took place, important changes occurred: while at the beginning of the century, the aim was to explain what living beings were within a general picture of the living world, fifty years later, the aim was to observe how living beings behaved and how they were composed. Consequently, while, for Schelling, activity in the living world means a primitive dynamic force, for Haeckel, living activity means being capable of nutrition, reproduction, sensibility, and minimal motility. Similarly, in a nineteenth-century methodological framework into which Darwin bursts, life is necessarily understood in relation to heredity and adaption. In this sense, for Haeckel, plastidules are the ultimate living particles because they have the ability, with their chemical composition, to take into account the environmental influence, to memorise it and to respond to it. In other words, each plastidule has its own specific chemical layout which adapts, passes on its information and so constitutes the specificity of heredity. In this sense, the chemical layout of plastidules is the specificity of vital activity. As Haeckel explains:

> However, those modifications are due to chemical changes in the atomistic composition and, consequently, in the molecular movement of plastidules and these changes are the result of diverse influences from the external environment, namely the external conditions of existence, and of the extraordinary mobility and the complex preservation of constitutive atoms. Plastidules do not forget these experiences. Rather, they transfer them to their progeny in the form of a modification of the primitive movement of the plastidule. In this sense, heredity can essentially be explained as a transmission of the individual movement of the plastidule which is necessarily linked to any reproductive process. [...] Hereditary transmis-

---

**40** "Die wahren activen Factoren des Lebens-Processes" (Haeckel 1876, p. 77). My translation.
**41** "Der Lebenstheilchen" (Haeckel 1876, p. 65). My translation.

sion is the plastidules' memory, variability is the plastidules' receptivity.⁴² (Haeckel 1876, pp. 68–69)

Thus, the Haeckelian moneron bears the marks of both contemporary biological research and of speculative monadic research. Here, the monera are the simplest living beings and the simple building blocks of life; for Haeckel, they complete the cell theory and connect it to Darwinian interrogations with the conception of plastidules. And yet, at the same time, monera are the basic living beings that contain the true active factors of the vital process. By articulating the notions of activity, life and living beings within a derived dynamic atomism, the moneron remains indirectly connected to constellation of problems posed by the metaphysical-speculative monadic approach.

## Conclusion

Between the seventeenth and nineteenth centuries, the methodological approach and the scientific tools for understanding life and living beings deeply evolved. The general scientific paradigm profoundly changed and, in our research into the conceptual basis of the basic vital entity, the path from monads to monera appears quite sinuous. And yet, we can draw a clear conceptual line from Leibniz to Haeckel regarding how the understanding of the first vital and active entity evolved.

Leibniz establishes metaphysical entities at the basis of the living world. Characterised by their perception, monads are principles of unity, of activity and of life in living beings. By understanding Leibniz through a Spinozist lens, Schelling uses the perceptive nature of monads and transforms them into driving forces. For him, natural monads as natural intensive drives have to be postulated at the basis of the living world; they are the authentic active and in-

---

42 "Diese letzteren aber sind zurückzuführen auf chemische Veränderungen in der atomistischen Zusammensetzung und demgemäss in der Molekular-Bewegung der Plastidule, welche bei der außerordentlichen Beweglichkeit und verwickelten Lagerung der constituirenden Atome unmittelbar durch die veränderten Einflüsse der umgebenden Aussenwelt oder der äusseren "Existenz-Bedingungen" herbeigeführt werden. Diese Erfahrungen vergessen die Plastidule nicht. Sie übertragen vielmehr dieselben als Modification der ursprünglichen Plastidul-Bewegung auf die Nachkommen. So erklärt sich die Vererbung wesentlich als die Uebertragung der individuellen Plastidul-Bewegung, welche mit jedem Processe der Fortpflanzung nothwendig verknüpft ist. [...] Die Erblichkeit ist das Gedächtniss der Plastidule, die Variabilität ist die Fassungskraft der Plastidule." (Haeckel 1876, pp. 68–69). My translation.

tensive basis that found the productive force in nature. With Schelling, even though they are still intensive, natural monads have windows, they conflict with one another, and constitute the active basis of living beings. They therefore belong to the natural world. Hence, this way of embracing this natural reading of Leibniz is structuring for understanding the *Naturphilosophie* of the young Schelling and this *Naturphilosoph* use of monads played a major part in the requalification of monads. In this sense, this speculative meaning of the monad remains current throughout the nineteenth century, while a more positive, taxonomic meaning of "monad" develops. During this period, the monad refers either to an intensive unit at the basis of the living world, a soul-like entity, the smallest living building blocks or the primitive living beings and this kind of contradiction between different meanings makes the conceptual legibility gradually fade away. Facing this lack of structuration in the monadic meaning, Haeckel rejects the term "monad" in favour of his own "moneron". Haeckel's refusal to apply the term "monad" allows him to reject the legacy of *Naturphilosophie* and to claim a connection with Darwin. And yet, while presenting monera as the simplest living beings in a taxonomic way and an evolutionary way, Haeckel also gives to monera a kind of soul and establishes that within the protoplasm there exist plastidules which are like the true active factors of the vital process. In doing so, Haeckel appears as the heir of the *Naturphilosophie*'s dynamic atomism.

Consequently, in the same way the Leibnizian monad is not a naïve metaphysical entity completely detached from the living world, the Haeckelian moneron is not only a morphological entity existing within a strict empirical taxonomy but also the descendant of a speculative tradition. In a new Darwinian age, something from older paradigms carries on and these conceptual shifts make the Haeckelian moneron the result of a combination of different legacies. From monads to monera, we face a powerful process of naturalisation of the understanding of how the basic living entity must be in order to be alive, and this process directly impacts the definition of life and living beings in the nineteenth century.

# Abbreviations

GP   Leibniz, Gottfried Wilhelm (1875–1890): Die philosophischen Schriften von Leibniz, VII Bde. Berlin: C. I. Gerhardt.

# Bibliography

Beiser, Frederick C. (2008): *German Idealism. The Struggle against Subjectivism 1781–1801.* Cambridge: Harvard University Press.
Du Bois-Reymond, Emil (1870): "Öffentliche Sitzung der Akademie zur Feier des Leibnizischen Jahrestages". In J. E. Strick (2004): *The Origin of Life Debate Molecules, Cells and Generation,* vol. 6. Bristol: Thoemmes.
Fichant, Michel (2005): "La constitution du concept de monade". In: Enrico Pasini (Ed.): *La monadologie de Leibniz, Genèse et contexte.* Paris, Milan: Mimesis, pp. 31–54.
Haeckel, Ernst (1876): *Die Perigenesis der Plastidule oder die Wellenerzeugung der Lebenstheilchen.* Berlin: Verlag von Georg Reimer.
Haeckel, Ernst (1878): *Das Protistenreich : eine populäre Übersicht über das Formengebiet des niedersten Lebewesen.* Leipzig: E. Günther.
Haeckel, Ernst (1904): *Die Lebenswunder, ergänzungsband zu dem Buche über die Welträtsel.* Stuttgart: Alfred Kröner Verlag.
Haeckel, Ernst (1921): *Die Welträtsel, Gemeinverständliche Studien über monistische Philosophie.* Stuttgart: Alfred Kröner Verlag.
Leibniz, Gottfried Wilhelm (1952): *Theodicy.* Ed. by Austin Farrer and trans. by E.M. Huggard. New Haven: Yale University Press. EBook of The Project Gutenberg.
Leibniz, Gottfried Wilhelm (1989): *Philosophical Essays.* Ed. and trans. by Roger Ariew and Daniel Garber. Indianapolis, Cambridge: Hackett Publishing Company.
Leibniz, Gottfried Wilhelm (2006): *The Shorter Leibniz Texts: A Collection of New Translations.* Ed. and trans. by Lloyd Strickland. London: Continuum.
Levey, Samuel (2007): "On Unity and Simple Substance in Leibniz". In: *The Leibniz Review* 17, pp. 61–97.
Müller, Otto Friedrich (1773): *Vermium Terrestrium et Fluviatilium, seu Animalium infusoriorum, Helminthicorum et Testaceorum, Non Marinorum, Succincta Historia.* Heidelberg and Leipzig: Heineck et Faber.
Nachtomy, Ohad (2011): "Leibniz on Artificial and Natural Machines: On What It Means to Remain a Machine to the Least of Its Parts". In: Smith Justin E. H. and Nachtomy Ohad (Ed.): *Machines of Nature and Corporeal Substances in Leibniz.* Amsterdam: Springer Netherlands, pp. 61–80.
Oken, Lorenz (1805): *Die Zeugung.* Bamberg, Wurzburg: J.A. Goebhardt.
Oken, Lorenz (1844): *Elements of Physiophilosophie.* Trans. by Alfred Tulm, London: The Ray Society.
Pasini, Enrico (2011): "The Organic Versus Living in the Light of Leibniz's Aristotelianisms". In: Smith Justin E. H. and Nachtomy Ohad (Ed.): *Machines of Nature and Corporeal Substances in Leibniz.* Amsterdam: Springer Netherlands, pp. 81–94.
Phemister, Pauline (2005): *Leibniz and the Natural World, Activity, Passivity and Corporeal Substances in Leibniz's Philosophy.* Amsterdam: Springer Netherlands.
Phemister, Pauline (2016): *Leibniz and the Environment.* London, New York: Routledge.
Rey, Anne-Lise (2009a): "L'ambivalence de la notion d'action dans la Dynamique de Leibniz. La correspondance entre Leibniz et De Volder (1ère Partie)". In: *Studia Leibnitiana* 41, pp. 47–66.
Richards, Robert (2008): *The Tragic Sense of Life: Ernst Haeckel and the Struggle Over Evolutionary Thought.* Chicago: University of Chicago Press.

Schelling, Friedrich (1803): *Ideen zu einer Philosophie der Natur.* Landshut: Philip Krüll.
Schelling, Friedrich (1858): *Sämmtliche Werke*, III Bd. Stuttgart, Augsburg: J.G. Cotta.
Schelling, Friedrich (1861): *Sämmtliche Werke*, X Bd. Stuttgart, Augsburg: J.G. Cotta.
Schelling, Friedrich (1988): *Ideas for a Philosophy of Nature.* Trans. by Errol E. Harris and Peter Heath, Cambridge: Cambridge University Press.
Schelling, Friedrich (2004): *First Outline of a System of the Philosophy of Nature.* Trans. by Keith R. Peterson, New York: State University of New York Press.
Wilson, Catherine (2010): "Leibniz's Reputation in the Eighteenth Century". In: John Roger, Tom Sorell and Jill Kraye (Ed.): *Insiders and Outsiders in Seventeenth-Century Philosophy.* New York, London: Routledge, pp. 294–308.
Zammito, John H. (2017): *The Gestation of German Biology, Philosophy and Physiology from Stahl to Schelling.* Chicago: University of Chicago Press.

Robert Kocis
# Idealism and Darwin – Rejection, Accommodation, Appropriation: James Hutchison Stirling and David George Ritchie

**Abstract:** Darwin's materialism and rejection of teleology (or divine plan) seemed to contradict the most central tenets of British Idealism. Not surprisingly, though, the idealists responded in roughly three ways: some, like Whewell and Stirling rejected evolution; others, as Boucher argued, sought an accommodation between British Idealism and Darwinism; Ritchie, though, sought to appropriate Darwin's revolution as an instance of idealism at work. To do this, Ritchie needed to invent a fledgling philosophy of science, based on Kant's science of cognition. Each of these responses is examined critically. Ritchie's invention of a fledgling philosophy of science, based on Kant's science of cognition, is found to be a significant philosophical achievement.

## I Introduction

Even philosophical traditions with strong internal coherence – like British Idealism – contain tensions. For example, Bosanquet praised Herbert Spencer's writings ("cannot be too strongly urged", 2001, p. 97) while Ritchie (1896) sought to refute Spencer's *errors*. It should be no surprise, then, that idealists varied in responding to the *zeitgeist*-shaking ideas of Charles Darwin. Darwin's fundamental challenge evoked engagement along (roughly) three lines. 1) Obvious (and profound) differences between the two traditions did lead some to oppose him; Whewell and Stirling, for instance, *rejected* "Darwinianism." 2) Obvious (but not necessarily profound) differences may permit an *"accommodation"* if there

---

Earlier versions of portions of this paper – the quotations and expositions of Ritchie's Hegelian and Kantian arguments – appeared in the *History of Political Thought*, XXXIX: 1, Spring 2018, pp. 156–180.

---

**Robert Kocis,** The University of Scranton

https://doi.org/10.1515/9783110604665-009

is no inherent contradiction; a prominent researcher in the field, David Boucher[1] (2014), has argued that this strategy was largely successful. 3) Obvious differences did inspire proponents to hold that idealism is logically prior to Darwinism, so that Darwin is *appropriated* as idealism at work.

Part II articulates Stirling's *rejection* of evolution; Part III summarizes efforts at *accommodation*, including Boucher's and Ritchie's; and Part IV examines Ritchie's efforts to *appropriate* Darwin. Our principal foci will be the writings of Stirling and Ritchie.

## II Stirling's *Rejection* of Darwin: Design and Naturalism

### A Preliminaries: Non-Trivial Disagreements

Ambiguities about "species" and the lack of known causation for variations bedeviled Darwin. Stirling suggested that Darwin was well summarized by Huxley:

> That new species may result from the selective action of external conditions upon the variations from their specific type which individuals present and which we call 'spontaneous' because we are ignorant of their causation – that suggestion is the central idea of the *Origin of Species* and contains the quintessence of Darwinism (GLE).

Stirling detected five problems. First, organisms which vary "from their specific type" may be only variants, not species. Second, "we are just simply ignorant of [variation's] causation" (GLE). These objections were more significant in the nineteenth century because the idea of a species was less fixed than it is now and the significance of Mendel's work on the genetics of peas (1865) was not fully appreciated until the work of DeVries (1900; 1904).

Darwin himself complained that "No one definition has satisfied all naturalists; yet every naturalist knows vaguely what he means when he speaks of a species ... The term "variety" is almost equally difficult to define ..." (OS, p. 58; pp. 63–65). This ambiguity persisted until consensus emerged around Mayr's (1944) definition: species are groups of morphologically and behaviorally similar organisms "that can breed only among themselves, excluding all oth-

---

[1] In addition to being a world-renowned expert in British Idealism studies, Boucher is a professor of political philosophy and international relations at Cardiff University; he has served repeatedly as "convener" for the British Idealism interest group of the Political Studies Association; and he is editor of *Collingwood and British Idealism Studies*.

ers."² Darwin could have clinched his argument if he had shown that the new varieties of finches could not interbreed. Stirling sensed this deficiency:

> Because he found in these islands so many finches³ and in the different islands different ones Mr. Darwin was led to speculate on their possible origin. There was a common analogy in all of them ... only on a certain South American type. The obvious inference accordingly was that all these finches ... had been actually modified and all of them out of a single characteristic type ... As one sees it is at once assumed here that the thirteen different finches constitute or represent thirteen different species; and consequently the first thing it occurs to us to ask is What *is* a species? ... We really should like to know if they cannot pair together." (GLT)

Second, Stirling deplored our ignorance of the causes of "spontaneous" variations. After DeVries (1900), we can *now* say that in the processes (meiosis and mitosis) of cell divisions, genetic material, which almost always creates an accurate "carbon copy" of itself in the new cells, sometimes randomly mutates. While this objection is *no longer* damaging, Stirling was right to criticize this gap in the theory.

Third, the Galapagos seemed a strange place to stumble upon the "struggle for existence."

> We have seen that Mr. Darwin speaks of the struggle for existence as an essential element of the theory ...; what countenance then does the very feeding ground and breeding ground and originating ground of natural selection show it? Why none – absolutely none! Through-

---

2 Horses and donkeys, different species for morphological and behavioral differences, also cannot *successfully* reproduce; the hybrid attempts (mules) are sterile. This is consistent with Mayr's definition. However, to complicate matters even for today's understandings, today's wolves (apparently a different species from the one that gave rise to today's dogs) are seen as different species from today's dogs for morphological and behavioral reasons. But dogs and wolves *can* interbreed and the hybrids seem to be fertile, creating legal dilemmas for jurisdictions having statutes prohibiting wolves for pets; are the hybrids wolves or dogs for legal purposes? Although Mayr's definition is now taken as authoritative, there are still gray areas at the edges.

3 Interestingly, the fifteen types of finches that Darwin identified are still largely taken to be distinct species because they tend not to interbreed (although hybridization *can* happen) and because of morphological and behavioral differences. When researchers sequenced their genomes, they found that a single gene (ALX1) was largely but not exclusively responsible for variations in the shapes of beaks. [https://www.bbc.com/news/science-environment-31425720] Perhaps genomic variations are a bit like dialect areas (or continua): with settlements spread across significant distances, individuals from any two adjacent villages are mutually intelligible while speakers from the extreme ends of the variations are not (Bloomfield 1935, p. 51 and Chambers and Trudgill 1998, *passim*). Similarly, variants from island A may be able to interbreed with those from island B, but not with those from island K.

out the whole of the Galapagos archipelago there is not a vestige of the struggle for existence – not a trace! (GLE)

In Darwin's own words, the native fauna exhibited no fear of him or of any predator. We are left with Darwin's explanation that he found this idea in the writings of Malthus and Spencer (OS, p. 74) and that they are of general validity (OS, Ch. 3).

Fourth, in *Darwinianism: Workmen and Work*,[4] Stirling (1894) hypothesized that Darwin conjured a false dichotomy between creation *or* variation. He returned to this theme: Darwin "came to his idea of ... 'Creation *or* Modification ...'." (GLE) Darwin, it seemed, could not quite get beyond the view that acts of creation had to be singular acts like the way that "workmen," say, carve a duck. But the frontispiece of OS features a quotation from Whewell: "But with regard to the material world, we can at least go so far as this – we can perceive that events are brought about not by insulated interpositions of Divine power, exerted in each particular case, but by the establishment of general laws" (Darwin cites Whewell's *Bridgewater Treatise III*; this could not be independently verified). God *could* have created humanity by setting in place the laws of variation and natural selection and letting them operate. Darwin's view of divine creation was less cramped than the "workman's" view.

Finally, he questioned if Darwin overreacted to the exotic nature of the area. What about the visit to the Galapagos evoked so strange a theory as godless evolution? "One cannot wonder that such a region as this went to the heart of Mr. Darwin," Stirling wrote; "one cannot wonder that it was here he found the motive for his peculiar theory. This spot was solitary and remote; and what life there was upon it seemed to have for him only a strange unnatural and old-world look."(GLN) In short, Darwin may have been impressionable.

## B Teleology

And that brings us to the [central] question that is between Mr. Darwin and ourselves – the question of design ... How was the woodpecker for instance so wonderfully formed for the climbing of trees [Darwin] asked himself; and he could not quiet himself by the answer it

---

[4] *Darwinianism: Workmen and Work* is a unique and almost peculiar work. From the title one might expect a detailed, philosophical examination of the central tenets of Darwin's theory. But Part I ("The Workmen") is a psychological examination of the characters of Darwin, his father, and his grandfather, Erasmus. Meanwhile, Part II ("The Work") is a detailed examination of Darwin's intent, given data drawn from his extensive correspondence.

has been just so made. That was a supernatural explanation and he for his part could only be satisfied with a natural one. (GLE)

Stirling believed in Divine creation and that everything in nature has a purpose. For Stirling, "The whole effort of nature in its zoology is to get to man; and it is a long ascent to get to him through sponge and mollusc[,] fish and reptile[,] bird and beast" (GLE).[5]

Stirling detects a divine plan behind what seems contingent (random or not-necessary):

> Nature scatters its living products abroad as the sea its shells upon the strand ...Contingency is the world; he that cannot put himself at home with contingency as philosophically understood will never philosophize this world. Mr. Darwin's inherited individual differences will never prove a match for the contingency that *is* (GLE).

So, Darwin's contingency pales in comparison to divine or natural contingency.

However, for Darwin the contingent ways that *differences* are randomly distributed through populations are what is significant. Stirling understood this clearly:

> Mr. Darwin said to himself 'Children resemble their parents; but they also differ from them.' Evidently therefore they are as likely to propagate differences as to propagate resemblances ... Now any given difference may be an advantage or it may be a disadvantage. That is[,] the animal by reason of the difference propagated and inherited may be obstructed in the exercise of its functions and the use of its conditions ... The ultimate of obstruction can only be extinction. But in the case of furtherance... say for incalculable periods the ultimate can only be something perfectly new – can only be a new organism in fact that is tantamount to a new species (GLE).

To the extent that an organism resembles its parents, it has neither more nor less survival value; but (random) differences result either in extinction or a new evolutionary advantage. Stirling seems to have understood this part of Darwin

---

[5] Curiously, Alfred Russell Wallace, co-founder of natural selection with Darwin, and whose work on the subject was co-published with Darwin's, became a spiritualist later in life. From this new perspective, divine intervention was required at three points in human evolution. First, only God could have created the first life; second, divine intervention would have been required for the emergence of the consciousness in the higher forms of animal life; and finally, only divine intervention could have created the excellences of the human mind (Wallace 1914). Lyell split with Darwin and joined Wallace, Whewell apparently allied with Stirling, while Huxley and Hooker held firm with Darwin.

quite well, but never accepted it because it lacks divine design.[6] Perhaps a divine plan is such a self-evident truth that denying it is an error.

## C Materialistic Evils

Some of Stirling's most interesting arguments against "materialism" are found in his response to Huxley's lecture on protoplasm (1869).[7] Huxley reasoned that "there is some one kind of matter which is common to all living beings" (p. 1): the protoplasm within cell walls. While organic diversity may appear to be striking – "What, truly, can seem to be more obviously different from one another... than the various kinds of living beings?" (p. 3) – behind obvious variety lies a fundamental similarity. "A single physical basis of life [underlies] all the diversities of vital existence." (p. 4). Further "What has been said of the animal world is no less true of plants" (p. 7).

To explain his view that "Protoplasm, simple or nucleated, is the formal basis of all life," Huxley conjured a metaphor: "It is the clay of the potter: which, bake it and paint it as he will, it remains clay, separated by artifice, and not by nature, from the commonest brick or sun-dried clod" (p. 7). More prosaically, it is a fact "that all the forms of protoplasm which have yet been examined contain the four elements, carbon, hydrogen, oxygen, and nitrogen ..." (p. 8). The basic building blocks of life have been discovered: the smallest organism is a cell and the protoplasm in all cells is bio-chemically similar. "But it will be observed, that the existence of the matter of life depends upon the pre-existence of certain compounds"; remove "any one of these from the world, and all vital phenomena come to an end" (p. 11).

Huxley conceded that:

> Experience leads me to be tolerably certain that, when the propositions I have just placed before you are accessible to public comment and criticism, they will be condemned ... I should not wonder if 'gross and brutal materialism' were the mildest phrase applied to them in certain quarters (p. 12).

No doubt, "the terms of the propositions are distinctly materialistic." Nevertheless, two things are certain: "the one, that I hold the statements to be substan-

---

[6] Stirling's position is usually somewhat clear; less clear is *why* he disagrees with Darwin so intensely.
[7] Fascinatingly, Huxley denies believing in the doctrine of materialism, but Stirling would not see the difference between ontological and methodological materialism.

tially true; the other, that I, individually, am no materialist, but, on the contrary, believe materialism to involve grave philosophical error" (p. 12).

This requires explanation. Huxley professed to have learned from Hume "the limits of philosophical inquiry" (p. 12): it is impossible to find absolute certainty in answers to philosophical questions; advocates of any of the ontological "isms," whether spiritualism or materialism, err in believing they have found such an impossible truth. While Huxley is methodologically a materialist, he would not embrace *any* ontological doctrine, even materialism. Considering potential progress of science, he writes:

> materialistic terminology is in every way to be preferred... But the man of science, who, forgetting the limits of philosophical inquiry, slides from these formula and symbols into what is commonly understood by materialism, seems to me to place himself on a level with the mathematician, who should mistake the $x$'s and $y$'s, with which he works his problems, for real entities – and with this further disadvantage, as compared with the mathematician, that the blunders of the latter are of no practical consequence, while the errors of systematic materialism may paralyse the energies and destroy the beauty of life (p. 16).

Stirling was having none of this ontological skepticism. Once one begins with materialistic terms, assumptions, and propositions, there is no avoiding reductionism: all life and all consciousness are mere matter. "Nor is there any logical halting-place between" Huxley's contention that life is based upon combinations of compounds "and the further and final one: That all vital action whatever, intellectual included, is but the result of the molecular forces of the protoplasm" (Stirling 1872, pp. 22–23).

Stirling then raises the question as to how we could know if the protoplasm of a living cell is the same as that of a dead cell – after all, we kill the cell to conduct the chemical analysis. Hence,

> Chemically, dead protoplasm is to Mr. Huxley quite as good as living protoplasm ... [I]t must be pointed out that ... living protoplasm ... is unlike dead protoplasm. Living protoplasm, namely, is identical with dead protoplasm only so far as its chemistry is concerned ... and it is quite evident, consequently, that the difference between the two cannot depend on that in which they are identical – cannot depend upon the chemistry. Life, then, is no affair of chemical and physical structure, and must find its explanation in something else (p. 40).

What is missing in dead protoplasm is exactly the "life" that we are trying to understand. Since a clump of dead chemicals exhibits no life, chemicals alone cannot produce life. To create life is to create "a new world – a new and higher world, the world of a self-realizing thought, the world of an entelechy" (p. 40). Stirling sees here a key tenet of Absolute Idealism: "But this [living] force is a rational unity, and that is an idea ..." (p. 43).

This implies a duty to "resist the extravagant assertion that all organized tissue, from the lichen to Leibnitz, is alike in faculty, and again the equally extravagant assertion that life and thought are but ordinary products of molecular chemistry" (p. 54). Stirling concludes: "I need go no farther ... the case is now complete" (GLT). For him, idealism is totally incompatible with Darwinian materialism.

## III Accommodating Darwin

### A Boucher's Overview

The central tenets of Boucher's approach are a recognition of the tensions between idealism and Darwin, coupled with a belief that there are no *fatal* contradictions. "Darwin's scientific method was inspired by Newton's astronomy ... and developed in a neo-Kantian direction" under the influence of William Whewell[8] (Boucher 2014, p. 307). Darwin's scientific methodology thus was akin to idealism. Thus his achievement in applying evolution to the natural world:

> was widely applauded by the British Idealists. It was a hypothesis that gave credence to the fundamental starting point of the philosophy of the British Idealists, that is, the unity of existence, in that evolution posited a continuity between nature and spirit. Principally, it asserted the unity of life... (p. 307).

There were other methodological similarities. Idealists, like scientists, viewed hypotheses as central to expanding our knowledge; "Evolution as a hypothetical conjecture constituted, for the British Idealists, an absolute postulate, or what R. G. Collingwood was later to call an 'absolute presupposition' [...], exercising 'subtle dominion' in all aspects of experience" (p. 307). Boucher notes similarities to Kuhn's "paradigms," overarching views of reality rooted deeply in human cognition and language that help form scientific communities and shape "normal" research (p. 307).

Further, they subscribed to complementary views of truth. Scientists believe that hypotheses have truth value only in context; idealists believe that propositions have truth only as part of a coherent whole. For scientists, "$e=mc^2$" is without truth value – unless it is contextualized in Einsteinian theories of relativity. For idealists, such propositions are found to be true only if the whole is coherent.

---

[8] Whewell, a Cambridge Idealist, was a friend and mentor to Darwin. Boucher is suggesting that the idealism of Whewell's *Philosophy of the Inductive Sciences* likely influenced Darwin.

"In essence, then, the logic of [both] the empirical sciences and of British Idealism is propositional. They differ only ... over the criterion of truth" (p. 308). A difference so subtle constitutes no stumbling block; idealists resemble working scientists more than do empiricists – who view propositions in isolation.

The "theory of evolution was particularly conducive to" idealism's "manner of philosophizing" (pp. 308–309). The idealists had a theory of evolution and drew on similarities between Darwinian and Hegelian evolution. "Many of the British Idealists consciously conflate Hegelian philosophy and the theory of evolution in order to appropriate its favorable evaluative force..." Others were seizing evolution for "all sorts of political pretexts, and therefore they had to adapt it to their own ends, while remaining faithful to the spirit of Hegelian philosophy" (p. 311).

The heart of Boucher's accommodation thesis is this idealist conviction that "evolution was the key to understanding the ultimate character of the universe and... all of the separate forms of knowledge or experience were converging on the universal form of explanation" (p. 311). They believed that "the categories of biological evolution, heredity, inheritance, natural selection, and the struggle for existence, had to be accommodated in idealist social explanation" (p. 311). Given that strategy, "The tactic of the British Idealists was to argue that Hegel was a far better evolutionary theorist than Darwin, Spencer, or Huxley. Evolution was indeed completely compatible with idealism because idealism was itself an evolutionary philosophy" (p. 311). William Wallace even "argued that what Darwin had done for the process of evolution in the organic world, Hegel did for the self-development of thought in philosophy" (p. 311).

Nevertheless idealism "was not a naturalistic philosophy and therefore felt compelled to challenge the naturalistic postulates of evolution and provide a more satisfactory theory based on Hegelian principles" (p. 315). Boucher was not daunted by these differences: "The great merit of Darwinian naturalistic theories of evolutionary ethics is that they explicitly recognize the unity of the cosmos..." Idealists further shared Darwin's belief that humans are not "so different in kind from the rest of nature that they require completely different forms of explanation" (p. 315). This permitted idealists to affirm "the unity of nature and spirit, and the operation of natural selection in nature and society." They argued "that natural selection accounts for both organic development and moral progress" (p. 318).

Boucher concludes: "I have tried to show that the idea and hypothesis of evolution so thoroughly permeated educated society in the Victorian era that all modes of thought had to show how they were compatible with it." Idealism cannot be properly "understood without locating its arguments within the context of evolutionary debates..." and idealists "believed their own philosophy to be a more sophisticated version of evolution" than "the naturalistic ideas of Darwin and Spencer ..." (pp. 319–320).

Nor is Boucher alone. Passmore (1959) observed that "it is nakedly apparent that there is more in common between Darwinism and Absolute Idealism than there is between Idealism and the orthodox Christianity which, it was at first expected, the Idealists would save from the onslaughts of Darwin" (p. 53). Similarly, Neill (2003) contended that "in some ways, the analysis given by these two groups was highly similar. Both argued that it made little sense to view humans as abstract individuals, since they had to be seen in the light of their involvement in (or indeed constitution by) the 'social organism' of the state" (p. 315). In summary, two of Stirling's concerns are not yet resolved; first, Darwinists believed the cosmos to be materialistic ("naturalism") while idealists did not; second, idealists were teleologists while Darwin was not.[9]

## B Ritchie's Accommodation

Ritchie's accommodation argument is slightly different. Boucher sensed that difference without articulating it fully; those human faculties, he notes, which contributed so much to our continued survival, are also what make science possible:

> Language, Ritchie argues, makes possible the transmission of experience which is not biologically inheritable. The possession of consciousness, the ability to reflect, and the use of language give human beings a tremendous advantage in the struggle for existence. The origins of these human powers or capacities are best explained, however, by the hypothesis of natural selection (Boucher 2014, p. 318).

This may be the weaker of Ritchie's two arguments, but it illustrates well his methods while also providing premises supportive of the *appropriation* argument.

Ritchie begins by noting that there is no antipathy between science and philosophy. "In every age philosophy has been affected by the sciences, i.e., the methods and conceptions which are used in the attempt to make some particular province or aspect of the Universe intelligible have exercised a fascination over those who are seeking to understand the universe as a whole" (CW II, p. 38). A kind of cross-pollination occurs, as disciplines borrow from one another;

---

[9] As Gould (1977) argued, Darwin's rejection of Aristotelian and Christian teleology gave him pause about publishing his findings because he knew that there would be resistance, especially from parts of the religious community.

and ideas permeate each *zeitgeist*. Notions related to evolution, like development, were common at the time. As early as 1794, Erasmus Darwin had posited a type of evolution; in addition, "Buffon, Geoffrey St. Hilaire, Lamarck, had all attacked the orthodox dogma of immutable species ..." (p. 42). "Evolution is in every one's mouth now ... But nothing grows up quite suddenly. During the latter half of last century many isolated thinkers had, in this or that department of science, come to apply the idea of development" (p. 42).

Idealists, in particular, were aware of scientific discoveries and even contributed to them. "Though in Kant as a philosopher the idea of evolution ... is conspicuously absent, yet the same Kant, as a man of science, was the author of the nebular hypothesis ...," a belief that stars "evolved" from nebula or amorphous gases (p. 42). It is significant that Kant did *not* believe that employing physical assumptions and discovering material evolution contradicted any part of his philosophy.

So, Ritchie, like Boucher's other idealists, seeks initially to reconcile evolution with Hegel's notion of *emanation*. "Thus, Hegel grew up in an intellectual atmosphere in which the conception of evolution, and especially of biological evolution, was no inconsiderable element" (p. 43). Similarly, Hegel embraced the view that organisms are more or less-highly developed, being drawn to Goethe's law that:

> The more imperfect a being is the more do its individual parts resemble each other, and the more do these parts resemble the whole. The more perfect the being is, the more dissimilar are its parts. In the former case the parts are more or less a repetition of the whole; in the latter case they are totally unlike the whole. The more the parts resemble each other, the less subordination is there of one to the other. Subordination of parts indicates high grade of organization (p. 44)[10]

Such teleological considerations led Hegel to prefer *Entwicklung* (or *emanation*) over evolution. Ritchie quotes Hegel: "The two forms in which the series of stages in nature have been apprehended are Evolution and Emanation" (p. 44).[11] So "Evolution was thus familiar to Hegel, both the theory and the word. Everywhere in Hegel we read about Entwickelung (sic); but of Evolution he does not speak in so friendly a manner" (p. 44). Why the preference? Ritchie explained that emanation "is meant the process from the less perfect to the more perfect" while evolution is "the process from the more perfect to the less perfect"

---

**10** Ritchie attributes this expression of Goethe's "law" to Lewes, *Life of Goethe*, p. 358; apparently the second edition (London, 1864). Lewes in turn refers to Goethe's *Zur Morphologie*, from *Werke*, XXXVI, p. 7.

**11** Ritchie cites Hegel, *Naturphil.*, p. 34.

(p. 45). Of the two Hegel "prefers the conception of Emanation, because it explains the lower from the point of view of the higher, whereas Evolution carries one back 'into the darkness of the past,' and only gives us a series of stages following one another in time" (p. 45). Put differently, Hegel would have preferred *The Ascent of Man* to Darwin's choice of *The Descent of Man*.

However, Ritchie believed that our task is not to dwell on such "details in Hegel" but to articulate "his general method and spirit of philosophising" (p. 55) So he shifts our attention from evolution: "Now it is 'natural selection' which seems to me the really epoch-making scientific theory: it is that that has produced that 'change of categories' which, as Hegel says,[12] is the essential thing in all revolutions, whether in the sciences or in human history" (p. 55). Since Darwin's natural selection *transforms* our way "of looking at nature," Hegel's "method of philosophising Nature could adjust itself quite easily to the new scientific theory..." (p. 56). Here is *what is most important* of this Hegelian argument for accommodation: scientific revolutions involve "changes of categories."

Unfortunately, Ritchie permits this promising argument to veer off topic. Natural selection operates through struggle, he reasoned, so it is a process of negativity. "In the stage of mere Nature this negativity is mechanical and external" (p. 56). In higher stages, negation works at the level of morality and becomes conscious: "Morality, to begin with, means those feelings and acts and habits which are advantageous to the welfare of the community" (p. 63). So, we begin with nature and end with morality. Ritchie concludes that "This seems to me a type of interpretation of human evolution which is in entire accordance with Darwin's theory of natural selection and which yet admits of what is most valuable in Hegel's dialectical method" (p. 66).

The argument however does not succeed in its purpose: reconciling evolution (a-telic and materialistic) with emanation (telic and ideal). At the most basic level, if one strips "emanation" of its ideal and teleological aspects, and "evolution" of its material and teleological aspects, all that remains of each is "a process of change." Everything distinctive about each has been removed; this is not a reconciliation of the two ideas but a gutting of the essential characteristics of each.[13]

---

[12] Ritchie cites Hegel, *Naturphil.*, p. 19.
[13] Apparently, I stand alone among Ritchie scholars on this issue. Nicholson's "Introduction" contained contemporaneous commentary that, in the context of his ruminations on Hegel, quotes (approvingly) a reviewer who believed that Ritchie had "captured" the idea of evolution for idealism (CW I, p. xviii). He goes on: "In the striking article 'Darwin and Hegel' Ritchie claimed that Darwinian ideas of evolution and Hegelian ideas of the development of reason in the world are compatible, and indeed when rightly understood can be seen to support one another."

## IV Ritchie's *Appropriation* of Darwin

Ritchie discerned that science is only possible if the Kantian account of human cognition is true. From this he developed a still under-appreciated account of science that roots it solidly in Kantian cognition and posits the existence of a "social" dimension (a community of knowers) for a science and, by extension, for all knowing.

Ritchie's intention was not to invent a philosophy of science; rather, he sought to protect Green's liberal commitments against Spencer's harsh political economy. "Survival of the fittest" was the Spencerian term that Darwin borrowed to describe the competition among individuals, varieties, and species over millennia (OS, p. 74). If one were to apply the mechanism to societies over short periods of time without the mitigation of compassion, one might be tempted to prefer Spencer to Adam Smith (for whom morality was a pre-condition to a successful capitalist society; see his *Theory of Moral Sentiments*). But Darwin was more akin to Smith than Spencer. He wrote:

> Important as the struggle for existence has been and even still is, yet as far as the highest part of man's nature is concerned there are other agencies more important. For the moral qualities are advanced... much more through the effects of habit, the reasoning powers, instruction, religion, etc., than through natural selection; though to this latter agency the social instincts, which afforded the basis for the development of the moral sense, may be safely attributed" (Darwin 1871, vol. II, p. 386).

Biology does not doom humans to viciousness; nature has selected "the social instincts" as the basis for "development of the moral sense," which is "more important."

So Ritchie is not wrong to "seek to prove that *The theory of Natural Selection (in the form in which alone it can properly be applied to human society) lends no support to the political dogma of laissez-faire*" (CW I, p. iii). This placed him at odds with the view that a nature "red in tooth and claw"[14] required economic arrangements based on units that need to outperform one another in order to

---

However, this "created something of a sensation" among philosophers "because Darwinism had been regarded as a natural enemy of Hegelians" (p. xviii).

**14** This has sometimes been attributed, I think wrongfully, to Spencer. Berlin has attributed the idea to de Maistre (Berlin 1990 [2013], pp. 111–112); others to Tennyson, *In Memoriam*, Canto 56. I could independently verify only the Tennyson source.

survive. To save Green, Ritchie needed a philosophy of science. He therefore directs our attention to epistemological concerns.[15]

Traditionally, epistemological antagonists were 1) empiricists – all knowledge is *only* sense data and its implications – and 2) rationalists – all knowledge is the product of deductions. The balance between the two was disturbed when Hume famously awakened Kant from his "dogmatic slumber" (Kant 1783 [1912], "Preface") by questioning the existence of causes. Hume's account of "causes" was disturbing because, as an empiricist, only sensation could be a valid basis for an existence-claim. In his billiards metaphor, he questioned whether there were any *perceptible* causes at work. Clearly one could see: 1) the cue ball, 2) the stroke of the stick which moved the cue ball, and 3) the trajectory of the cue ball – all these existed. But we have no similar perceptual warrant for saying that a *cause* was extant. Clearly, there was no visible entity moving across the surface of the table, carrying the ball into the pocket. Without a sensual warrant for asserting the existence of causes, Hume was left with the skeptical conclusion that causes did not "exist" *qua* entities in the material world. Causation was (merely?) a relationship between entities. So a new epistemological crisis emerged.

That the Enlightenment ideal of "Reason" could not be sustained was, then, an implication of Hume – and a delight to Romantics, along with grist for Kant.[16] Where the *philosophes* of the Enlightenment had unlimited confidence that Rea-

---

[15] It is possible that Ritchie was prepared to sacrifice a part of the ontology of idealism – the belief in the Absolute – for the sake of saving the remainder. Certain of the idealists, most prominently Hegel, had believed that reality, including what the ordinary citizen would call the physical or material world, is merely "Idea" seeking to develop its potential in a dialectical process. But if Darwin and Ritchie are right about the nature of science, those ontological commitments are no longer necessary. One implication of Ritchie's argument is that the only way to have a cogent science of the material world is to assume that it *is* material and to work out fully the implications of that assumption. Ritchie's epistemological solution to the problems posed by Darwin might be seen as the compromise that saved the remainder of the idealist project, making possible the recent revival. If abandoning a belief in "the Absolute" makes the rest possible, it may be a wise choice.

[16] Isaiah Berlin addressed the (primarily French) Enlightenment and the (Germanic) Romantic reaction against it in *The Roots of Romanticism* (Berlin 1999). Where Berlin differs from this account is that he sees idealism in general (including Kant, Fichte, Hegel, Bradley and Bosanquet) as arising from Romanticism and consequently fraught with a "metaphysical division of the person;" but idealism can be seen as an alternative to both the Enlightenment rationalism and the Romantics' emotionalism. In tying Kant so tightly to Hamann, Berlin overestimated the importance of Romanticism and "darkness" in Kant's philosophy. The problem for Berlin's interpretation is that, despite his friendship with Hamann, Kant denounced Romanticism and cannot be seen as a disciple of Hamann's "darkness."

son could provide the one, true answer to *all* of life's pressing questions, the Romantics rejected it wholesale and celebrated a mysticism that could evoke more (moral) purity from humanity than could "cold" rational calculation. A decade later, Kant sought to transcend the divide between empiricism and rationalism by shifting the conversation away from the *knowledge* – that comes from reasoning or sensual data – to a science of *erkenntnis* or *cognition*. (See the addendum for the German text.) Cognition, and not merely knowledge, becomes the mystery to be explored; with this, cognitive science was born (Brook, 1987).

Kant delineated the limits of reason, penning *A Critique of Pure Reason* that both celebrated its successes within narrow spheres, like mathematics, while attending carefully to its limits for the rest of life's questions. In this new Kantian science, cognition is a series of mental functions that are "embodied" in ways that form, order, and shape the manner that we conceive of – and perceive – the world. Our cognitive frameworks, then, are partly a function of that world – a reality external to us – and partly a function of the minds (and bodies) becoming cognizant of it. Given the ways that our bodies are shaped (bilaterally symmetrical, eyes forward, ears to the sides), we are inclined to employ certain cognitive strategies and categories to order sensations: "in front of or behind," "before or after," "to the left or right," "above or below," "because," and so on (Thagard, 2005). It is almost as if, given who we are physiologically and mentally, we *must* think in the ways that we do – or not think at all. Human cognition as we know it could not operate without thinking in terms of causes.[17]

Disciplines – fields of cognition – have their own ways of thinking, partly dependent upon the limits and strengths of cognition and partly dependent upon the subject being studied. In other words, there are communities of knowers and we join one by learning its language. Nicholson believed that the only type of "socialism" that Ritchie "accepted unconditionally was an epistemological socialism ..." (CW I, p. xxvii). One *cannot* become a cognitive being in isolation; one learns to know as a member of a community. Applied to the sciences, this implies that if one wants, say, to join the geometers, one must make their assumptions and learn their specialized language ("a point is a location without extension," "a line is the shortest distance between two points on a plane," etc.). Thus, Ritchie conceived of biology as a cognitive collective that made materialist assumptions in order for thought to proceed while still adhering to the view that

---

[17] One thing that Collingwood shared with Berlin was a belief that the "concepts and categories" with which we think and become cognizant of the world vary, changing over time and in response to circumstances. This similarity is most likely a result of Collingwood's suggesting the work of Vico to Berlin. So it is less than certain that we must think in *today's* categories. Still, without *some* categories, we could not think at all.

biological thinking is but one type of (idealist) cognition. We therefore conclude that cognition 1) is more than perception, 2) includes an objective element and a subjective element, 3) is tightly connected to language, 4) occurs within communities of knowers, and 5) is the basis for such communities.

Ritchie emphasized Hegel's discovery that when scientists have reason to alter or abandon a set of categories, a scientific revolution occurs. Since there can be no scientific *revolutions* without *changes* in categories, there can be no *science* without *categories of thought*. For example, in Newtonian physics, *mass* refers to material entities which can be acted upon (moved or changed) by *energy*; but in Einsteinian physics, *mass* refers to an entity in which *energy* is "stored" in ways permitting its release by fission or fusion. In the earlier conceptualization, *mass* and *energy* are distinct and irreconcilable concepts; in the latter, mass is (another "phase" of) energy; $e = mc^2$. This is true even though many (if not all) of the *referents* in the old and new sciences are the same and even though, in everyday processes, they respond in much the same way as they had before the scientific revolution. But the first conceptualization precludes fission and fusion while the latter reveals those possibilities.

### Ritchie's Argument to Appropriate Darwin

In prefacing this central argument, Ritchie raised expectations while acknowledging that the proof would not convince everyone: "I take the essence of the transcendental proof to be what I am going to state, and I cannot see that such proof admits of any refutation except from the consistent sceptic..." (CW II, pp. 9–10). He then offers a quick but deft refutation of the consistent sceptic: "he must be left to doubt his own scepticism and so to contradict himself" (pp. 7–8). How could this line of reasoning work? Consistent skeptics adhere to the proposition $\Xi$ that "no propositions can be known to be true." The dilemma moves swiftly: a) if one does *not* believe $\Xi$, then one is not a consistent skeptic; but b) if one does believe $\Xi$, then, since there is at least one proposition ($\Xi$ itself) that they profess to be true, *it* ($\Xi$) cannot be true. Skepticism is, for Ritchie, self-refuting.

We have already seen how important this Kantian argument was to Ritchie: It is meant to "convince anyone" and cannot be "disproved." So something fundamentally important to Ritchie *must be happening* in this argument and he italicizes every premise and the conclusion, which he did not do for any other argument. He introduced it by calling attention to the "empirical psychologist" who has "every right in saying that knowledge begins as sensation. That is true as a matter of mental history. He is only wrong when he goes on to say that knowledge is nothing but sensation and the products of sensation ..." (p. 9).[18]

Then the central argument is presented:[19]

- *If knowledge be altogether dependent on sensation, knowledge is impossible.*
- *But knowledge is possible, because the sciences exist.*
- *Therefore knowledge is not altogether dependent on sensation.* (p. 10).

Unfortunately, *as phrased*, Ritchie's argument cannot support the weight he placed upon it. If it is true at all, it may be little more than a tautology.

However, returning to Kant's phraseology enables us to reveal the true power of Ritchie's insight. In the *Prolegomena*, Kant writes of the possibility of a science of cognition (*Erkenntnis*); "If one wishes to present a body of cognition as a *science*, then one must first be able to determine precisely" its scope (Kant 1783, p. 14). Kant lists some possibilities: "Whether this distinguishing feature consists in a difference of the *object or the source of cognition*, or even of the *type of cognition*, or some if not all of these things together, the idea of the possible science and its territory depends first of all upon it" (p. 15). Clearly, the content of this new science is going to be human cognition; as Brook contended, cognitive science is born here. (Doubters are referred to the addendum.) Following Kant's lead, we can salvage Ritchie's argument. With one term changed, it can rise to the challenges he set for it:

- *If cognition be altogether dependent on sensation, cognition is impossible.*
- *But cognition is possible, because the sciences exist.*
- *Therefore cognition is not altogether dependent on sensation* (p. 10, revised).

What "the psychologist" got right was the history of human cognition; one must be cognizant of *something*; and experience gives us the *objects* of human cognition. But *experience* alone cannot create cognition. For beings to be cognizant, they must be able to do more than passively absorb sense data; cognitive beings *actively* employ a variety of cognitive skills, including ordering and categoriz-

---

**18** The target here is J. S. Mill: "Whatever we are capable of knowing must belong to the one class or to the other; must be in the number of the primitive data, or of the conclusions which can be drawn from these. (Mill 1950, p. 8).

**19** Note that the conclusion directly opposes a central belief of empiricists: that perception depends totally and exclusively upon sensation. For example, the possibility of perception without conceptualization was at the root of the controversy between J. S. Mill (the empiricist) and William Whewell (the idealist).

ing.[20] We humans routinely perform these cognitive functions just to perceive the world. Whatever else may be true of today's robots and computers – even super computers – which can passively receive sensory input and store it in memory – no one has yet ascribed cognition to them. By contrast, cognition is typically said to emerge in humans late in the first year (Rochat, 2014).

Kant drew the parameters of his new science of cognition:

> First, concerning the sources of metaphysical cognition, it already lies in the concept of metaphysics that they cannot be empirical. The principles of such cognition (which include not only its fundamental propositions or basic principles, but also its fundamental concepts) must therefore never be taken from experience; for cognition ... [lies] beyond experience. Therefore it will be based upon neither outer experience, which constitutes the source of physics proper, nor inner, which provides the foundation of empirical psychology. It is therefore cognition *a priori*, or from pure understanding and pure reason. (Kant 1784, p. 15)

In short, Kant believed human cognition requires the *a priori* categories of human thought, derived from "pure reason." So: when one is trying to understand "metaphysical cognition," one may legitimately make whatever metaphysical assumptions advance our sciences; similarly, when trying to understand machines, one may legitimately make mechanical assumptions; and, in the case of Darwin, one may make materialist assumptions about life forms.

The *form* of Ritchie's argument is an instance of *modus tollens: if P, then Q; not Q; therefore not P*. Since *modus tollens* is deductive, we can be certain of the conclusion whenever the premises are true. The first premise – if cognition depends totally and exclusively upon sensation, then cognition is impossible – has already been established. But three additional lines of argument remain. First, if cognition were nothing more than sensation, then merely perceiving old acquaintances would be enough to recognize them (to cognize again, to re-cognize); but re-perception does not always result in re-cognition. Second there must be a consciousness (or an *I* or *ego*) unifying sensations or they are little more than a cacophony of data. One must be cognizant of oneself (logically and maybe chronologically) before one can gather sense data about the world. Third, we understand things as members of categories; I might have the sensory input necessary for knowing that a thing in front of me is a yellow umbrella, but if I have no apparatus for conceiving "umbrella" and "yellow," I cannot see it *as a yellow umbrella* – that is, I cannot understand it.

---

[20] "Categories" is used here rather broadly, as Hegel used it, not in the narrow technical sense employed by some Kantians when they speak of the categories of human thought.

The second premise, that cognition is possible, can be verified in at least two ways. First, per Ritchie, cognition is possible "because the sciences exist." The facts that sciences and scientific revolutions exist are proof that cognition, with its conceptualizations and re-conceptualizations, is real. Second, a direct intuition of the activities of one's own mind (*cogito, ergo* cogitation exists) can validate the premise; one can only be cognizant of oneself as a cognitive being if cognition exists.

We may conclude, then, that "cognition is not altogether dependent on sensation." This is a *tour de force,* proving one of philosophy's most contentious propositions. But it also requires that Darwin's revolution, as a cognitive revolution, can only be understood in terms that are fundamentally idealist; empiricism is not adequate. We know, for instance, that Darwin's thoughts upset thinking in many fields; but empiricists like Mill believed that scientists merely add another true proposition to the scientific pile without disturbing anything else in our understanding of what it means to be human and what could be, and should be, our place in our material world. In short, Ritchie's work refutes crass empiricism and provides an anticipation of the work of Karl Popper (1934) and Thomas Kuhn (1962).

From these considerations Ritchie drew two conclusions: the Kantian project is sound and Kantian insights into cognition enable us to understand the Darwinian revolution. In short, Ritchie has not simply "saved" idealism from the threats of materialism, but he has also shown that materialist assumptions about the material world make sense without threatening idealism.

## V Summary

Idealists coped with the challenge of Darwinism in at least three different ways: some, like Stirling and Whewell, chose to reject Darwin because of his materialism and his rejection of teleology. Others, including those covered by Boucher and Ritchie in his argument from emanation, sought an accommodation. Finally, Ritchie appropriated the Darwinian revolution for the idealist cause by showing Darwin's revolution was possible only if the idealist theory of cognition is true. Ritchie's appropriation of Darwin represents a philosophical achievement of the highest order. Whatever we may still discover and come to understand about the origins and nature of organisms will make sense only if we are cognitive beings, operating within this idealistic and dialectical context.

## Addendum

Kant's *Prolegomena:*
"Wenn man eine Erkenntnis als Wisenschaft darstellen will, so muß man zuvor das Unterscheidende, was sie mit keiner andaren gemein hat und was ihr also eigentumlich ist, genau bestimmen könnnen; widrigenfalls die Grenzen aller Wissenschaften ineinanderlaufen, und keine derselben ihrer Natur nach grünlich abgehandeit werden kann." (Kant 1784, p. 13; § 265)[21]

A fairly strict translation: "When one wants to (or should one want to) create a science of cognition, one must first define it, by determining accurately its peculiar features which constitute it properly; otherwise, the boundaries of all sciences become indistinct and none of them can be dealt with appropriately to its unique nature." This passage – and several subsequent passages – concern efforts to create a science of cognition [*Erkenntnis*]. Subsequently Kant writes the source of metaphysical cognition must be metaphysical and not physical. Further, since the science of cognition is metaphysical, it must consist of a priori (and not a posteriori) judgments. Kant cannot be understood as doing anything different from a science of cognition.

## Abbreviations

CW   Ritchie, David George (1998): *Collected Works of D. G. Ritchie*, Peter Nicholson, ed. London: Thoemmes.
GLE   Stirling, James Hutchison (1890): Gifford Lecture the Eighteenth (GLE) https://www.giffordlectures.org/books/philosophy-and-theology/gifford-lecture-eighteenth visited on 4 June 2018.
GLN   Stirling, James Hutchison (1890): Gifford Lecture the Nineteenth (GLN) https://www.giffordlectures.org/books/philosophy-and-theology/gifford-lecture-nineteenth visited on 4 June 2018.

---

[21] There is a great variety of minor differences in translation here: "If it becomes desirable to formulate," "if one wants to offer," "when one seeks to," and so forth. There is also a bit of variation as to what is to be presented, formulated, or offered: "a body of cognition," "any cognition as science," etc. Given what is presented in the next paragraphs, perhaps the best current translation might be: "If one is to formulate any *science* of cognition." The 1903 translation by Carus comes very close to this: "If it becomes desirable to formulate any cognition as science, it will be necessary first to determine accurately those peculiar features which no other science has in common with it…" (p. 12). In short, Kant was offering a genus-species definition of a science of cognition.

GLT  Stirling, James Hutchison. (1890): *Gifford Lecture the Twentieth* (GLT) https://www.giffordlectures.org/books/philosophy-and-theology/gifford-lecture-twentieth visited on 4 June 2018.
OS   Darwin, Charles (1859 [1958]): *The Origin of Species*. London: John Murray; New York: New American Library.

# References

Berlin, Sir Isaiah (1999): *The Roots of Romanticism*. London: Chatto and Windus and Princeton: Princeton University Press.
Berlin, Sir Isaiah (1976): *Vico and Herder*. New York: Viking Press.
Berlin, Sir Isaiah. (1990 [2013]): "Joseph de Maistre and the Origins of Fascism," *The New York Review of Books*, Sept. 1990; also in *The Crooked Timber of Humanity: Chapters in the History of Ideas*. Princeton: Princeton University Press. pp. 95–177.
Bloomfield, Leonard (1935): *Language*. London: George Allen & Unwin.
Bosanquet, Bernard (1899 [2001]): *The Philosophical Theory of the State*, Sweet and Gaus, (Eds.). South Bend: St. Augustine's Press.
Boucher, David (2014): "British Idealism and Evolution" in *Oxford Handbook of British Philosophy in the Nineteenth Century*. W.J.Mander (Ed.). Oxford: Oxford University Press. Also at: https://books.google.com/books?hl=en&lr=&id=UDeAgAAQBAJ&oi=fnd&pg=PA306&dq=William+Wallace+Darwin+Hegel&ots=ukw_y6E9e1&sig=_VqDOW6KQh2a06RKTsvRR1xSgY#v=onepage&q=William%20Wallace%20Darwin%20Hegel&f=false visited on 15 May 2018.
Brook, Andrew (1987): "Kant and Cognitive Science" in *The Prehistory of Cognitive Science*. Andrew Brook, ed. Basingstoke: Palgrave Macmillan, pp. 117–136.
Chambers, J. K. and Peter Trudgill (1998): *Dialectology*. 2$^{nd}$ edn. Cambridge: Cambridge University Press.
Collingwood, R. G. (2005): *The Idea of History*. Oxford: Oxford University Press.
Darwin, Erasmus (1794): *Zoonomia*. London: J. Johnson.
de Maistre, Joseph (1821): *Les Soirées de Saint Pétersbourg*. Paris: La Librairie Ecclesiastique de Rusand. Later editions at: https://archive.org/details/lessoiresdesai02maisuoft visited on 15 June 2018.
DeVries, Hugo (1900): *Species and Varieties: Their Origin by Mutation*. Chicago: Open Court.
Gould, Stephen J. (1977): "Darwin's Delay" in *Ever Since Darwin*. New York: W.W. Norton; 21–28.
Huxley, T. H. (1869): *The Physical Basis of Life*. New Haven: The College Courant. Also at: http://nzetc.victoria.ac.nz/tm/scholarly/tei-Stout18-t1-g1-t13.html visited on 14 June 2018.
Kant, Immanuel (1783 [1912]): *Prolegomena To Any Future Metaphysic*. Leipzig: Verlag von Felix Meiner and Chicago: Open Court Publishing Co. Also at: https://archive.org/details/kantsprolegomen00carugoog visited on 1 July 2018.
Kocis, Robert (2018): "Idealism *with* Materialism: Darwin Through the Eyes of David George Ritchie," *History of Political Thought*, XXXIX: 1; 156–180.
Kuhn, Thomas (1962): *The Structure of Scientific Revolutions* Chicago:Chicago University Press.

Lewes, G. H. (1864): *The Life of Goethe*. London: Smith, Elder & Co.
Mayr, Ernst (1944): *Systematics and the Origins of Species*. New York: Columbia University Press.
Mill, J. S. (1950): *System of Logic*. Ernest Nagel, ed. New York: Hafner Publishing. Also at: https://archive.org/stream/johnstuartmillsp012451mbp/johnstuartmillsp012451mbp_djvu.txt visited on 28 June 2018.
Neill, Edmund (2003): "Evolutionary theory and British idealism: the case of David George Ritchie," *History of European Ideas*, 29; pp. 313–338.
Passmore, John. (1959): "Darwin's Impact on British Metaphysics." *Victorian Studies*, 3:1, pp. 41–54.
Popper, Karl (1934 [1959]): *The Logic of Scientific Discovery*. Vienna: Verlag von Julius Springer, and New York: Routledge.
Ritchie, David George (1893): *Darwin and Hegel with Other Philosophical Studies*. London: Swan Sonneschein. Also at: https://archive.org/details/darwinhegelwitho00ritcrich visited on 28 June 2018.
Ritchie, David George (1896): *The Principles of State Interference: Essays on the Political Philosophy of Mr. Herbert Spencer, J. S. Mill, and T. H. Green* 2nd Edn. London: Sonneschein
Ritchie, David George (1901): *Darwinism and Politics*. London: Swan Sonneschein. Also at: https://archive.org/details/darwinismandpoli028701mbp visited on 28 June 2018.
Rochat, Philippe (2014): *Early Social Cognition: Understanding Others in the First Months of Life*. New York: Lawrence Erlbaum.
Smith, Adam (1759): *The Theory of Moral Sentiments*. London: Henry O. Bohn. Also at: http://oll.libertyfund.org/titles/smith-the-theory-of-moral-sentiments-and-on-the-origins-of-lanquages-stewart-ed visited on 28 June 2018.
Stirling, James Hutchison. (1872): "As Regards Protoplasm" London: Longmans, Green, & Co. Also at: https://archive.org/stream/a622876900stiruoft#page/10/mode/2up visited on 13 June 2018.
Stirling, James Hutchison (1894): *Darwinianism: Workmen and Work*. Edinburgh: T. & T. Clark. https://play.google.com/books/reader?id=BKElAAAAMAAJ&printsec=frontcover&output=reader&hl=en&pg=GBS.PR5 visited on14 June 2018.
Tennyson, Lord Alfred (1849 [1895]): *In Memorium A. H. H.* Boston: Houghton Mifflin. Also at: https://archive.org/stream/inmemoriambyalfr00tennuoft/inmemoriambyalfr00tennuoft_djvu.txt visited on 29 June 2018.
Thagard, Paul (2005) *Mind: Introduction to Cognitive Science*, 2nd Edn. Cambridge: MIT Press.
Wallace, William, (Ed). (1874): *Hegel's The Science of Logic*. Oxford: Clarendon Press. Also at: https://archive.org/details/logicofhegeltran00hegeiala visited on 28 June 2018.
Wallace, Alfred Russell (1914): *The World of Life*. London: Chapman and Hall, Ltd.. Also at: https://archive.org/details/worldoflifemanif00walliala visited on 28 June 2018.
Whewell, William (1833): *Bridgewater Treatise* London: Pickering. Also at: https://archive.org/details/astronomygeneral00whew_3 visited on 28 June 2018.
Whewell, William (1847): *Philosophy of the Inductive Sciences*. London: John W. Parker. Also at: https://archive.org/details/philosophyofindu01whewrich visited on 18 June 2018.

Yūjin Itabashi
# Biology and the Philosophy of History: Nishida Kitarō and the Philosophy of "Necessity that Includes Freedom"

**Abstract:** Nishida Kitarō, the philosopher known as the founder of the Kyoto School, is a representative of early Showa period Japanese thought. In one of his most famous articles, "Logic and Life" (*Ronri to Seimei* 論理と生命, 1936), Nishida actively incorporated the theory put forward in the biologist J. S. Haldane's The Philosophical Basis of Biology into his own work. On this note, I would like to ask: what was it about Haldane's theory of life that was so interesting to Nishida? Why was Nishida so intent on actively incorporating Haldane's theory into his own philosophy? Moreover, we can also ask why he was so interested in the field of biology in the first place. To discuss these questions, a further question must be raised: What kind of problems did Nishida believe that the field of "biology" needed to clarify? Nishida incorporated Haldane's biology into his philosophy insofar as he believed it to be relevant to the fundamental problem of the philosophy of history – the core of his later philosophy –, which has an inevitable connection to his theory about the nature of biology.

## Introduction

Nishida Kitarō, the philosopher known as the founder of the Kyoto School, is a representative of early Showa period Japanese thought. In one of his most famous articles, "Logic and Life" (*Ronri to Seimei* 論理と生命, 1936), Nishida actively incorporated the theory put forward in the biologist J. S. Haldane's *The Philosophical Basis of Biology* into his own work.[1] Furthermore, Nishida remained

---

Translated by Yukiko Kuwayama and Richard Stone

**1** I have used *Nishida Kitarō Zenshū* (Complete works of Nishida in Japanese) (『西田幾多郎全集』 second print) from the year of 1965–66 (Iwanami Edition). All citations of Nishida's works are abbreviated to NKZ (including the volume and page number). For more detailed research on the generative process and the content of thought between Nishida's introductory work, *An Inquiry into the Good* and his later thought, especially his conceptualizations of "the lived body",

**Yūjin Itabashi,** Rissho University

https://doi.org/10.1515/9783110604665-010

sympathetic to Haldane's position all the way until the writing of his article "Life (*Seimei*, 生命)" in 1944. On this note, I would like to ask: what was it about Haldane's theory of life that was so interesting to Nishida? Why was Nishida so intent on actively incorporating Haldane's theory into his own philosophy? Moreover, while it is well known that Nishida was quite interested in the results found in the work of his contemporary biologists, we can also ask why he was so interested in the field of biology in the first place. To discuss these questions, a further question must be raised: What kind of problems did Nishida believe that the field of "biology" needed to clarify? Otherwise, what kind of philosophical interests or issues were in the background of Nishida's foray into biology? To state my conclusion in advance, I believe that Nishida incorporated Haldane's biology into his philosophy insofar as he believed it to be relevant to the fundamental problem of the philosophy of history. Thus, Nishida's philosophy of history – the core of his later philosophy – has an inevitable connection to his theory about the nature of biology. In this contribution, I would like to examine this inevitable connection between biology and Nishida's philosophy of history.[2]

# 1 Dialogue with Haldane's Biological Theory of Life

From the very first time Nishida referred to Haldane's *The Philosophical Basis of Biology* in "Logic and Life", one of Nishida's most important works, Haldane's biological (physiological) theory of life became an indispensable tool for Nishida's work.[3] As the story goes, Nishida learned of Haldane's *The Philosophical Basis of Biology* through his disciple Iwao Takayama in 1936.[4] Nishida, who had long been interested in biology, quickly tried to respond to Haldane's biological theory in the aforementioned essay "Logic and Life". Nishida thus attempted to give a sketch of Haldane's ideas while simultaneously relating them to his own philosophy.

---

"the continuity of discontinuity", "contradictory self-identity", acting intuition and historical necessity, refer to Itabashi Yūjin (2004).
**2** It should be kept in mind that this paper may cover many of the same points given in the description of Nishida's philosophy of work I have previously published in Japanese (Itabashi: 2011).
**3** See Noé (2002), about the meaning of Nishida's reference to Haldane's life theory.
**4** See Gülberg (2005).

> True life, however, is established as the self-determination of the expressive world. And we can regard the world of biological life in light of this as well. What the physiologist Haldane says about life becomes comprehensible in light of the above as well. Life possesses an environment not only outside of the organic body but also within. Life involves a normative structure unique to the specific species and the active maintenance (*nōdōteki iji* 能動的維持) of its environment. Further, it cannot be a merely physical or chemical conglomeration but rather must be a persisting unity. It would have to be an individual expression of nature itself. Life has no spatial boundaries. Whether we start from the vitalist's idea of the organic body as separate from the environment or from the mechanist's idea that takes the organic body as a part of matter, we would be unable to construct the biologist's notion of life. (NKZ 8, 287–88).[5]

He further states, "An organic body does not only possess an environment on the outside but also possesses an environment within. Life, according to Haldane, is the active preservation of a particular normative structure with its environment. It is neither a mechanistic process nor a vitality" (NKZ 8, 293). [6] In *Philosophical Essays II* (1938), the collection of essays in which we find "Logic and Life", Nishida argues that "[t]he environment is both external and internal to life. Life has both an external environment and an internal environment (as Haldane claims)" (NKZ 8, 508). In the following pages, I will consider both the meaning and significance of these passages while summarizing Nishida's thoughts in *Philosophical Essays II* (1938).

Let us first look at the part of Haldane's *The Philosophical Basis of Biology* to which Nishida is referring here. Haldane defines life as the maintenance of an "internal environment", or the physiological environment inside the body. In other words, physiological functions inside the blood (such as the circulation of blood) keep the body in a consistent *normal and specific* state relative to the external environment.[7] Therefore, the structure, form, and function of living organisms are all mutually dependent on one another according to this concept of active maintenance. That is to say, the structure and form of an organism express the maintenance of functions, and the function expresses the maintenance of the structure and form. Furthermore, since the active maintenance of an organism means the maintenance of both the organism and environment as a whole, life cannot be explained merely mechanistically nor in terms of vitality. Life is rather an expression of the structure and function, i.e., the appearance of a unity between the organism and its environment. In this sense, the organism and its environment cannot be separated from one another. We may even say that

---

[5] Translation from Nishida (2012), p. 112.
[6] Translation from Nishida (2012), p. 115.
[7] See Haldane (1931), p.16

the organism (life-form) permeates the environment. As Haldane further argues, life is nature expressing itself as a specific whole without any spatial boundaries.[8]

According to Nishida, all living organisms are only able to exist insofar as they have an environment to which they have adapted. However, living organisms must also be capable of *living by themselves*. That is, organisms exist insofar as they have an inherent 'independence' or "subjectivity" from their environment. This means that organisms exist as organisms only as long as they continue interacting with their own environment/creating new environments (NKZ 8, 195). Therefore, we are able to find the existence of the activity in which "as life alters the environment, the environment alters life"[9] (NKZ 8, 283). In other words, the environment and the organism neither exist as independent and separate entities nor do they merely merge into one activity (*hataraki*働き). On the one hand, the organism and the environment will always be differentiated from each other and they can never merge together. On the other hand, the two can only exist insofar as they exist in a relationship of mutual negation with one another. As long as the environment and the organism, i.e., life, are negating, acting upon, and forming each other, they will remain individuated. "True life exists in [a relationship] of negative mutual determination with the environment" (NKZ 8, 508). In other words, the environment and organic life forms exist insofar as they determine each other through mutual negation and, in a sense, express themselves within one another. Neither one is anything more than one aspect or moment of this mutual expressive-formative act through negation. From Nishida's point of view, the environment and the organism are established in the self-formation of the "expressive world" (i.e., the whole of a mutual expressive-formative operation existing between organism and environment).

In order to demonstrate his own theory of organic life, Nishida argues, "Life, according to Haldane, is the active preservation of a particular normative structure with its environment."[10] (NKZ 8, 293). Indeed, even for Haldane, organic life-forms (*seimeitai*生命体) and the environment are inseparable and exist in so far as they maintain their consistency and uniformity as a whole. This is expressed and manifested through the very way in which living beings actively maintain their normative and specific structures. Moreover, Nishida continues to cite Haldane when claiming that the simultaneously mutually negative/mutually expressive-formative operation between the organism and the environment can be seen

---

[8] See Haldane (1931), p.74
[9] Nishida (2012), p. 109.
[10] Nishida (2012), p. 115.

in the structure and function of the organism. As Haldane informs us, as long as we believe that the structure and function of an organism are mutually expressive, and thus that all organisms are equipped with this "structure=function", then there is no way the existence of this structure=function can precede the whole of the mutually formative/mutually expressive operation occurring between organism and environment. Nishida believes that Haldane's theory is capable of demonstrating this point. The formation of life in self-determination and the self-formation of the "expressive world" mentioned above refer to the way in which the whole world forms and expresses itself in the "structure=function" of the organic living being. As long as this is valid, the world is established as the world.

## 2 "Form" of Life

In this way, we can assume that the "structure=function" of the environment expresses itself in the "structure=function" of the organism. However, this does not mean that Nishida considers life forms (or their relation to the environment) simply as "organic" i.e., harmonious. Nishida argues, "Normally, life is understood as organic unity. However, true life must also include death [within itself]" (NKZ 8, 528). Nishida thus distinguishes his theory of life from a harmonious, organic theory of totality and system theory. Nishida also states as follows. "The environment is both internal and external to life. Life has both an external environment and an internal environment (as Haldane claims). Thus, we can say that life includes negation [within itself]; we can say that life includes death [within itself]." (NKZ 8, 508)

Here, Nishida problematizes the negation that consistently accompanies one's life, which is not directly attributable to Haldane. This "structure=function" itself is established through the negation of the environment, or – if we were to rephrase this point in the terms of Haldane's philosophy – organic living beings express the external environment as an internal (environment). From Nishida's point of view, this is possible because the self-identity of organic living beings is established through contradiction and self-negation. In other words, living beings are established through the "negation" of life, i.e., they are established with "death" as their "internal condition". However, this sense of "death" is not death in the sense of being the limit [i.e., the end] of the life's continuous flow (for Nishida, this would not be death in the true sense). It is rather something which is not derived from life at all. Death understood in this way is something altogether other to life, and is a result of the environment acting upon the organism. Since the active self-maintenance of life is the constant "maintenance" of mutual formations and mutual expressions through the negation of living beings and the environment, it is not possible for the organism to carry out this maintenance by itself. The active self-maintenance of life has a contradictory self-identical way of being constructed through the identity formed between the continuous self-maintenance and the self-perpetuation of the life on one side and the negation of this "continuity" (i.e., discontinuity) on the other. The self-formation of the "expressive world" described above is precisely the self-formation of the "discontinuous-continuous world". In other words, it is the self-formation of the "contradictory self-identical world".

In addition, Nishida continued to think of the concrete concept of the "self-formation of contradictory self-identical world" as the entirety of the interaction achieved through the negation of both (organic) life and the environment. Nishida regards the "normative structure" he found in Haldane's work as "a type of form"[11] (NKZ 8, 287–88) and thus interprets it as "the life of species"[12] (NKZ 8, 290–91). This positioning is especially apparent in his translation of Haldane, where the expression "to a certain genus" is added by Nishida himself: "Life […] is the active preservation of a particular normative structure with its environment."[13] Still, even when it seems that Haldane is thinking about the active maintenance of the structure of an organism as an individual, maintaining the individual's normal and standardized structure (=function) cannot exceed the range of what is normal or normative for the "species". Therefore, Nishida's interpretation cannot be seen as problematic. Moreover, if this interpretation is possible, then for Nishida the organic life of an individual can be maintained while also maintaining the normal and standardized characteristic structure i.e., function (structure=function) of the "species." Nishida thus concludes that "species" are "shapes" or "paradigms" in the sense of the ground-form (*konpontai* 根本態) of structure, form, and function, of which the organic life of the individual is a part. Additionally, looking from an alternative viewpoint, the "species" can also be taken as the "shape" (*katachi* 形), or in other words, the "mold" (*kata* 型) of the working life of the individual. As a result, "species" are precisely the "shape" or "type/mold" of the self-formation (self-determination) of the "contradictory self-identical world" where organisms as individuals are established.

Yet this does not immediately mean that the (organic) life of an individual is unilaterally constrained and subordinate to the life of the species. The

---

[11] Nishida (2012), p. 112.
[12] Nishida (2012), p. 114.
[13] Nishida (2012), p. 115.

"shape" of the "contradictory self-identical world" in which life holds and expresses itself, is also the "shape" established in the "world of the discontinuous continuity". Therefore, "[l]ife is, as the self-identity of contradictions, formative. Life is found at the junction between having a determinate shape as species and constantly breaking through this determinate shape [...] a living species is only a living species so long as it includes its own negation within itself. Without this negation, it would be nothing more than a dead shape." (NKZ 8, 451). This means, the "contradictory self-identical world" as "the expressive world" now refers not only to individuals and the environment, but also to the mutual-formation through negation (mutual-determination through mutual-negation) occurring between individuals, species, and environment. "The species has a [certain] shape and determines itself by itself. It is both that which determines the individual and that which is determined by the individual. There is no such thing as an individual that exists completely separately from its species. The individual is borne from the species. It is both thoroughly determined by the species as well as that which determines the species." (NKZ 8, 451) It is in the individual organism's normative structure=function that this "contradictory self-identical world" forms and expresses itself. Insofar as this is the case, we can say that this "structure=function" expresses the "shape" of the species as the grounding form (*konpontai*根本態) of the individual organism's structure (form)=function.

However, for Nishida, when the normative (standard) structure, i.e., the function of life, is not able to go beyond mere physical instinct, "the individual is [but] a slave to the species" (NKZ 8, 450). This means that such individual living organisms are restricted to the life of their species. That is to say, in this situation individuals have no adequate positive significance beyond obeying the physiological instincts prescribed by the life of their own species. They cannot have any positive meaningfulness and there is no individuality which differentiates one individual from another in a positive sense. Nishida assumes that organic beings, with the exception of human beings (we will talk more about his definition of human beings soon), correspond to this definition. In other words, in organisms other than human beings, active life with independency and subjectivity (*shutaisei* 主体性) that is capable of acting directly on the environment is not realized. According to Nishida, in biological life, the activity of "creation" towards the future or creating itself in contrast with the environment as something already "shaped" or "fixed" in the past, is not realized.

By contrast, the normal and standardized individual structure i.e., function (structure=function) of individual life forms in "human life" (*ningenteki seimei* 人間的生命) is understood as the "body". Yet, Nishida assumes that this "body" is not merely a physiological internal environment. He instead explains his position in the following way: "We are borne from the species, but we also actively form the species" (NKZ 8, 450). Thus, in contrast to the environment that has already been formed, we find the realization of a subject that can create new forms and the subject's body. But in what way can this be the case?

## 3 Historical Body

Nishida describes the act of "making things with tools" as the distinguishing feature of human life. This behavior makes it possible for each individual human actively to do work upon the environment using tools (from simple tools to complex advanced machines) in order to change it. As human beings living in an environment in which our medium is tools, we recognize tools as tools, i.e., something replaceable or general. This also means we are conscious of the purpose of using tools and we are able to imagine (i.e., cognize) a future which has yet to be realized. This act furthermore shows our capacity to go beyond the pre-formed environment of the past and cognize and will an intention to create something new. Furthermore, this behavior indicates that humans recognize themselves as the active subject which creates new things. This means that the individual is realized as an independent entity capable of distancing itself from a pre-formed environment. In comparison to this, Nishida claims that even if there are cases in which they may seem to be using tools at first glance, "[a]nimals do not truly have tools" (NKZ 8, 450). The relation between the individual, the species, and the environment in human beings is qualitatively different from that of purposive and instinctive ones, seen in organisms, dependent on the environment. Individuals have inherent subjectivity (shutaisei 主体性) which cannot be reduced to the species. This means, (one's organic) life itself has its own independent and unique subjectivity that cannot be reduced to the environment. It is here that we find the actualization of the world of "contradictory self-identity" as a "discontinuous continuity"; i.e., the world in which life is established as life.

The individual life of the human is, therefore, defined as the active maintenance of its normative, standard and specific structure i.e., function, as formed and expressed in "mutual determination through negativity" between individuals, species, and environment. The "structure=function" of the human's individual life is precisely the "lived body". According to Nishida, the active maintenance introduced by Haldane is not simply maintaining a physiological structure i.e., the function in human life. It is rather the maintenance of structure i.e., function that enables practice that, in turn, further enables us to use tools and thus realize technology (although Haldane does not mention this directly).

Hence, this already means the maintenance of a form or mold (*kata*型) of an action. The "lived body" is a "structure=function" which actively maintains the act of its newly forming the environment by using tools, while recognizing tools as tools, and being formed from the environment which we exist in, all at the same time. Therefore, the "life of the species", which is understood as the fundamental form or mold (*kata* 型) of the "structure=function" that individual organisms must actively maintain, must also be recognized as the form or mold of social and cultural actions in "human life".[14]

In that case, how does the normative "structure=function" known as the "lived body" actually exist in reality? As an example, consider a case in which I use tools to make something new in order to change the environment and improve my quality of life. Here, my changing of the environment is enabled through the actions I perform upon pre-existing entities. This includes my use of tools and my mastery of various techniques or methods. In this way, I am affected by the various customs and cultures that have conditioned how these tools are used. Moreover, it is through reforming these customs or techniques and taking them to a more advanced state than when I initially picked up the tools that I am able to make new improvements. Therefore, on the one hand, this act is a free and creative act. On the other hand, though, it is an act which is made possible thanks to the various properties of already existing things. It is enabled through our actions upon the properties of things, which thus allow for as-of-yet undiscovered properties inherent to the object to be realized through our actions. We can only view this as us *necessarily* having to act upon objects in some determinate way. Here, the following two aspects are realized as one event in which one simultaneously causes the other: on the one hand, we find my own free and creative act, and, on the other hand, we see the properties of the things that do not allow for me to move in a purely arbitrary way, and thus necessarily realize themselves in a certain way. On the contrary, this means that only when I follow the necessity (inevitability) inherent to the properties of things that I know what I want to do, and that it is only here in these sorts of acts that I can realize my freedom.

As we can see in these examples, one's free actions can be realized only in the form of acting upon things in a set way determined by their properties. Free

---

**14** From Nishida's standpoint, the dualistic scheme between material as mechanical moving bodies and consciousness as a conceptualizing act is a scheme which is conceived of through the individualization of the two mutually exclusive dimensions in the form of life, i.e., through individualizing and objectifying the dimension of being formed by the environment (the former) and active forming of the conceptual consciousness.

actions can only be realized as a conception of necessity that includes freedom.[15] In other words, all facts formed from the already shaped to its new form to be shaped in the future, evoke and create one another. And it is exactly through this, that the whole of individuals is linked to each individual in itself as the whole. Therefore, it means realizing the connection of the necessity (inevitability) that should be and in which each individuality of oneself and things are emphasized and made lively (*ikasu* 生かす). Nishida describes this state of affairs in the following way: The world is "free yet necessary" (*jiyū ni shite hitsuzen narumono* 自由にして必然なるもの) and also "has an inevitable direction (*hitsuzenteki hōkō wo motsu* 必然的方向を持つ)" (NKZ 8, 440).

However, even while it may be true that this "free yet necessary" interaction is realized in our own actions, we are not its origin (i.e., we are not the reason it exists). One's actions are always mediated by something not found within oneself, i.e., the inevitable and normative acts of things that negate the will of the self. It is through the act of becoming one with something alien to one's own self that action becomes free (and necessary) for the first time. In other words, this act is realized by denying the realization of the self by oneself (which is continuous and harmonious self-derivation). This means that the negation of existence, the negation of the fact that we live on our own (i.e., death), is actually what enables us to realize our own intrinsic/individual life itself.[16]

Overall, the normal and normative criterion of the "structure=function" of human life, in which life is simultaneously death, is none other than the "body". The human body expresses freedom that includes necessity, i.e., the way in which life is both free and pre-determined at the same time. This remains to be the case while simultaneously maintaining "continuous discontinuity" or "contradictory self-identity". Nishida refers to this "free yet necessary (*jiyū ni shite hitsuzen* 自由にして必然)" state-of-affairs as the "historical".[17] This means that the human body is, in this sense, a "historical body" (*rekishiteki shintai* 歴史的身体). It is not merely the body of "biological" life as a mere harmonic continuity between the body and the environment. It is rather a "body" established through a negation of this harmonious continuation.

---

**15** As is well known, Nishida calls this way of acting of the self "acting intuition".
**16** While I could not treat the topic in great detail in this paper, we can note that, for Nishida, the fact that death is actually life is most profoundly realized in the realm of religious experience. For more, refer to Ueda (2002) and Itabashi (2008).
**17** According to Nishida, history is nearly a movement which develops from "the formed thing to the active forming." It is a movement which cannot be grasped through the mechanical causality or purposive causality and is thought to be grasped through the "self-determination of the eternal now." For further analysis, refer to: Itabashi (2008).

It should be noted, however, that the sense of the word "necessary" I have described here with the phrase "free yet necessary" is neither necessary in the sense that it is based on the results in the past that have led up to this present moment nor is it necessary in the sense of being oriented towards some goal in the future. Moreover, it is rather "historical" necessity, as can be seen in the following quotation: "reality is both entirely determinate and constantly fluctuating" (*genjitsu wa doko made mo ketteiteki de arinagara itsumo dōyoteki de aru* 現実は何処までも決定的でありながら、いつも動揺的である) (NKZ 8, 481). From Nishida's point of view, "historical" necessity is not given prior to the new formation of the presence which is discontinuous from the past. It is rather something that can only act and be proven through the realization of a new formation of things, and only in the midst of a new formation does this become clear.

# 4 The Philosophy of "Historical Necessity" and Biology

Now, keeping all of our considerations up to this point in mind, let us tackle our initial question of why Nishida was so convinced of the importance of Haldane's theory of life and what kind of science "biology" was for Nishida. The primary meaning we can take away from Nishida's reference to Haldane's biology is this: neither the mechanical theory (in which Haldane distinguishes the environment from individual life forms, and in which the environment is thought of as the principle of the existence of the individual life forms) nor the vitalist theory (in which individual life forms as the principle of life) can accurately grasp life. Life is assumed, rather, to be the active maintenance of a normal and standardized specific and special structure (i.e., function) in the interrelation between environment and life forms. To conclude, for Nishida biology is neither mechanistic nor vitalist. It is rather a science that aims to clarify the manner in which life-forms exist as the active maintenance of their own normative "structure=-function" and, furthermore, as the active maintenance of the environment and individual life forms. This scientific field is hugely significant in virtue of its capacity to clarify how life forms are established through the interactions between the environment and particular organisms.

From Nishida's viewpoint, these biological conclusions lead us to a world view about life in which the "shape/form" of the normative structure-as-function of the life is "discontinuous-continuous" or "contradictory self-identical." Furthermore, it is precisely something that leads us to "historical life" or to the historicity of life, that the "shape/form" of the real, "free yet necessary" world is simultaneously "entirely determined yet always fluctuating". This is why Nishida believes biology to be so important.

Nishida's insights about historical life do not directly result from Haldane's theory of biological life. Rather, Nishida tries to reveal the true value of "negation" – which he believes Haldane failed to clarify – when looking at the interaction between the environment and life forms. From Nishida's viewpoint, the "organic" structure understood by biologists as the inter-relation between the environment and the living organism is precisely the realization of the structure of historical life as one phase of the continuity of discontinuities.

Then, what philosophical problems or aspirations lie hidden in the background of this evaluation of biology? Through our considerations up to this point, we can understand that the problem Nishida addressed was the following: how can one understand the historical necessity that realizes itself while still including the freedom of the acting subject. The key to this question was actually the exact question raised by two contemporary critics of Nishida's philosophy, Tosaka Jun and Tanabe Hajime.[18] Upon receiving their critical remarks about accurately grasping life as active maintenance of the normative "structure=function", Nishida proposed a way of being "free" (to be precise, "free yet necessary") in the world of "historical life" which has a "necessary direction". The mutual formation of the environment and organic lives as a "mutual determination of (through) negation" or "discontinuous continuity" has a "form" that cannot be derived continuously from the environment and organic lives, and in this sense, has an objective reality and necessity. Nishida shows through analyzing the "form" of life that the realization of the current free self-creation is the realization of the indispensable linkage (connection) of all historical facts, and that realizing one's own freedom thus means the realizing the necessity which includes freedom within itself.

# Conclusion

In our investigation up to this point, we have seen how Nishida integrated the results of biological research into his philosophy in order to grasp the process of historical necessity. Thanks to this effort, the "entirely determined yet constantly fluctuating" manner of historical progression, which seems to have

---

[18] See Itabashi (2011), in which I take up Tosaka's and Tanabe's attitudes towards biology and thus investigate the discourse that took place between them and Nishida.

been overlooked for a long time, can finally be clarified.[19] Thus Nishida postulates a contribution of biology to logic that clarifies historical necessity, including freedom. This approach seems to be quite distinct from current perspectives on this topic. However, we should remember that Nishida's approach to biology was hardly a rarity. Indeed, similar approaches to biology or physiology can be seen in the work of thinkers contemporary to Nishida, such as Hans Driesch or Haldane himself. In this sense, Nishida's work was very much in line with the trends of his time period. Driesch's and Haldane's theories of life tried to find a particular law of life that can be distinguished from the laws of mechanical causality we apply to inorganic matter. These laws of mechanical causality in inorganic matter are certainly thought to be capable of demonstrating the laws of life from the perspective of universal natural scientific law. However, the goal of these theories of life is to clarify the autonomous and active self-formation of life activities. Therefore, their goal was not to explore any deterministic principle that would unambiguously derive the activity of life from its preceding state. Several biological theories on life during this time period attempted to analyze the activity of the autonomous subject and active self-formation, along with its lawful necessity. With this in mind, Nishida saw an opening in biology that could contribute to an elucidation of the necessity included in human beings' freedom, subjective acts and the necessity which includes both of them.

However, in the biology of his time, unlike Nishida's point of view, there was already a tendency to reduce even the law of human social life into the law of the activity of general biology. In the modern sciences, or more specifically, in modern natural sciences (including biology), the tendency to ignore fundamental differences between human beings and other creatures in the law of activity was even more intense. In comparison with Nishida's thought that developed the logic of "historical" necessity, the confidence in the existence of human "history" was overlooked. As a result, it can be seen as proof that the reality of subjective free acts, i.e., the creation of a new one-time event, not in the past, is also being overlooked. This alone is a good enough reason to admit the significance and range of Nishida's philosophy and the need for current philosophers to re-evaluate his work.

---

**19** In Nishida's article "Logic and Life" (「論理と生命」), Nishida makes some thinly-veiled critical remarks on Tanabe's article on "The Logic of Species and the World Schema" (「種の論理と世界図式」). Although not explicit, we can see that Nishida did react to Tanabe's logic and, in this sense, Nishida analyzed the normal and normative structure i.e. function of the life and through this, he tried to explicate the form as the form of species.

# References

Driesch, H. (1909) *Philosophie des Organischen*, Leipzig: W. Engelmann.
Driesch, H. (1905) *Der Vitalismus als Geschichte und als Lehre*, Leipzig: Johann Ambrosius Barth.
Gülberg, Niels (2005). "Nishida to Haldane" in *Nishida Kitarō Zenshū Geppō 14. Vol. 11.* (西田幾多郎全集月報). Tokyo: Iwanami.
Haldane, J.S. (1931) *The Philosophical Basis of Biology*, London: Hodder and Stoughton, Ltd.
Haldane, J.S. (1936) *The Philosophy of a Biologist*, Oxford: Clarendon.
Itabashi, Yūjin (2004). *Nishida tetsugaku no ronri to hōhō* (西田哲学の論理と方法), Tokyo: Hōsei University Press.
Itabashi, Yūjin (2008). *Rekishiteki genjitsu to nishida tetsugaku* (歴史的現実と西田哲学), Tokyo: Hōsei University Press.
Itabashi, Yūjin (2011). "Seibutsugaku to rekishitetsugaku" (生物学と歴史哲学)、editied by Kanamori, Osamu: *Shōwazenki no kagaku shisōshi* (昭和前期の科学思想史) P. 341–412. Tokyo: Keisou Shobō.
*Nishida Kitarō Zenshū* (Complete works of Nishida in Japanese) (『西田幾多郎全集』second print) from the year of 1965–66, Tokyo: Iwanami.
Noé, Keiichi (2002) "Shutai to kankyō no seimeiron" (主体と環境の生命論). In: *Nihon no tetsugaku* (日本の哲学). No. 3, P. 29–51. Kyoto: Shōwadō.
Ueda, Shizuteru (2002). "Shi no tetsugaku to zettaimu" (死の哲学と絶対無) in: *Ueda Shizuteru shū* (collections of Ueda Shizuteru, 上田閑照集) No. 11. P. 107–163. Tokyo: Iwanami.

# Cited Translation

Nishida, Kitarō (2012). *Place and Dialectic: Two Essays by Nishida Kitarō*. Trans. by John W. M. Krummel and Shigenori Nagatomo. Introduction by John W. M. Krummel. Oxford and New York: Oxford University Press.

Takeshi Morisato
# Tanabe Hajime and the Concept of Species: Approaching Nature as a Missing Shade in the *Logic of Species*

**Abstract:** This chapter aims at providing a methodological reflection on the possibility of a philosophy of nature in Tanabe Hajime's "logic of species." It will firstly highlight the challenges involved in discussing the notion of "organism" and in providing a substantial contribution to the ongoing discussion on the significance of natural beings in Tanabe's system. Secondly, without discounting these challenges concerning Tanabean philosophy, it will highlight the possibility of applying the logic of species to our approach towards the sphere of the natural; and further hypothesize Tanabe's philosophy of nature by contextualizing his concepts in reference to basic terms in biology. By taking a dialectical approach, this chapter will demonstrate how the logic of species in Tanabe's account can be relevant to the field of philosophy of nature; and also, how its application to the biological concept of species can generate a particular understanding of the natural world that fits within Tanabean metaphysics.

## Preamble

A founding member of the Kyoto School, Tanabe Hajime (1885–1962), distinguished himself as an original thinker by developing his own philosophical system, called the "logic of species" (*shu no ronri* 種の論理). It took approximately fifteen years, from the early 1930s, for the philosopher to articulate his new metaphysical system in thirteen voluminous essays.[1] During his impressive career as a professor of philosophy at Kyoto Imperial University, Tanabe's students and major publishers encouraged him to publish these writings as a single monograph (Tanabe 1963h p. 523). However, it was only towards the end of his life that he agreed to allow this work to be posthumously included in the *Complete Works of Tanabe Hajime* (hereafter *THZ*) both as a record of his life's achieve-

---

[1] The list of the thirteen essays on the logic of species in vol. 6 and 7 of *Tanabe Hajime Zenshū* (plus two commentaries in their appendixes) is provided in the bibliography.

**Takeshi Morisato,** Sun Yat-sen University (Zhuhai) and Université libre de Bruxelles

https://doi.org/10.1515/9783110604665-011

ment and, more importantly, as a unified whole, a monument of his philosophical thinking (Tanabe 1963h, 524).

At the outset, given the highly abstract rendering of the logic, saturated with a plethora of metaphysical vocabularies that demonstrate a considerable influence of the thinker's earlier engagement with mathematics and theoretical physics, many readers may expect that the "logic of species" could promise Tanabe's serious engagement with the discipline of biology or at least succeeds in providing a robust theory of nature. It is clear that Darwin's use of the term in *The Origin of Species* was not unfamiliar to him.[2] However, it becomes clear that Tanabe's own interests led him away from reflection on the philosophical significance of nature or the ontological structure of natural life.

What I would like to do in this chapter, therefore, is to develop a dialectical approach to the logic of species. First, I would like to highlight the challenges involved in discussing the notion of the organism or providing any substantial discussion of the significance of natural beings in Tanabe's system. It will become apparent that Tanabe's rendering of ontology through the metaphysical notion of species is unlikely to help us account for the significance of any natural being in its concrete singularity. Nor does it seem to help us develop any positive notion of nature as a whole.

Without discounting these negative aspects of Tanabean philosophy, I would, however, highlight the possibility of applying the logic of species to our approach to the sphere of the natural. The dialectical structure of Tanabe's thinking is such that the lack of an explicit discussion of a theme does not necessarily mean that it cannot be addressed. This is because Tanabe's thinking has the potential to point us beyond itself: that is to say, his logic can – and must be – developed so as to allow it to become more comprehensive, such that it can illuminate every aspect of concrete reality, thereby demonstrating its dialectical capacity to fully account for the negative. It is only by means of this dialectical advancement, so to speak, that we will grasp the way in which Tanabe's logic of species can shed light on nature and the significance of natural beings (i.e., "natural born monads").

---

[2] Kōyama Iwao (1905–1993) recalls his conversation with Tanabe in Kyoto in the early 1930s (Tanabe 2016, p. 470). Immediately after joining the afternoon walk, Tanabe suddenly asked his friend, "Have you ever thought about something we call *shu?*" At first, Kōyama could not understand what word he was referring to. As if anticipating the confusion, however, Tanabe quickly responded in English, "the 'species' of *On the Origin of Species.*" Kōyama warns us not to equate Tanabe's account of species with Darwin's theory. Nevertheless, I think this account shows that the biological notion of species was not entirely absence in Tanabe's rendering of it.

## Social Ontology and the Immediacy of Nature: Why Can't We Talk About Biological Species when We Talk About the Logic of Species?

It is undeniable that Tanabe's primary concern in developing the logic of species is the metaphysical question of "mediation" (*baikai* 媒介).

While commenting on Heidegger's *Kant and the Problem of Metaphysics*, the first essay, *From the Time-Scheme to the World-Scheme*, immediately tackles the question of the relation between categories of understanding and intuition in Kant's critical philosophy (Tanabe 1963a, p. 3).

The second essay, "The Logic of Social Existence: An Essay on Philosophical Sociology" (hereafter LSE), refers to the Aristotelian notion of species in the same vein and begins to show Tanabe's own take on the idea of mediation:

> From the idea of democracy to the theory of social contract, the natural law theory (which regards individuals as being subject to the rule of the state) is based on the relation of the universal and the particular. Perhaps, the individual is the limit of the particular's particularization; and we can think of a determination vis-à-vis the universal as arriving at the individual through the particular. At the same time, if we look at it from the side of the individual, the particular is a kind of the universal and the ultimate stage of universalization is humanity as the whole. Accordingly, the conflict between the whole and the individual can be relativized to the relation of the universal and the particular... In this case, the whole and the individual, while being in conflict with each other, are unified through the relative mediation of the particular. As a result, we cannot help but lose sight of the particular's unique function to serve as the concrete mediation [between these opposing terms]. (Tanabe 1963b, pp. 53–54)

In reference to Aristotle's logic, Tanabe identifies the middle point between the universal and the individual as "species" and recognizes that the "limitation of the particular [i.e., species], which restricts the universal that unifies the whole of humanity in its process of grouping multiple individuals, forms the specific state in reality or a particular society" (Tanabe 1963b, 54) just as "this mediatory system of genus, species, and individuals, makes sense only in relation to [our understanding of] organic beings" (Tanabe 1963b, 55). In so far as "species" is conceived of as a way to speak about the notion of mediation, it seems reasonable to think that there is some ambivalence in Tanabe's use of the term. In any case, it provides an opening to the ontology of nature just as much as to that of human existence.

However, as the titles of his essays indicate, the Japanese philosopher repeatedly claims that his primary concern in writing these texts has much more to do with social ontology than with advocating a new metaphysics of nature. For that reason, his ontological investigation tends to focus on what pertains to human existence. In his "Responding to Criticisms of the Logic of Species," Tanabe declares that "the first motivation [for working on the logic of species for several years] was to think about the origin of the state's obligatory power towards individuals and then to search for the rational ground of the very obligation" (Tanabe 1963f, 399). The overture to "Clarifying the Meaning of the Logic of Species" (hereafter CMLS) identifies two reasons for the development of this very logic. The first of them, Tanabe argues, derives from his thought that it would be impossible to think of the unity of an ethnic group or of state's governing power (that was taken quite seriously in his society and in other countries at that time) from the standpoint that understands society merely as an interrelation of individuals (Tanabe 1963g, p. 449). The introduction to the last essay written in 1946 also looks back on the author's effort to develop a new dialectical logic, based on the notion of species developed between 1934 and 1940, thereby consistently claiming that the logic was meant to elucidate the rational foundation of state in relation to human existence (Tanabe 1963n, p. 253).[3] There are some indications in his writings that the primary focus of social ontology entails more than the author initially expected. However, precisely because the authorial intent clearly focuses on the significance of species in the context of human (societal) existence, it is not easy to see its meaningful implications in the field of the philosophy of nature or the philosophy of biology.

There are several occasions on which Tanabe refers to the concept of nature and makes a passing remark on the importance of elucidating the ontological status of natural beings in reference to his social ontology. However, the brevity of his remarks discourages many critics of Tanabe from according them any real significance. Additionally, the philosopher repeatedly argues that the biological concept of species fails to capture the mediatory function of the term that he is trying to elucidate through these essays. The LSE, for instance, recognizes the ef-

---

[3] "From 1934 to 1940, I conducted a study of a dialectical logic that I called "logic of species" and through it, I tried to elucidate the concrete structure of state. The motivation behind this investigation was based on the fact that I tried to take up democracy that was beginning to come to the forefront of the political and the intellectual scene at that time as a philosophical problem and further by rejecting totalitarianism that finds its ground merely on democracy and criticizing liberalist philosophy that governed us, tried to discover the rational foundation of state as the practical unity of reality and ideality at the standpoint of absolute mediation that represents "substrate-qua-subject" and "subject-qua-substrate" through the intermediation of the individual (which gives the subject of liberalism) and an ethnic group (which serves as the substrate of the individual)."

ficacy of using the triadic system of genus, species, and individual for expressing the classification of living beings. However, it immediately warns us that the structure of society cannot be understood through the external organization of classes that the biological understanding attributes to these terms. Moreover, Tanabe continues, the distinction between genus and species, which is developed out of the biological classifications, tends to dismiss their negative opposition, a crucial component in his logic of species that unpacks the dialectical structure of (human) existence (Tanabe 1963c, pp. 198–199).[4] The small number of references to nature or natural beings, in addition to the critical distinction drawn between the philosophical and the biological use of the relevant terms, makes it quite difficult for us to discern the relevance of the logic to our understanding of nature.

A leading Tanabe scholar, James Heisig, argues that the logic of species (at least at the outset) "looks like vintage Hegel" (Tanabe 2016, p. 18). The triadic structure of the logic, as well as Tanabe's continuous emphasis on the intermediation of any opposing metaphysical terms through the principle of (self-)negation, clearly shows that the Kyoto school philosopher took Hegel as the starting point for developing his own framework of thinking. What makes this similarity between Hegel and Tanabe problematic in our attempt to read the logic of species for our understanding of nature? Tanabe's philosophy inherits the initial distinction between the positive term (e.g., infinite, necessary, rational, true, etc.) and the negative term (e.g., finite, contingent, natural, untrue, etc.) from Hegel (tacitly à la Descartes). Like Hegel, Tanabe spent considerable time in articulating the proper form of the intermediation between these terms. However, unlike Hegel, he says little about the way in which their mutual implication will shape a comprehensive study of the negative. Nature, in Tanabe, in other words, is explained as that which is more than the initial distinction (mis-)conveys. However, the question of how the reconfiguration of these terms through their mutual implications, in line with his version of dialectic, will lead us to another revision of the philosophy of nature is left unanswered.

The authorial intent and the main thesis of the work clearly show that these texts on the logic of species are primarily concerned with the societal existence of human beings. A lack of sufficient reflection on, or any reference to, contemporary studies of nature becomes more troublesome for the ongoing discussion of this book once we see the philosopher distinguishing his use of the central term, species, from the way in which it (according to him) is used in the domain

---

4 Cf. Heisig 2016a, pp. 271–274. In this section, Heisig clearly indicates that the notion of species is mainly seen as the "socio-cultural substratum of historical peoples."

of natural science. Lastly, since Tanabe did not further articulate his philosophy of nature, based on his reformulation of dialectical logic, the concept of nature in the logic of species ultimately remains too abstract to deliver any clear insight into our ontological understanding of it. For these reasons, many readers may hesitate to read Tanabe's *Logic of Species* for comprehending the structure and the content of natural existence.

## Nature Must Be Accounted For: Reading Tanabe's *Logic of Species* with a Tanabean Spirit

The lack of evidence for a philosophy of nature in Tanabe's texts on species is the strongest indication that it was not his primary concern. What is at stake is not the biological notion of species and its significance for social ontology. Some readers may wonder, at this point, if a philosophical account of natural beings is not a missing shade in the *Logic of Species* but is simply missing altogether.[5] The authoritative reference work, *Japanese Philosophy: A Sourcebook*, implicitly shares this view since it translates the title of the work as the *Logic of the Specific* (Heisig et al. 2011, p. 670).[6] It succeeds in showing the Japanese term, *shu* 種, as having nothing to do with the biological concept of species in Tanabe's account and highlights the fact that it only denotes a metaphysical category. With this translation, we can defuse the tension between Tanabe's dialectical notion of species and the notion of species in the philosophy of nature, by showing that they are irrelevant to each other. That being said, I would like to point out in what follows that there are, however, some passages in Tanabe's text where we can decisively suggest the possibility (if not the necessity) of extending his social ontology to the domain of the philosophy of nature.

On more than one occasion, Tanabe identifies the notion of species as the "will to life" (*seimei ishi* 生命意志), while contrasting it with the individual as

---

[5] It highlights the fact that it should not be translated as "species" (which implies a biological concept) but only as the logical category of the specific. In this case, there is not tension between what I have described in the previous section and what I am going to show in the rest of this paper. If we are going to recognize some tension between Tanabe's dialectical notion of species and the notion of species in philosophy of nature, it is better to transliterate the title as the *Logic of Species*.

[6] See also (Heisig 2016a, p. 261). Heisig explicitly explains the reason for substituting with "species" with the "specific" to remove any ambiguity from the term. However, my contention is that the very ambiguity is the key for elucidating what is missing at the surface of the logic of species. For my critique of Heisig's translations of Tanabean terms, see (Tanabe 2016, pp. 469–473).

representing the "will to power" (*kenryoku ishi* 権力意志) and genus as the "will to salvation" (*kyūsai ishi* 救済意志) (Tanabe 1963b, p. 141).[7] On the basis of these definitions, we can anticipate how the concept of species comes to signify a substratum of all living beings in Tanabe's mind. Moreover, in LSE, species is developed into that which is comparable to nature as genus is to the divine absolute, thereby signaling a strong connection between the notion of species and the realm of nature (Tanabe 1963c, p. 212). This is probably because Tanabe acknowledges that the concept of species, as the substrate of all living beings, originates from the Aristotelian understanding of the notion, which involves appropriate attentiveness to the characteristics of organic existence (Tanabe 1963d, p. 272–273). Tanabe remains critical of Aristotle's metaphysics on the basis of the fact that it emphasizes the ultimate primacy of form over matter, and construes an undifferentiated continuity between human/societal existence and nature.[8] However, as far as the aforementioned passages are concerned, the notion of species has an indisputable connection with the notion of life: hence, Tanabe's social ontology must address the notion of natural beings since it aims at comprehending human existence *in* nature.

The dialectical nature of the logic of species, moreover, guides Tanabe to recognize the significance of social ontology – or his conception of human existence (i.e., self) in society – as maintaining a comprehensive relation to its opposing term, "nature," regardless of the fact that his logic must maintain their mutually irreducible difference. This is in line with a kind of development that Tanabe recognizes in working on the *Logic of Species*. We have seen that the CMLS identifies two reasons why he developed this logic, and the second of them, as Tanabe argues, "… pushed me beyond the practical meaning tied to the current socio-political circumstance but it came to possess more general meaning that relates to the method of philosophy" (Tanabe 1963g, p. 466).

Two interrelated concepts that Tanabe develops as the general "method of philosophy" in these thirteen essays are (1) absolute mediation and (2) history. The former focuses on the proper way in which we can conceive of the intermediation between any opposing metaphysical terms, while the latter sets forth the ultimate standpoint from which we can understand reality in concrete terms, which is supposedly comprehensive of both the human and the non-human

---

**7** Cf. Tanabe 1963e, p. 303: "The continuous whole of species should precisely give the continuous whole of life. The individual living being takes species as the ground of its life and yet as a product of the will to love that does not reflect on the sacrifices of numerous individuals for its preservation, it remains irreducible to species." See also Tanabe 1963e, p. 358: "An individual living being is an individualized species."
**8** Cf. Tanabe 1963k, p. 155.

standpoint (which he calls the "world-scheme" or "space-and-time"). I believe that Tanabe's reflections on these complex ideas provide us with some room for extending his social ontology to the domain of the philosophy of nature.

In the "Logic of Species and World-Scheme" (hereafter, LSWS), Tanabe argues that absolute mediation is the core structure of the dialectical logic of species, and further claims that we have to attend to the way in which it intermediates both positive and negative terms. First, he claims that the logic of species, as absolute mediation, cannot be a kind of thinking that "abstract[s] the non-rational and include[s] what is rational alone as its content" (Tanabe 1963c, 176). Rather, as proper dialectical reasoning"

> The logic that denies immediacy must suspend it (i.e., the negation of self) as the mediatory moment of itself and deny [the immediacy] by affirming it under the condition of absolute negation (which is a negation of negation); otherwise, the logic cannot truly deny the immediacy. Thus, [the] non-rationality that immediacy possesses should not be thrown out but must be maintained as the mediator of the logic. If we say that the logic that denies this [immediacy] is rational, it should not be that which abandons the non-rational by merely opposing it. But rather, it should constitute non-rationality-qua-rationality that affirms it through negating it. With regard to this sense of absolute rationality, [the] non-rationality of immediacy must be an indispensable mediatory moment in the process of establishing itself. This standpoint is the absolute mediation that turns the non-rational into the medium of its own self through maintaining [the non-rational]. (Tanabe 1963c, 176)

This intermediation of self and other, based on the notion of "absolute mediation," continuously surfaces in Tanabe's writings, especially when he expresses his critical stance towards other thinkers. In "Three Stages of Ontology," for instance, the relation of nature and self becomes the standard model for outlining the structure of his dialectic, and it is laid out in relation to his critique of Hegel's philosophy of nature (Tanabe, 1963d, pp. 296–297).[9] Additionally, in "Ethics and Logic," while explaining the mediatory relation of existence and ethical ideals, the philosopher weaves his understanding of absolute mediation into Schelling's view of nature. He argues:

---

[9] "Hegel's philosophy ignores nature's independent, negative opposition to mind and that is why it tends to be interpreted as painting nature as a result of mind's self-alienation within itself. In this case, when claiming nature as existing externally to mind, this transcendence cannot indicate that which is negative to immanence as a moment of the absolute mediation. But it can only signify ideal transcendence within mind. Here, the peculiar meaning of the transcendence of space, discussed in overture to philosophy of nature is lost, whereby nature is reduced to ideality. However, nature that ultimately remains in immanence as that which is emanated from absolute spirit cannot be the mediatory moment as the substrate of existence."

> Nature that stands over against human action as an obstacle rather serves as the mediator that affirms the human action as the negative: so, to overcome nature through human action, in turn, must mean to follow nature. In this case, however, nature loses the meaning of "standing over against action." Rather it takes the human action as the mediator of its formation. In short, nature simultaneously comes to mean what human action produces. It is no longer nature that self must overcome as non-self but rather it turns itself into self as the mediator of self. Hence, it must be trans-natural but at the same time trans-self because self follows the nature and so long as that takes place, nature turns self that is [initially] denied therein into an affirmation of it. The absolute mediation of self-qua-non-self or nature-qua-self makes self and nature as two moments of the absolute unity of subject-qua-object. (Tanabe 1963l, p. 179)

In "Eternity, History, and Act" (hereafter EHA), Tanabe begins to label this transformative unification of nature and self as the work of "absolute nothingness," thereby adding a far more metaphysical (and even religious) tone to the notion of absolute mediation.[10] Once the notion of the transformative and metanoetic absolute is introduced to the logic of species, Tanabe's social ontology comes to mean something more than a metaphysical understanding of human existence because the concept of the absolute necessarily implies the transformed sense of the whole reality (including nature and all sentient beings) in relation to self's self-negating self-fulfillment in its social existence.[11] Since, in this context, "self-awareness" based on absolute mediation comes to "represent reflections on the transformative unity of nature and self" (Tanabe 1963m, p. 233), the logic of species contains a line of thinking that extends far beyond the initial intention to examine the contradiction between the state's authority and individual freedom. This metaphysical advancement in Tanabe's writings can only make us wonder what the renewed sense of nature under the logic of absolute mediation would look like, and invites us to apply the very logic to the study of living beings.

---

**10** Cf. Tanabe 1963k, p. 156: "In action, self does not remain to be self alone but while entering nature beyond itself and working on it, nature goes through formative change through the mediation of self and takes the self into itself. The act is established through the transformative mediation where self's work becomes nature's and vice versa. In short, in act, self as the acting subject and nature as the object that is worked on are transformed through absolute nothingness and thereby self becomes nature and nature becomes self. That is to say, through the transformation of absolute nothingness, both self and nature negatively become one. This serves as the ground of comparative studies between nature and human action."

**11** Tanabe identifies this renewed sense of nature as "historical nature" (in opposition to an ordinary sense of it as "material nature"). Cf. Tanabe 1963k, p. 164: "The concepts of nature, causation, and technology can have a broad meaning of the negative moment of historical and ethical act in general... Historical nature is more concrete than physical nature. The concept of material itself comes to be historicized and comes to have historical meaning."

There are some methodological suggestions that we can draw from Tanabe's writings. The EHA states that, "through the transformation of [absolute mediation or] absolute nothingness, both self and nature negatively become one; and this serves as the ground of comparative studies between nature and human action" (Tanabe 1963k, p. 156). This statement alone proves that we may apply his logic to our study of nature or relate it to the biological concept of "species." In fact, the philosopher himself often applies his dialectical logic to the field of mathematics and theoretical physics. In this sense, our application of it to the philosophy of biology should be a welcome addition to his philosophical findings.[12]

This interdisciplinary practice, however, is further supported by Tanabe's metaphysical insight. He insists that his intention in making mathematical inferences or offering philosophical insights in matters of theoretical physics is not designed "to prove the structure of the historical world through the world understood in terms of physics as if it were the ground of existence" (Tanabe, 1963e, p. 364). However, he continues:

> Existence in physics is not anything that we can directly grasp through sense perceptions because it requires a negative mediation of the perceptual content and mathematical concepts. Hence, mathematics that belongs to a historical culture enters the world understood in terms of physics as mind's formal self-awareness; and thereby, the world picture that pertains to physics is not a mere reproduction of existence. Rather, it includes the historical fruit of mathematical thinking as its negative moment. In order for the historical world to reach the level of concrete self-awareness, therefore, it must include the structure of the world understood in terms of physics as a mediatory moment [for its establishment]; and if the logical structure of the latter is not understood as the abstract aspect of the logical structure in the former, then we would have to say that the understanding pertaining to the former is incomplete. (Tanabe 1963e, p. 364)

This is to say that the logic of species, based on the notion of absolute mediation (i.e., nothingness), does not only serve as the foundation for the comparative studies of philosophy and biology (among other disciplines in the field of natural and social sciences), but also, without accounting for the significance of the world-view obtained through a certain structure of understanding in the specific field of sciences, the philosophical world-view (or what he calls "historical world") that we can access through the logic of absolute mediation remains an unfinished business. To judge the validity and soundness of Tanabe's insight,

---

[12] Tanabe classifies biology as one of the disciplines in sciences that become a component of the history. See Tanabe 1963n, p. 341.

therefore, it is crucial that it be compared with findings in other academic disciplines.

The task of teasing out a philosophy of nature, understood as a missing dimension in the *Logic of Species*, is quite novel, and requires familiarity with the basic concepts or the conventional interpretations of the system. Only about twenty pages of the *Logic of Species* have been translated into English, and given the size of the original (roughly 900 pages in two volumes of the *THZ*), it is not surprising that it has not been adequately treated in English-speaking academia.

However, Tanabe's style of doing philosophy should set our minds at ease. More particularly, in the texts written from the 1930s onwards, Tanabe tends to perform a critical study of other contemporary thinkers (including Heidegger, Jaspers, Nishida, Bergson, Durkheim, et al.) in such a way that it always benefits the formation of his own philosophical ideas. What he often does with the works of these thinkers (many of whom were not translated into Japanese or fully studied among scholars at that time in Japan) is to provide a very concise summary of their basic ideas. Moreover, instead of devoting himself to lengthy philological work, he presents their theories in terms of various "philosophical types" (*shisō no kata* 思想の型) (Tanabe 1963c, p. 225). By doing so, he is able to set forth his standpoint in relation to these authors clearly, and to clarify his own, original, contribution to the field of philosophy.

The logic of species, for instance, is known to have taken up Bergson's *Two Origins of Morality and Religion* soon after its publication. Tanabe was undoubtedly one of the first thinkers to cite this work in Japanese and his engagement with Bergson's text in the LSE indeed predates the first Japanese translation of it, which was only published in 1936.[13] After expressing his deepest respect for this monograph, he claims that, with his dialectical manner of delivering his points, "he cannot hope for the clear, intuitive, and concrete impression and persuasiveness that the philosophical genius of Bergson is able to provide through his philosophy of intuition" (Tanabe 1963b, p. 73). Nevertheless, he continues, "it is more convenient for me to take his description of the philosopher's ideas as a medium [of establishing his own standpoint] and to further advance my own ideas through critiquing it" (Tanabe 1963b, p. 73). According to Tanabe, a philosopher does not have to wait for the ideal scholarly environment where a number of translations and critical studies of the primary sources are available before

---

**13** The first Japanese translation of Bergson's *Two Origins of Morality and Religion* was done by Hirayama Takaji 平山高次 in 1936 through the publisher, Shibashoten, and later reprinted through Iwanami Shoten.

making original claims based on the ideas provided by them. What is important is to take the basic description of what we can find in any under-represented work as a medium for forming and transforming one's own ideas. While striving to remain faithful to what we can find within the bounds of Tanabe's logic of species, this article will take the same methodological stance to his essays as he did to Bergson's, among others. The spirit of Tanabean philosophy, therefore, fully supports our venture to elucidate the logic of species and explore its implications for the philosophy of nature.

## The Logic of Species: Defining the Terms and Their Inter-Mediatory Relations

The threefold distinction of genus, species, and individual corresponds to Tanabe's rendering of the universal, the particular, and the singular. Given that the history of philosophy, in its ongoing task of achieving a comprehensive understanding of reality (including human existence), exhibits a constant struggle between emphasizing the primacy of the universal and maintaining faithfulness to the irreducibility of the single individual, Tanabe proposes "species" as the middle term that serves as the foundation for establishing a proper intermediation between these. It is quite challenging, however, to define one of these terms in a determinate and coherent fashion, precisely because each definition presupposes its relation to others. Simply put, we cannot define one term without defining the others.

Given that one term could mean a number of different things depending on how we formulate its relation to other terms, it is a matter of course that some parts of the explanations of the single term will have to wait for further elucidation vis-à-vis other terms and vice versa. In what follows, therefore, I will provide a preliminary outline of what Tanabe means by each term, in the awareness that this does not exhaust Tanabe's understanding of absolute mediation. This exercise is necessary if we are to demonstrate how Tanabe sees their mediatory relations to each other and why he speaks of the "logic of species."

With *genus*, Tanabe indicates the notion of the universal. It is abstract or mere ideality when it is understood apart from other terms, that is to say, it is non-existent in and of itself. In the context of social ontology, he refers to it as the totality of all human beings (i.e., humanity) and further attributes the notions of the "state" and the "(divine) absolute" to it. In the context of metaphysics and the philosophy of religion, he argues that it represents the notion of nothingness as a polyvocal, open unity of all that is, and describes it as a "will to salvation" or the "city of bodhisattva" (from which all sentient beings are derived and to which they all return) (Tanabe 1963b, pp. 127; 131–132; 141). It, therefore, represents a sort of principle of unity and rationality that is at work in other terms such as species and individual. However, if we interpret genus merely as the totality of all beings in immanent terms, Tanabe argues, we must be mistaking it for an unmediated sense of species or reducing genus to a bad (i.e., unmediated) sense of the latter. To understand genus properly, as what it is, we must first conceive of it as "absolute negation" or negative unity that constitutes the inseparable unity of all terms without reducing their differences to an unmediated sameness. In this sense, the concept of nothingness is indispensable for properly interpreting Tanabe's rendering of genus.

With *species*, the Japanese philosopher indicates the notion of the particular, that is to say, that which is more universal than the singular and more specific than the universal. It is described as the substrate (*kitai* 基体) of all living beings and, unlike genus, it can be seen as being both immediate and irrational in itself. Species in its immediacy, in other words, constitutes a continuous and particular whole that compiles a multitude of individuals into a relatively determinate group, and this specific group can serve as the foundation of individual lives. The prime example of this relation between species and individual would be the relation of a nation state and its citizens. If I reflect on my existence as a singular person, I can easily think of my nationality, i.e., Japanese, as a specific characteristic of it. The socio-political identification of the nation-state is usually more or less given immediately (that is to say, it is usually easy to identify the nationality of an individual) and my citizenship in this particular country has provided me with certain elements that have enabled me to live my life as a free individual in this world.[14]

What is interesting about Tanabe's argument is that rationality ultimately belongs to genus. In view of this fact, a species can provide no rational ground for legitimizing its identity over the others. Its specificity is simply beyond reason. As I will explain later, Tanabe comes to argue that species can serve as the ground for manifesting the ideality of genus in concrete reality, thereby highlighting the possibility that various forms of species can be placed in a hierarchical order. In other words, he does not deny that one species or a small group of

---

[14] In some cases, it is much more complicated to identify the nationality of an individual. Additionally, citizenship of a nation-state that is going through a violent socio-political-economic turmoil is almost always detrimental to its holder. However, this does not deny the fact that her belonging to the specific nation state is foundational to the fate of her life as an individual. Regardless of the positive or the negative outcome of the belonging, the socio-political substratum itself is undeniably significant for the life of an individual.

them can come to play a leading role in manifesting the ideality and rationality of genus vis-à-vis other species. However, even in this process of relativizing species by means of genus, this realization of ideality is made through the free act of individuals, and whether its formal specificity should be defined in one way or another in relation to the manifestation of the rational whole, remains completely contingent. In this sense, Tanabe concludes that the "species is the grounding source of [the] individual's immediate life and what ultimately ought to be denied [since] it only gives the expedient being that exists as the negative mediator of absolute nothingness's manifestation in this world" (Tanabe 1963n, p. 267).

If we revisit the example of my nationality, the specificity of being Japanese can manifest its rational ground only when its citizens live up to the ideality of humanity (which Tanabe explains as practicing the act of nothingness) and transform their socio-political belonging in the nation-state into an enabling ground for an open community of free individuals. But this concrete manifestation of the universal ideal of humanity (genus) can be made through any specific group of individuals, and there is no reason why it has to be done by means of one form more than by some other, as regards their specific way of living their lives. When a group of individuals identifies the source of its individual life in a specific socio-political milieu (species) as the ultimate ground of all things (genus) and fails to recognize the contingency and irrationality of the former in its negative relation to the latter, then we witness the totalitarian dictum of nationalism where the participants erroneously reduce genus to a bad (i.e., unmediated) sense of species. Tanabe was clearly against this reduction of the notion of genus to the unmediated sense of species.

To ground the contingency and irrationality of the specific, Tanabe comes to realize, the notion of species must represent something more than the given irrational that we can immediately talk about as something irreducible to our individual existence *in concreto* or to the ideality of human existence *in abstracto*. In response to a criticism that Tanabe's rendering of species seems to lack any mediation regardless of the fact that it should demonstrate an axis of absolute mediation, the "Social Ontological Structure of Logic" strives to articulate species' in-between status in relation to genus and individual as the principle of self-negation or, more precisely, an endless movement of auto-generated oppositions that has its source in nothing other than itself:

> Species does not possess the negative outside itself but within itself carries that which denies itself. This self-negation has nothing beyond itself that grounds it and it is the self-negation that is entirely specific to species. The reason why species is not universal but particular is because it is simply self-negating and does not give any absolute negation [that pertains to genus]. When we say that the particularization of the universal is universal's self-alienation, what we mean is that absolute negation [of genus] is reduced to self-nega-

tion [of species]. But without this self-negation, we cannot talk about species as species. Nor can we know its existence. The fact that we have already known and been talking about it [as the immediately given] presupposes some negation of, and opposition to, immediacy. We just have to understand species as that kind of self-negation. (Tanabe 1963e, p. 313)

This insight certainly allows Tanabe to emphasize the contingent characteristics of species. There is no external reason why the series of divisions between diverse species is made in the way it is, nor can we reach any teleological viewpoint from which we can place a number of species in a hierarchical order based on their specificities. Species, as the principle of self-negation, is just that which divides itself: hence, it is simply marked with contingent finitude, whereby no rationality can be ascribed to the ways in which it divides itself.

However, this insight also leads Tanabe to revise the initial definition of species as having a primitive unity or a continuous whole in its immediacy.

[The reason why we have to make this modification] is because, so long as we think that the essence of species is to divide itself through the principle of self-negation, whichever part of species we look at, we will recognize a conflict between the positive and the negative power; and thereby, based on this conflict, the division towards the negative opposition and the unification towards the whole that opposes the very division accompany each other. Species represents this unceasing movement that contains the double negation that tries to divide itself through the struggle between opposing powers while, in turn, trying to preserve its unity through opposing this division. We cannot possibly think of it as a fixed unity at rest. (Tanabe 1963e, p. 315)[15]

It is undeniable that we can articulate species as something that is directly involved with the life of the individual. We can see, for instance, that a specific nation state, an ethnic group, or a particular society can serve as an enabling condition for the possibility of one's existence as a free individual. Tanabe's qualification of species as a principle of self-negation, here, does not contradict this point. However, he qualifies that the determinate phase of species as the immediately given is only half of the whole picture. The specific socio-political, linguistic, or cultural milieu is contingent and subject to change (and in some cases to extinction) regardless of the fact that it constantly strives to preserve its unity. That is because it carries within itself the seed of self-negation as species. It divides itself as it strives to preserve its particular identity and as a contingent particular, it is always subject to endless change. Thus, the notion of species repre-

---

[15] For the "self-negating structure of species," see Tanabe 1963e, p. 321; and for the "absolute-negative transformation of species," see Tanabe 1963g, pp. 484–485.

sents the principle of dynamic self-negation that constitutes two opposing phases of immediate positivity (i.e., unity) and mediated negativity (i.e., disunity).

Species, in Tanabe's account, therefore, manifests itself in two types: (1) the positive type which opens itself to ideality and the rationality of genus while recognizing its relativity via-à-vis other species, and (2) the negative and unmediated type which strives to preserve its relatively unified identity in immediacy and tempts the individuals that belong to it to regard it as the absolute ground of all things (i.e., genus).[16] The question, then, is as follows: Where precisely is the basis on which genus is (1) manifested through species without compromising its specificity, or (2) reduced to the partial picture of species as the immediately given (thereby disregarding its essential relativity vis-à-vis other species)? The answer is the following: it lies in free choice and the actions of the individual.

If the notion of species is to serve as the foundation of individual, it cannot be a simple amalgamation of individuals but must represent that which is both irreducible to and constitutive of individual's concrete existence. When it is conceived as the negative and unmediated whole, moreover, its immediate unity has a propensity to "swallow up the individual" (as every form of nationalism tends to do) (Tanabe 1963e, p. 306).[17] Species constitutes the continuous whole that denies the individual but at the same time gives the societal substrate as the foundation through which the individual can come to enjoy its existence. In this sense, Tanabe argues, "in their negative relation, individual and species presuppose that ... the individual can discard species through its own power while simultaneously species can bring the individual to extinction" (Tanabe 1963c, p. 196). In short, "species gives birth to individual as it kills the individual" (Tanabe 1963e, p. 306).

The flipside of this dialectical argument is that the individual must also stand in an oppositional relation to species and, in that sense, Tanabe defines the individual as having the "will to power." What he means by this is "freedom" or the "ability to deny the essence of determinate identity (species) in itself; and [to] take this freedom as its essence" (Tanabe 1963b, p. 109).[18] This sounds quite

---

**16** These two types roughly correspond with Heisig's interpretation of "open" and "closed species." See his rendering of it in relation to the political discourse in Heisig 2016b, pp. 329–331.
**17** Cf. Tanabe 1963e, pp. 302–303. Tanabe means here that, in the process of establishing itself a continuous whole, [the] species requires a number of individuals to dedicate themselves cross-generationally to its preservation. In this case, many individuals must die to maintain the whole as it provides a foundation for sustaining the life of individuals. The preservation of species, in this sense, paradoxically provides a ground for the life of individuals and takes them for maintaining its determinate identity.
**18** Cf. Tanabe 1963b, p. 119.

natural to our ears if we apply it once again to the relation between a nation state and its citizen. A single individual owes much to her country for enjoying her life but what grants her the status of the singular is her ability to transcend her national identity and to make her own decisions which may support or go against what is expected of her as her civic duty in that particular socio-political climate. The examples of this "will to power" could range from an extreme case of high treason to a milder instance of naturalization in another state. Humanitarian aid to citizens of other countries would also qualify as the exercise of this freedom. The point is that the individual has the ability to suspend species when their interests are in conflict with each other.

In the LSE, Tanabe examines the source of the individual's radical freedom. His argument is that the individual's differentiation from species through the will to power cannot be self-given but must come from somewhere else. In the process of responding to a critique of his earlier engagement with Durkeim's totemic principle and Tönnies's *Gemeinschaft–Gesellschaft* distinction, Tanabe traces the rise of individual freedom in the conflict between particular societies.

> The differentiation (分立) of the individual [from species] finds an opportunity in the mutual conflict between various species. Species is the womb of the individual and simultaneously unifies them all [into itself] as a whole. Nevertheless, its conflict and struggle with other species serves as an opportunity for the individual to differentiate itself from its species. The demand of the individual for its autonomy is awakened by the existence of other species (that it does not belong to) and begins to put itself in motion through the struggle among them. The external conflict and the internal opposition, in this sense, are always relative to each other. (Tanabe 1963j, p. 80)

What takes place in this moment of conflict between species is that the individual "usurps" the self-negating power of species. In the LSE, Tanabe defines the condition for the possibility of exercising freedom:

> Only an individual who usurps the power of species that both determines and unifies all individuals can obtain the will to power through the mediation of this governing force pertaining to species. Stated succinctly, the individual can be established through its oppositional unity vis-à-vis species and because of that, its will to power represents that which is mediated through species as the will to life. (Tanabe 1963b, p. 124)

The essential determination of the individual, therefore, is that it can stand in its opposition to species as it exercises its "spontaneity that transforms the determination of species into a medium of its establishment" (Tanabe 1963b, p. 112). This self-differentiation of individual from species as radical freedom maintains a paradoxical relation between the two terms. On the one hand, the individual is completely in opposition to species, thereby claiming its irreducible difference to, and independence on, species. Yet, on the other, since the source of its division from species comes from the self-negating principle of the latter, there is a significant degree of indebtedness to what the individual is departing from.

This leads us to an interesting outlook on the relation between individual and species in reference to the two types of the genus–species relation. The negative sense of species generates a reduction of genus to species where the latter mistakes itself for the absolute, while disregarding its specificity.[19] In this case, this species loses sight of itself since, on the basis of its finite specificity, it maintains its quality in its relativity to other species. Under the reign of this "closed" species, the individual will come to experience their conflict with other species. Additionally, as the species aims to preserve its identity, it fails to exercise its full potential as the principle of self-negation and makes a self-contradictory demand of unwanted service on the individual. In this case, the individual will be awakened to its freedom and bring its "closed" species to a transformation (if not extinction).

This transformative change of the closed species only leads us to the positive sense of it that gives an open relation between genus and species where the irrational substrate of species can manifest rationality of genus without losing its specificity that keeps its relativity with other species. The principle of self-negation with regard to this species is fully realized in this "open" species: hence, the individual is fully exercising its freedom and making possible the manifestation of genus in species. This is what Tanabe means by the absolute mediation of genus, species, and individual, or the triadic logic of species. What is important in relation to the nature of the individual is that it plays the crucial role for realizing the open mediation between genus and species, thereby fulfilling the promise of its freedom as derived from the self-negating principle of species.

## Absolute Mediation as the Logic of Species

We have seen that the rationality of genus cannot exist in and of itself but only manifests itself through species that serves as the substrate of life. Species represents the contradictory principle of self-negation: hence, it has a double directionality in its way of (re- or misre-)presenting the genus. On the one hand, it strives to preserve its continuous unity by imposing its immediate and determinate identity on the individual as the absolute whole (i.e., a counterfeit genus).

---

[19] This point concurs with Heisig's analysis of the "absolutizing of the specific." See Heisig 2016a, p. 280.

In this case, the species mistakes itself as genus while disregarding its finite specificity and equality with other species. On the other hand, it also strives to deconstruct its determinate identity and constantly calls for a transformation that brings itself in its relativity to other species. This self-contradictory nature of species serves as the foundation of freedom in the individual, and only through exercising this freedom in a particular way, can the single individual concretize genus in and through species.

Tanabe describes this formal structure of the logic of species as absolute mediation and an examination of this notion can help us sketch out the content of the intermediation between the metaphysical terms in question. In the LSE, he argues:

> Absolute mediation means that in order to establish one, it is always mediated through another. Thus, since one and the other mutually negate each other, it means that absolute mediation cannot be carried out unless every affirmation is made through negation as a medium of its establishment. As the so-called negation-qua-affirmation, absolute mediation requires that an affirmation is always an affirmation through negation. Hence, it excludes all immediacy. Even for what we call the absolute, we cannot allow it to be established directly unless it takes any relative that negates it as the medium of its establishment (Tanabe 1963b, p. 59).

The double directionality in mutual affirmation-through-negation among three terms in the logic of species is much easier to picture in reference to the relation of species and individual. Species is conflicted between its self-preservation and self-negation and because of that, this inner split ultimately characterizes its nature as the self-negating principle. The individual, then, takes on this principle as the basis of its freedom, and can choose to contribute to the preservation of the species or the transformation of it into a different kind. However, the similarity of self-negation between the individual and the species does not necessarily exhibit what kind of action from the side of individual will contribute to the conservative or the innovative formation of its species. The question remains what kind of action constitutes the full realization of freedom by the individual and how it contributes to the intermediation of other terms.

In the LSWS, Tanabe describes what the intermediation of three terms would look like with an emphasis on genus as the unity of species and individual (Tanabe 1963c, p. 218). He argues that the individual does not stand in an oppositional relation to genus, but genus enables the individual to recognize itself in its negative relation to other individuals. Also, since it negates species through itself as absolute mediation (or what Tanabe calls affirmation-qua-negation):

> Species does not disappear simply as species but at the same time it is preserved in the unity of genus that transcends species. Accordingly, the special characteristics of species that is to stand in an opposition to other species is suspended and thereby comes to demonstrate the relation of difference-qua-sameness or otherness-qua-self. Genus accepts an infinite number of species and transcends them. In here, genus exhibits the meaning of the absolute whole. ... In this whole, not only the individuals in the same species, but also the individuals that belong to other species maintain the relationship of equality, as they constitute the unity of difference-qua-sameness in the same genus (while belonging to different species), and they stand in the relation of self-qua-other whether or not they belong to the same species. In short, the reason why genus realizes the whole of absolute mediation is because [in it] all conflicts between species and individual; between single individuals; and between multiple species are negatively unified, whereby they enter into their mediatory relations to each other. (Tanabe 1963c, p. 218)

Genus, as the principle of absolute mediation, therefore, embodies an open whole that both unifies all opposing terms and preserves their differences. Once again, it grants them their equality with each other without disregarding their specificities. The content of the absolute mediation, in this sense, is a kind of agapeic release that non-insistently disposes autonomous power to what is other to itself and enables them to be what they are in their oppositional relation to each other. For this reason, Tanabe comes to describe this "open unity of species and individual in genus" as the notion of nothingness, and declares that individual freedom must carry out the act of nothingness so that, through its intermediation with other individuals, it can manifest the ideality of such universal genus in the substrate of species.

It takes a while for Tanabe, in his essays on the logic of species, to clarify what exactly the act of nothingness should look like. Given that genus provides some space for both species and individual to determine themselves, and also that species represents the principle of self-negation at work in the heart of individual existence, Tanabe comes to argue that the individual must practice a kind of self-sacrifice as a negative mediator of one's own self-realization in its relativity to the self-negating nature of genus and species (Tanabe 1963f, p. 402; Tanabe 1963g, p. 506). "The act of [the] individual as the subject," he further elucidates, "represents the transformation of self-negation pertaining to species into the absolute negation of genus and thereby [the] individual is established through the transformation (*tenkan* 転換) into genus via the mediation of self-negation in species instead of simply by standing in an opposition to species" (Tanabe 1963g, p. 404).

What he means by the term transformation or conversion (*tenkan*) becomes clear only when he ties the practical notion of "metanoesis" (*zange* 懺悔) to it in the 1940s. The last essay, the "Dialectic of the Logic of Species," which is a revision of an essay published in 1946, and was published after the appearance of

*Philosophy as Metanoetics*, clearly indicates that the act of nothingness must represent a kenotic movement of self-negation, an agapeic selflessness that brings itself to nothing so that what is other to itself becomes everything. To embody and signify the absolute nothingness of universal genus in the particular species, therefore, [the] individual must be actively engaged in the act of constituting the inter-subjective (net)works of love with other individuals, and this community of selfless individuals gives a particular manifestation of genus in their species.[20] What makes this logic triadic is that this open species of free individuals, as the concrete manifestation of nothingness, consists of multiple communities of individuals that recognize the source of their freedom in other relevant metaphysical terms. With three terms giving space for each other through their self-negation, individual freedom can be seen to complete the triadic logic of species.

# The Intermediation of Species Through the Notion of Absolute Mediation: A Self-Negating Theory of Evolution?

The application of the triadic logic of species to our philosophical understanding of natural beings gives an interesting insight into the philosophy of nature. In the context of social ontology, Tanabe regards a nation-state or any particular sociolinguistic, cultural, ethnic or even racial grouping of human individuals as species. This is clearly incompatible with the biological understanding of humans as *homo sapiens* (which applies to all human beings regardless of their particular differences pertaining to the categories just mentioned). However, what we need to keep in mind in applying Tanabe's logic of species to the philosophy of nature is that it is essentially concerned with the structure of the ways in which a specific group of living beings organizes itself and manifests its ideality through its interrelation with other species. The ontology of species, in other words, is abstract enough to give us a formal definition of the term in its relativity to other terms, such as genus and individual, thereby enabling us to identify what counts as a biological species according to Tanabe. In the following, we will hypothesize Tanabe's philosophy of nature by contextualizing his concepts in reference to the basic terms in biology.

---

20 Tanabe remains inconclusive in the *Logic of Species* regarding the source of the self-negating self-transcendence of individuals as the concrete manifestation of genus in open species. In *Philosophy as Metanoetics*, however, he will clearly state that it should originate from the nature of nothingness.

According to Tanabe, species represents an irrational substrate of life for individual beings and, more importantly, it carries within itself an irresolvable energy of contradiction. As that which gives life to the individual, species suffers from a double directionality towards self-preservation and self-negation. Since it is incapable of bringing this mark of finitude within itself to rest, Tanabe has emphasized it as constituting the principle of self-negation. There is no ground outside species to account for the contradictory nature of its specific existence; and in fact, because of that, its specificity is radically contingent and, precisely in that sense, it is irrational (cf., Heisig 2016b, pp. 275–276). Genus represents the ideality or rationality of existence; and yet since it cannot exist in and of itself, it must be revealed through various forms of species as the intermediation of individuals. Species serves as a place in which the interrelation of individuals can concretize the ideality of genus. Lastly, the individual takes on the double directionality of species, thereby attaining freedom. On the one hand, it can choose to affirm itself over against all the others (including both individual and species) and, on the other, it can negate itself to fully realize its nature as a contradictory existence. What Tanabe has in mind as the act of self-negation in [the] individual is a kind of selfless giving. This leaves room for other individuals to freely determine themselves and, through this agapeic kind of self-negation, it constitutes an open community of selfless individuals as the ideal manifestation of genus in species.

Let us see, then, how this triadic logic of species might guide our understanding of a lion as an instance of *panthera leo*. Tanabe's framework of thinking does not frame species (*leo*) as a constitutive form (i.e., lioness) that gives a unifying intelligibility to the individual being (whereby it serves as the ontological ground of its existence as a lion). Rather, it represents the irrational substrate of life that transmits its ontological energy to the individual, the power of existence that is torn between two contradictory movements towards its continuation and discontinuation. That is to say, every species has a natural propensity to encourage procreation and to preserve its continuous whole beyond the death of individuals. Yet what makes it a specific whole is the principle of self-negation that limits itself in relation to other species and ultimately negates its existence. It is ultimately finite and there is no species that can escape extinction.

Additionally, in reference to other species in the same genus (e.g., jaguar or *panthera onca*), there is no reason why the individual has to be a lion rather than a jaguar. The notion of species in Tanabe's account identifies radical contingency with a specific way of being vis-à-vis all the other possible ways. Not only could a lion have been a jaguar or any other species of *panthera*, we cannot give any rational ground or teleological reasons for prioritizing one species over the others. When they are properly understood as various manifestations of the same genus, we ought to see them as being equal to each other (without disregarding their differences).

The notion of *panthera* as genus cannot refer to any concrete existence in nature but demonstrates an idea that ought to be manifested through different species. In this case, the idea of genus enjoys rationality and necessity, unlike species: hence, this seems to imply, at least in Tanabe's account, that its extinction (as we are now seeing, for instance, in the case of *Beatragus*) should be avoided at all cost. However, note that there cannot be any direct preservation of genus and that there can only be proper or improper manifestations of it through various forms of species. This implies a paradox: viz., a series of improper manifestations of genus through species that primarily focuses on its self-affirmation or self-preservation (at the cost of engendering other species) could lead to an extinction of the genus.

The triadic logic of species, moreover, indicates that genus represents the kenotic principle of absolute negation. This is to say, genus has an inexhaustible wealth of potentiality and, through emptying itself, manifests its ontological richness through various forms of species in reality. What makes this kenotic movement of genus possible is also the inter-subjective behaviors of individuals that have a visible impact on the (trans)formation of their species. A species, according to Tanabe, cannot continue its existence when each of its members puts itself above the other members, or when the species as a whole dominates other species with the same principle of self-determination. In both cases, the delicate balance of the ecosystem that generally guarantees a sustainable supply to the species' demand for its survival is put under enormous pressure. For the species to fulfill its mediatory function as the place in which genus can manifest its rich content through the inter-relational behaviors of individual animals, then, each individual animal must exhibit a kind of self-less behavior and dedicate itself to the well-being of others. As a result, the interrelation of multiple species must also exhibit the kind of behavior that mirror the self-negating self-realization of genus and individual.

Finally, this leads us to the nature of the intermediation between multiple species and enables us to provide a Tanabean theory of evolution. Tanabe refers to Ranke's theory of history in contrast with the Hegelian notion of it in the EHA. In this essay, he discusses the nature of each age (i.e., species) in relation to the notion of eternity (i.e., genus). Tanabe concurs with Ranke's criticism that the Hegelian model of history as "rational progress" has a problematic tendency to relativize each age as a moment for the development of the next one, thereby reducing the significance of each age to an instrumental value for the self-affirmation of the whole of history (Tanabe 1963k, pp. 104–105). The rational theory of history, in other words, comprises a teleological view of development in which each age serves as a particular means for the self-determining whole of universal humanity (i.e., the absolute self-knowing of the Spirit and/or the State).

Contra Hegel, Ranke provides a communal picture of world history where the particular (i.e., each age) has access to the absolute (i.e., eternity) without instrumentalizing the other specific age by determining itself to be the absolute (von Ranke, 1906, pp. 15 – 17). Tanabe builds on Ranke's philosophy of history in the following passage:

> With regard to the interrelation of particular ages, if each age is truly in touch with divine eternity, it does not remain as a mere particular but since it has to be the particular in divine universality, there is no longer the struggle [among the particular ages] for the power… But rather there must govern peace and harmony in which each age respects the particularity of the other and acknowledges the dignity of the other for the other. Also, since one age necessarily undergoes its transition to another, as each of them maintains its absolute and independent status, the new age should be able to hold its superiority to the old one in terms of order not through the mechanical relations of power, but by transcending this relation. This means that we can avoid simply placing the particular ages in parallel relation to each other, but place them in order of succession. Here we find the characteristics of the relations among ages. We will be able to find the path to acquire the concept of development, which consistently maintains the absolute status of each age as the relative relation of the ages that follow each other, without including the propensity to instrumentalize each of them for the ultimate purpose in the concept of progress. (Tanabe 1963k, 109)[21]

What we find here is more than a stark contrast between the history of progress in the Hegelian model that affirms the worth and value of one age at the expense of all the previous ones and an alternative theory (*à la* Ranke) that sees history as a multivocal development affirming the worth and value of each age for its own sake.[22] Tanabe offers the reflection that genus must consist of the transformative intermediation of species, and that this intermediation must be a kind of kenotic service that grounds the self-negating self-transcendence among multiple species. It certainly surpasses the limit of an instrumental relation between them.

---

[21] For Tanabe's view of history, consult also the essay published in June 1940, "Historical Reality" (Tanabe 1964). In pp. 161–163, he clearly gives up the notion of history as a linear development and describes it as a series of concentric circles that never close on themselves but are constantly getting wider. For the clearest exposition of this essay and Tanabe's notion of history, see also Goto-Jones 2008, p. 11.

[22] This reflection alone should enable us to conceive of the history of philosophy as the history of world philosophies where Tanabe's framework of thinking must be placed in an open community with other philosophical frameworks of thinking from various intellectual traditions.

As we have seen above, each species is radically contingent in the sense that we cannot rationally prioritize the value of one species over the other, while, at the same time, the logic of species (inherent in Tanabe's understanding of history as the open mediation of various ages) demands that they be respected as having equal dignity and equal ontological value. The transformation of species from one to another in the same genus, moreover, is no longer seen as a sort of progress by means of which the most recent conquers the previous. Rather, they are seen as equal manifestations of the ontological richness inherent in the notion of the genus. Tanabe makes the qualification that one species can play a leading role in this business of unpacking the inexhaustible ideality of genus. But, strictly speaking, in reference to the logic of species, it only means that the leading species (which is often the latest one) carries more potentiality to demonstrate the wealth of genus in reference to all the other species. This is not equivalent to the sense of advancement which we see in the case of the notion of "rational progress."

In order to realize the transformative change from one species to another and/or embody the open mediation of multiple species as the particular manifestation of genus, the individual must practice a kind of self-negation by acknowledging that genus and species are always related to each other. An individual animal, in this case, is no longer seen merely as dominating other individuals for its own survival or insisting on the survival of its own species at the cost of other species. Except for the purpose of fulfilling its nature as a singular, the individual animal ought to be dedicating its life selflessly to the well-being of other individuals and constituting an open community of individuals. Only in that case, can they fully manifest the ideality of genus through their specificity in species.

Picturing the selfless behaviors of animals is quite problematic. In terms of social ontology, Tanabe uses the term "agape" and ascribes the "act of love" to the notion of genus or, more particularly, the absolute mediation of nothingness. Examples are much more available in relation to human activities than to animal behavior. The best philosophical reflection on animality that we can use here, I think, would be George Bataille's analysis in *The Accursed Share: An Essay on General Economy* (1949) and the *Theory of Religion* (1973). In both texts, Bataille argues that animals are much more in rapport with the ways in which the world *both* brings forth *and* consumes life as excessive energy than humans. Bataille, like Tanabe, understands "existence as a contradiction" (Tanabe 1963n, p. 262). Bataille believes that the world is constituted by energy that both enables us to be, and to be "squandered without reciprocation" (Bataille 1988, p. 38). Human beings, with their "sovereignty in the living world" (Bataille 1988, p. 23), according to the French philosopher, constitute a strange and self-alienated apex in the global economy of excessive energy. Stated otherwise, human beings are equipped with self-consciousness that allows them to make object-subject distinctions in the face of given reality; and, because of that, they cannot fully immerse themselves in the world (Bataille 1989, p. 45). By constantly ascribing (supposedly undying) meanings to all that is (including their own existence), they often fail to live their lives in accordance with the irrational and self-contradictory principle of life and must willy-nilly follow in the end the destiny of its self-negation.

Unlike humans, Bataille argues, every "animal is *in the world like water in water*" (Bataille 1989, p. 19).[23] It does not distinguish itself from other individuals. Nor are animals capable of reflecting on their belonging to a specific species (Bataille 1989, pp. 18–19). As a result, an animal never breaks off from the world by positing things in front of it as the objects standing over against itself as the subject. Animals, in this sense, just eat, sleep, procreate, and die. Without pondering why or questioning the significance of their finitude, they gloriously follow the order of the living world. Like water in water, they are in union with the world that generously gives being to all that is and takes it away from everything without reciprocation. Tanabe's account of animals would be a bit more nuanced than Bataille's in the sense that, by defining species as the self-contradictory substrate of life, we can acknowledge a degree of self-negating freedom that we can ascribe to different species.[24] Nonetheless, just as it is difficult to ascribe the same degree of freedom to animals as to humans, it is difficult to imagine animals as purposefully setting out to dominate other species or singling themselves out from the rest of their species beyond the basic need of their individual survival. Perhaps, as Bataille puts it, animals are much more capable of letting go of their ego than humans, such that they can gracefully fall into the earth to die and give room for others to bear much fruit.

---

[23] See also Bataille 1989, pp. 23–25.

[24] Given that species represents the self-negating principle of life, we cannot accept in a Tanabean philosophy of nature any animals that are completely free from the double directionality of existence. Having said that, it is clear that one species is much more capable of granting freedom to its individual members than another and it is reasonable to acknowledge a degree of freedom that we can ascribe to different species.

## Concluding Remarks: Future Challenges for the Tanabean Philosophy of Nature

This paper has provided a methodological reflection on the possibility of a philosophy of nature in Tanabe's logic of species, and applied it to our basic understanding of the terms genus, species, and individual in reference to nature. I have demonstrated that the logic of species in Tanabe's account is relevant to the field of philosophy of nature and, moreover, that its application to the biological concepts of species can allow us to generate a particular understanding of the natural world that fits with Tanabe's metaphysics. I do not wish here to repeat here what has been discussed at length in the preceding text. However, in this concluding section, I would like to point out what remains unanswered at the end of this exploration, and what opens the way to one possible future development of Tanabean philosophy.

We have seen how the logic of species in Tanabe could bring forth a distinctive understanding of the relevant terms in biology. However, we have not been able to discuss whether or not Tanabe's ontology brings any structural change to the ways in which biologists formulate these terms and, if that were the case, how we would have to understand nature or natural beings differently from the way they are understood in natural science. Neither has our approach to Tanabe's thirteen essays on species provided any determinate solutions to the discussions surrounding "species" in the contemporary philosophy of science. The simple application of the logic of species to the biological terms used in the most general sense can test the insight provided in Tanabe's works. However, we cannot yet deny the possibility of an either/or: either that Tanabe's articulation of species will call for the necessity of overhauling the terms used in natural science, or that it only ontologically grounds their legitimacy. We have to test the legitimacy of the biological understanding in light of Tanabe's insight at some point, and further investigate whether or not this philosophical evaluation would provide any worthwhile approaches to the problems discussed in the philosophy of biology.

Given the lack of critical translations and the insufficiency of Tanabe scholarship on the theme of the logic of species, I believe that this chapter has sufficiently laid out the methodological foundation for the possibility of reading Tanabe in the context of philosophy of nature. It also provides an adequate introduction to the basic concepts of his logic of species. What we have seen in this chapter, however, is the possibility for a conversation between Tanabe and with philosophers of science (especially in the context of biology). I would like to close this chapter with an anticipation of this evolution in our reading of Tanabe's philosophical texts on metaphysics and the problems of social ontology.

# References

Bataille, Georges (1988): *The Accursed Share: An Essay on General Economy*. Vol. 1. Robert Hurley (Trans.) New York: Zone Books.
Bataille, Georges (1989): *Theory of Religion*. Robert Hurley (Trans.) New York: Zone Books.
Goto-Jones, Christopher (2008): "The Kyoto School and the History of Political Philosophy: Reconsidering the Methodological Domination of the Cambridge School". In *Re-Politicising The Kyoto School As Philosophy*. Christopher Goto-Jones (Ed.) London; New York: Routeledge, pp. 3–25.
Heisig, James W.; Kasulis,Thomas P.; Maraldo, John C. (2011): *Japanese Philosophy: A Sourcebook*. Honolulu: University of Hawai'i Press.
Heisig, James W. (2016a): "Tanabe's Logic of the Specific the Critique of the Global Village". In *Much Ado About Nothingness: Essays on Nishida and Tanabe*. Nagoya: Chisokudō Publications, pp. 261–293.
Heisig, James W. (2016b): "Tanabe's Logic of the Specific and the Spirit of Nationalism". In *Much Ado About Nothingness: Essays on Nishida and Tanabe*. Nagoya: Chisokudō Publications, pp. 295–340.
Tanabe, Hajime (1963a): "From the 'Time-Scheme' to the 'World-Scheme'". In *Tanabe Hajime Zenshū*. Vol. 6. Nishitani Keiji, Shimomura Toratarō, Karaki Junzō, Takeuchi Yoshinori, Ōshima Yasumasa (Ed.) Tokyo: Chikuma Shobō, pp. 1–50.
Tanabe, Hajime (1963b): "The Logic of Social Existence". In *Tanabe Hajime Zenshū*. Vol. 6. Nishitani Keiji, Shimomura Toratarō, Karaki Junzō, Takeuchi Yoshinori, Ōshima Yasumasa (Ed.) Tokyo: Chikuma Shobō, pp. 51–168.
Tanabe, Hajime (1963c): "The Logic of Spieces and the World-Scheme". In *Tanabe Hajime Zenshū*. Vol. 6. Nishitani Keiji, Shimomura Toratarō, Karaki Junzō, Takeuchi Yoshinori, Ōshima Yasumasa (Ed.) Tokyo: Chikuma Shobō, pp. 169–264.
Tanabe, Hajime (1963d): "Three Stages of Ontology". In *Tanabe Hajime Zenshū*. Vol. 6. Nishitani Keiji, Shimomura Toratarō, Karaki Junzō, Takeuchi Yoshinori, Ōshima Yasumasa (Ed.) Tokyo: Chikuma Shobō, pp. 265–298.
Tanabe, Hajime (1963e): "The Social-Ontological Structure of the Logic". In *Tanabe Hajime Zenshū*. Vol. 6. Nishitani Keiji, Shimomura Toratarō, Karaki Junzō, Takeuchi Yoshinori, Ōshima Yasumasa (Ed.) Tokyo: Chikuma Shobō, pp. 299–396.
Tanabe, Hajime (1963f): "Responding to Criticisms of the Logic of Species". In *Tanabe Hajime Zenshū*. Vol. 6. Nishitani Keiji, Shimomura Toratarō, Karaki Junzō, Takeuchi Yoshinori, Ōshima Yasumasa (Ed.) Tokyo: Chikuma Shobō, pp. 397–446.
Tanabe, Hajime (1963g): "Clarifying the Meaning of the Logic of Species". In *Tanabe Hajime Zenshū*. Vol. 6. Nishitani Keiji, Shimomura Toratarō, Karaki Junzō, Takeuchi Yoshinori, Ōshima Yasumasa (Ed.) Tokyo: Chikuma Shobō, pp. 447–522.
Tanabe, Hajime (1963h) "A Commentary". In *Tanabe Hajime Zenshū*. Vol. 6. Nishitani Keiji, Shimomura Toratarō, Karaki Junzō, Takeuchi Yoshinori, Ōshima Yasumasa (Ed.) Tokyo: Chikuma Shobō, pp. 523–536.

Tanabe, Hajime (1963i): "The Limit of Existential Philosophy". In *Tanabe Hajime Zenshū*. Vol. 7. Nishitani Keiji, Shimomura Toratarō, Karaki Junzō, Takeuchi Yoshinori, Ōshima Yasumasa (Ed.) Tokyo: Chikuma Shobō, pp. 1–24.
Tanabe, Hajime (1963j): "The Logic of State Existence". In *Tanabe Hajime Zenshū*. Vol. 7. Nishitani Keiji, Shimomura Toratarō, Karaki Junzō, Takeuchi Yoshinori, Ōshima Yasumasa (Ed.) Tokyo: Chikuma Shobō, pp. 25–100.
Tanabe, Hajime (1963k): "Eternity, History, and Act". In *Tanabe Hajime Zenshū*. Vol. 7. Nishitani Keiji, Shimomura Toratarō, Karaki Junzō, Takeuchi Yoshinori, Ōshima Yasumasa (Ed.) Tokyo: Chikuma Shobō, pp. 101–170.
Tanabe, Hajime (1963l): "Ethics and Logic". In *Tanabe Hajime Zenshū*. Vol. 7. Nishitani Keiji, Shimomura Toratarō, Karaki Junzō, Takeuchi Yoshinori, Ōshima Yasumasa (Ed.) Tokyo: Chikuma Shobō, pp. 171–210.
Tanabe, Hajime (1963m): "The Development of the Concept of Existence". In *Tanabe Hajime Zenshū*. Vol. 7. Nishitani Keiji, Shimomura Toratarō, Karaki Junzō, Takeuchi Yoshinori, Ōshima Yasumasa (Ed.) Tokyo: Chikuma Shobō, pp. 211–250.
Tanabe, Hajime (1963n): "The Dialectic of the Logic of Species". In *Tanabe Hajime Zenshū*. Vol. 7. Nishitani Keiji, Shimomura Toratarō, Karaki Junzō, Takeuchi Yoshinori, Ōshima Yasumasa (Ed.) Tokyo: Chikuma Shobō, pp. 251–372.
Tanabe, Hajime (1963o) "A Commentary". In *Tanabe Hajime Zenshū*. Vol. 7. Nishitani Keiji, Shimomura Toratarō, Karaki Junzō, Takeuchi Yoshinori, Ōshima Yasumasa (Ed.) Tokyo: Chikuma Shobō, pp. 373–386.
Tanabe, Hajime (1964): "Historical Reality". In Tanabe Hajime Zenshū. Vol. 8. Nishitani Keiji, Shimomura Toratarō, Karaki Junzō, Takeuchi Yoshinori, Ōshima Yasumasa (Ed.) Tokyo: Chikuma Shobō, pp. 117–170.
Tanabe, Hajime (2016), *Philosophy as Metanoetics*. Takeuchi Yoshinori, Valdo Viglielmo, and James W. Heisig (Trans.). Nagoya: Chisokudō Publications.
von Ranke, Leopold (1906): Über die Epochen der neueren Geschichte: Vorträge dem Konige Maximilian II. von Bayern im Herbst 1854 zu Berchtesgaden Gehalten. Leipzig: Duncker & Humblot.

Andrea Gambarotto
# Teleology, Life, and Cognition: Reconsidering Jonas' Legacy for a Theory of the Organism

**Abstract:** This chapter takes up Jonas' philosophical legacy as a site for developing a scientifically viable theory of the organism as a 'natural purpose.' It follows a suggestion by the late Francisco Varela that we need to move beyond the unstable position set out by Kant in the Critique of Judgement in order to provide a novel understanding of biological individuality. The chapter explores the theoretical potential of this claim by investigating the role the philosophy of Hans Jonas could play in helping us develop a naturalistic philosophy of biology that rejects both reductive physicalism and the metaphysical idea of teleology as intelligent design but still takes teleology seriously as a natural phenomenon.

> The lobster, in other words, behaves very much as you and I would behave if we were plunged into boiling water [...] Standing at the stove, it is hard to deny in any meaningful way that this is a living creature experiencing pain and wishing to avoid the painful experience. To my mind, the lobster's behavior in the kettle appears to be the expression of a *preference;* and it may be well be that an ability to form preferences is the decisive criterion for real suffering.
>
> David Foster Wallace, *Consider the Lobster*

## 1 Consider the Bacterium

This chapter takes up Jonas' philosophical legacy as grounds for developing a scientifically viable theory of the organism as a 'natural purpose,' following a suggestion of the late Francisco Varela. In his very last paper, published posthumously in collaboration with Andreas Weber, Varela argued for the "great need to bring to the fore the remarkable and recent convergence between the re-awakening of the philosophical discussion concerning natural purposes (with Jonas as the central figure), and an independent but convergent stream of thought concerning biological individuality and the organism (with the autopoiesis school

---

I would like to thank Luca Corti for reviewing the draft of this manuscript.

**Andrea Gambarotto,** UC Louvain

https://doi.org/10.1515/9783110604665-012

as the central figure)." The conceptual proximity of these developments led Weber and Varela to conclude that, "after two centuries, we can move *beyond* the unstable position set out by Kant in the *Critique of Judgement*, and therefore provide a fresh re-understanding of natural purpose and living individuality" (Weber and Varela 2002, pp. 97–98).

The 'unstable' Kantian position to which they refer is located in the tension inherent in Kant's understanding of teleology. On the one hand, Kant considered mechanism – i.e., the mere reference to efficient causes – insufficient to explain the peculiar features of living systems, notably their ability to self-organize and self-reproduce. This "mechanical inexplicability" (Ginsborg 2004) led Kant to the idea that, in order to make sense of vital organization, we must necessarily employ teleological concepts, i.e., consider "organized beings" as the product of a teleological principle that acts "for the sake of something." On the other hand, Kant was not inclined to ascribe teleological features to nature per se, and thus famously argued that natural purposiveness should not be understood as a constitutive feature of "organized beings" themselves but rather as a mere "regulative principle" of reason – a specific way that our power of judgment makes sense of the specific class of objects constituted by living organisms.[1]

This tension was rooted in Kant's Wolffian understanding of teleology (van den Berg 2013). For Christian Wolff, who coined the term 'teleology' in the early eighteenth century, the notion referred to the area of physics addressing the "ends of things." It was thus a "theologia experimentalis" whose function was to make the wisdom of God comprehensible through the study of nature. This kind of physico-theological argument was untenable to Kant and yet he could not find a way to conceive of natural-intrinsic purposiveness without ultimately referring to intentional-extrinsic purposiveness as a fundamental model, which led him to the "unstable position" criticized by Varela and Weber (among others). In a way, we could argue that the Wolffian understanding of teleology as "extrinsic purposiveness" that Kant albeit ambivalently subscribed to led him to conceive of teleology as necessarily dependent on the conscious cognition of a human-like agent, and therefore on something not found *in nature* or supported by objective scientific explanation as he defined it.

Varela's invitation to move beyond Kant stems largely from the need he felt later in life to "take teleology seriously" as a natural phenomenon.

---

[1] The literature on the topic is massive, as a general orientation cf. Breitenbach 2009; Ginsborg 2006; Goy and Watkins 2014; Huneman 2007, 2017; Kreines 2005; McLaughlin 1990; Richards 2000; Steigerwald 2006; Zammito 2006, 2012; Zuckert 2007; Zumbach 1984.

According to Varela and Weber, the general shift in our understanding of teleology has recently become possible due to the "remarkable convergence" between the biological theory of "autopoiesis" and the philosophical biology of Hans Jonas, rooted in the idea of a strong continuity between life and mind. The aim of this chapter is to explore the theoretical potential of this claim by asking the following question: *what role (if any) could the philosophy of Hans Jonas play in the development of a naturalistic philosophy of biology that would both reject reductive physicalism and the metaphysical idea of teleology as intelligent design but still "take teleology seriously" as a natural phenomenon?* The answer that I will sketch out is that many of the ideas Jonas formulated in the 1960s and 1970s easily lend themselves to a "naturalized" conception of organic nature, albeit one that diverges from the dominant framework of philosophical naturalism.

Naturalism is generally accepted as the dominant metaphysics of our time. As Barry Stroud explains, "naturalism seems to me in this and other respects rather like 'World Peace.' Almost everyone swears allegiance to it, and is willing to march under its banner. But disputes can still break about what it is appropriate or acceptable to do in the name of that slogan" (Stroud 2004, p. 22). Most important among such disputes is the question of what can legitimately be considered "nature" and how the realm of nature in general relates to what is specifically human, including phenomena such as intentionality or values.

These are far from being recent issues. One classical example is the Kantian distinction, formulated in the *Critique of the Power of Judgment*, between "concepts of nature" and "concepts of freedom", or rather, between that which is pure mechanism and that which pertains to human normativity and therefore responds to different (one would be tempted to say "super-natural") principles. This polarity has a long history in modern philosophy that continued throughout the twentieth century: one is reminded of Wilhelm Dilthey's distinction between *Naturwissenschaften* and *Geisteswissenschaften*, or the stark opposition Martin Heidegger established in his 1929/30 course on *The Fundamental Concepts of Metaphysics* between the animal who is "poor in world" (*Weltarm*) and humans who are "world-forming" (*Weltbildend*). This philosophical framework generates a strong antithesis between the natural and the human, perpetuating the old Cartesian distinction between machine-like, soul-deprived animals and human beings who are uniquely endowed with an immaterial principle of thought.

Interestingly, this dualism is not confined to continental philosophy and transverses the analytic tradition as well: the post-Sellarsian distinction between the "realm of laws" and the "space of reasons" is a case in point. Once again, this distinction institutes a strong opposition between nature, which works according to blind mechanical laws, and the world of human beings, characterized by something inaccessible to the methods and practices of the natural sciences.

Mario De Caro and David Macarthur have defined this issue as the "placement problem", which emerges from trying to determine where and how to locate "normative" phenomena usually associated solely with human beings within the scientific image offered by the natural sciences.

Two paths seem available to the naturalist philosopher seeking to resolve this dilemma: the first is to get rid of normative language altogether by showing that normative statements can be entirely reduced to non-normative ones: this is the position De Caro and Macarthur describe as "scientific" or "strong" naturalism. Another solution, the one Caro and Macarthur argue for, is a more "liberal" naturalism able to accommodate specifically human phenomena such as intentionality and values within the "scientific image" of the world: a "non-reductive form of naturalism and a more inclusive concept of nature than any provided by the natural sciences" (De Caro and Macarthur 2004, p. 1). As Lenny Moss and Daniel Nicholson have recently emphasized, the fundamental problem with such a philosophical enterprise is that "sailing this middle-course is easier said than done," because we still lack a coherent philosophical framework that would allow us to rethink "the relationship of normativity to nature *all the way down*" (Moss and Nicholson 2012, p. 88) to the simplest forms of biological organization. We thus lack the tools to ground the features we usually consider the exclusive prerogative of *human* beings with respect to those that make us "natural", and above all, *living* beings.

It is from this general background that I want to address the problem of natural teleology. Indeed, the question of whether something such as "intrinsic purposiveness" fits the philosophical framework of naturalism can be understood as an exemplary case of the general issue described above. We, much like Kant, are used to considering purposive activity an exclusive feature of human conscious action: to account for the fact that I am going to the kitchen, I provide the reason that, since I am hungry, I am aiming for the fridge *in order to* get a sandwich. Although the feeling of hunger may be identified as the *cause* of my going to the kitchen, it is not a cause in the same sense that gravity pushing a sphere down an inclined plane is a cause. Rather, it is a *drive*, mediated by a sense of need that induces me to make a specific decision through a form of conscious deliberation.

Even though the relation between the feeling of need and the deliberation to act in a certain way implies a lesser degree of mediation, it would not be entirely inappropriate to extend the same kind of reasoning to non-human animals: "if I observe my cat jumping down from the windowsill and going into the kitchen, and I know that the kitchen is where the milk bowl is located, then I may infer the reason why she went into the kitchen: namely, to get a drink of milk" (Barham 2012, 94). Or even further, in the now classic example: "consider a bacterium swimming upstream in a glucose gradient, its flagellar motor rotating. If we naively ask, 'What is it doing?' we unhesitatingly answer something like, 'It's going to get dinner.' That is, without attributing consciousness or conscious purpose, we view the bacterium as acting on its own behalf in an environment. The bacterium is swimming upstream in order to obtain the glucose it needs" (Kauffman 2000, p. 7; cf. also Varela 1991; 1997). In this sense, both the cat and the bacterium can be defined as characterized by a certain degree of *agency*, each conceived at least in a very general way as "acting on its own behalf".

This is precisely the main point of Jonas' philosophical biology, whose "statement of scope expresses no less than the contention that the organic even in its lowest forms prefigures mind, and that mind even on its higher reaches remains part of the organic" (Jonas 1966, p. 1). In other words, the main object of Jonas' philosophy of life is the reciprocal co-implication of life and cognition, which provides the theoretical grounding for a naturalized conception of biological purposiveness. It therefore overcomes the "antinomical" stance that has characterized our understanding of teleology at least since Kant, according to which proper scientific explanation of natural phenomena can be attributed to mechanical laws but with a specific class (living organisms) seeming to display features (self-organization, homeostasis, self-maintenance) that apparently exceed that form of explanation.

Contemporary philosophy of biology seems held hostage by a similar "unstable position": with, on the one hand, the "phenomenology" of living systems providing a rich array of cues that the behavior of living organisms should be understood as intrinsically goal-directed, and one the other a persistent commitment to Kant's rejection of goal-directedness as a constitutive feature of biological organization. This antinomy is best represented by J.B.S. Haldane's iconic characterization of teleology as a mistress a biologist cannot live without, but with whom he is unwilling to be seen in public (Mayr 1974, p. 115).

Jonas' philosophical biology can be understood as an attempt to find a way out of precisely this antinomy – not by supressing the "testimony of life" but rather by accepting it "as a call to a revision of the conventional model of reality inherited by a natural science which may be well passing beyond it" (Jonas 1966, 2). I believe this statement to be especially true today, when it can be fruitfully re-read against the background of recent calls for a more liberal, "normative" naturalism (De Caro and Macarthur 2010; Moss and Nicholson 2012), the current turn towards an "extended evolutionary synthesis" (Pigliucci and Müller 2010; Laland et al. 2014; Walsh 2015; Huneman and Walsh 2017), and the return of the organism as a key concept in biological research (Huneman and Wolfe 2010; Nicholson 2014).

This chapter is divided into four sections. In these introductory remarks, I have laid out the general stakes of my analysis of Jonas' philosophical program, namely the question of natural teleology in the context of contemporary naturalism's persistent Cartesian dualism between machine-like, world-poor, soul-deprived animals and human beings endowed with a "disembodied" principle of cognition. In Section 2, I outline Jonas' analysis of early-modern natural philosophy's materialist/mechanist account of life and disembodied account of cognition. Following Jonas' mobilization of the Leibniz-Stahl debate as the point of emergence for an alternative notion of soul, I elaborate how the concept of soul as a dynamic principle of vital organization developed in that exchange accounts for the forms of adaptivity and agency specific to biological systems. Taking up Jonas' analysis of materialism in early-modern natural philosophy, I address this issue from a historical point of view by asking the apparently metaphysical question "do animals have souls"?. The key figure in this historical analysis is the eighteenth-century physician Georg Ernst Stahl, who, though commonly portrayed disparagingly as a champion of immaterial vital principles, can be re-interpreted (on the background of Jonas' historical account) as the first endorser of a form of "vital-materialism" during the heyday of mechanical natural philosophy. I elaborate how Stahl framed "soul" as just another name for the dynamic organization of a living body, one that accounts for the typical autonomous features absent from non-organized bodies.

In Section 3, I develop this thread by focusing on Jonas' assessment of cybernetics and system theory. I argue that Jonas' philosophical biology serves as a useful springboard for overcoming a machine theory of the organism and challenging reductive physicalism, and thus opens the way for a naturalized account of allegedly "human" phenomena such as values, meaning, and normativity. Ultimately, however, it steps into "super-natural" territory, beyond the proper scope of natural science, through excessive emphasis on "inwardness" and on the first-person experience of "what it is like to be an organism". This turns Jonas' philosophical program into a scientific dead-end. Still, freed from such "biochauvinist" (Di Paolo 2009; Wolfe 2014) excesses, Jonas' legacy has much to offer those interested in a naturalized account of sense-making and intentionality.

## 2 Do Animals Have Souls?

In this section I address the question: "do animals have souls?" The answer I will try to develop in the following pages is that it depends on what we understand the notion of soul to mean and how we construe its relation to the phenomenon of life. More precisely, I will compare two models of the soul: the Cartesian conception of soul as an immaterial *res cogitans* – the critique of which motivated a large part of Jonas' research trajectory[2] – and the idea of soul as the principle of unity, or vital organization, of a living system allowing for a fundamental distinction between inert and living matter.

In a recent article, Charles T. Wolfe points out that "life is a controversial topic for the eighteenth century, not the seventeenth," and thus that life was not a major object of debate during the phase in the history of ideas we have come to define as the "scientific revolution".[3] In fact, it is not until the eighteenth century that "the existence of living beings suddenly again becomes an explanatory challenge or even a scandal" (Wolfe 2010, p. 189). This "scandal" was largely the result of the mechanistic scientific paradigm dominant at the end of the seventeenth century, which was particularly associated with figures such as Descartes and LaMettrie, and which considered organic bodies ultimately as "machines de la nature". This in fact was Leibniz's definition of an organism. Despite rejecting a simple (Cartesian) equation between living bodies and human-made machines – since, unlike the latter, living bodies are perfectly organized in their "most minute parts" – Leibniz nonetheless developed an account for living matter dependent on the heuristic of the machine.

It is my contention that what we tend to view as early-modern "mechanism" in fact implies at least two different (partially overlapping) theses: 1) an "ontological thesis" according to which the universe in general, and organized bodies in particular, must be understood as complex machines resulting from God's intelligent design, and 2) an "epistemological thesis", according to which legitimate explanation of nature, conceived as mere extended matter (*res extensa*), should exclusively rely on efficient causation that dispenses of all reference to "substantial forms" or "entelechies".[4] The exclusion of such Aristotelian con-

---

**2** Scholarship on Jonas usually divides this trajectory into three main phases: a first phase focused on historical studies of ancient gnosis, largely influenced by Heidegger's phenomenology of religion; a second phase dedicated to the development of a philosophy of the organism; and a third phase occupied by the question of the ethics of responsibility. The phase which will occupy our considerations in this chapter is the second one. For a detailed historical and philosophical reconstruction of Jonas' philosophy of the organism cf. Lories and Depré 2003; Franzini Tibaldeo 2009. For a Jonasian theoretical approach with an eye toward contemporary issues in philosophy of biology cf. Franzini Tibaldeo 2015; Michelini, Wunsch and Stederoth 2018.
**3** The term was notably introduced for the first time by Alexandre Koyré (1939) and later popularized by Herbert Butterfield (1959).
**4** These concepts play a complex and largely debated role in the history of Aristotelianism. In general terms, "substantial form" defines a sort of "platonic" universal embodied in nature, a principle that accounts for the overall organization of an entity and brings it towards the realization of its full potential, i.e., what that entity is supposed to be. An example of this is the

cepts of intrinsic purposiveness is not only compatible with an "extrinsic" teleology for which purposiveness (both natural and artificial) is necessarily the result of intelligent design. Rather, the ontological thesis according to which organisms are machines moreover seems to presuppose extrinsic teleology as a fundamental part of its theoretical foundations.[5]

In this sense, while on the one hand early-modern natural philosophy, the ostensible highpoint of the mechanistic paradigm, worked to expunge teleological language from its vocabulary, it did so only with regard to natural-intrinsic purposiveness and by (more or less explicitly) presuming an intentional-extrinsic purposiveness to underwrite that mechanism. As Jonas writes, "the maxim concerns teleology as a causal mode of nature itself, or immanent teleology, and not transcendent teleology such as might have been exercised by the creator of the existing system of nature in once creating it as it is" (Jonas 1966, p. 34). From this perspective, following Jonas' lead, we can even say that "extrinsic" teleology is a conceptual prerequisite for the "mechanistic" world-view: if organisms are "machines of nature" (ontological thesis) whose internal workings are the result of efficient causation among their component parts (epistemological thesis), then there must be a watchmaker who designed the machine and programmed its functioning (extrinsic teleology thesis).[6]

This historical reconstruction of the early-modern view of mechanism and its philosophical implications is largely aligned with Jonas' argument that:

---

adult organism as the entity realizing the final stage of human development. Since this process is per se intrinsically teleological, this final state can be defined as "entelechy", or the complete realization of a goal (*telos*).

5 Georges Canguilhem has expressed the same idea in different words: "it may thus be said that, in substituting mechanism for the organism, Descartes effaces teleology from life, but he does so only in appearance, for he reassembles it, in its entirety, at his point of departure. Anatomical form substitutes for dynamic formation, but as this form is a technical product, all possible teleology is contained within the technique of production. [...] In short, with the Cartesian explanation, in spite of appearances, it may seem that we have not taken a single step outside finalism. The reason is that mechanism can explain everything so long as we take machines as already granted, but it cannot account for the construction of machines" (Canguilhem 2008, pp. 86–87).

6 This watchmaker analogy – spanning over two centuries, from Descartes to Paley – provides the springboard for a general philosophical argument about the relationship between mechanism (in both the ontological and the epistemological sense) and teleology (in the sense of extrinsic purposiveness) that implies a co-implication evident, for example, in most 'adaptationist' accounts of design in nature as a form of optimal adaptation (notably criticized by Gould and Lewontin 1979) and in conceptions of evolution as a 'blind watchmaker' (Dawkins 1986).

> The conception of a divine engineer (of supreme skills but inscrutable ends) was actually a requisite for the mechanistic world-concept itself during its most vigorous phase of growth. The real problem was with final causes as *modi operandi* of and in nature itself. Historically their rejection was part of the great struggle with Aristotelianism which marked the birth of modern science, and in this setting it was closely connected with the attack on 'substantial forms.' Regarding final causes, we must observe that their rejection is a methodological principle guiding inquiry rather than a statement of ascertained facts issuing from inquiry (Jonas 1966, p. 34).

This erasure of intrinsic purposiveness from the discourse on nature explains why life became a problematic notion in the early eighteenth century: without the goal-directedness that the ancients (and above all Aristotle) ascribed to nature qua nature, nature itself was stripped of life – becoming no more than pure dead matter.

Jonas understands the relation between ancient and modern metaphysics as a form of dialectical opposition: if the ancients conceived of matter itself, in all of its forms, as endowed with life (hylozoism) – a statement which needs further scrutiny, but which we can still accept here as a general guideline – modern thought since the "scientific revolution" has instead been based on an ontology "whose model entity is pure matter, stripped of all features of life" (Jonas 1966, 9). In this sense, "l'*homme machine* signifies in the modern scheme what conversely hylozoism signified in the ancient scheme: the usurpation of one, disassembled realm by the other which enjoys an ontological monopoly" (Jonas 1966, 11), and it is thus "after Cartesian or LaMettrian concepts of *bêtes-machines* or *hommes-machines* that Life becomes a locus of a kind of ontological crisis" (Wolfe 2010, p. 190).

One of the loci in which this crisis emerges with most clarity is the famous *Negotium otiosum* debate over the notion of organism that took place between Leibniz and Georg Ernst Stahl (1659–1734), a medical doctor and professor at the University of Halle. Normally given a bad reputation, Stahl's physiology has been associated with the derogatory label of "animism" since not long after the publication of the *Theoria medica vera* (1708) – evidence can already be found in Albrecht von Haller (1708–1777) and Caspar Friedrich Wolff (1733–1794) – and was "rarely read...let alone discussed in detail" and rarely "reprinted after the 19th century, nor retranslated" (Huneman and Rey 2007, p. 226) – though the tide has been changing in recent years (Carvallo 2004; Nunziante 2011; Duchesneau and Smith 2016).

Stahl's work fundamentally complements Jonas' historical account of the erasure of life from the discourse on nature in early-modern materialism. In fact, what Stahl states shocked him above all in contemporaneous physical theories of the human body is that "*Life* was never mentioned nor defined, and

I could find no logical definition provided" (Stahl [1706] 1859, vol. 2, p. 224; tr. Wolfe 2010, p. 205). In his recent book, John Zammito points out how "Stahl insisted that the actual character of organic life – in humans as well as in other living things – entailed an *immanent purposiveness* that could not be derived from merely physical-chemical processes" (Zammito 2018, p. 24). This immanent purposiveness was evident in the fact that, although the fundamental elements of vital organization naturally tend toward chemical degeneration and physical dissolution, living matter seems to have an inherent tendency to self-maintenance. Stahl famously attributed this tendency to the "soul" (*anima*).

The way we interpret "soul" here is paramount, and in fact the *Negotium otiosum* dialogue between Leibniz and Stahl can be read as a debate on the very meaning of this notion. Leibniz ultimately remained Cartesian when it came to substance dualism, arguing that the soul cannot be the principle behind the movement and self-maintenance of organic bodies, since extended matter can be modified only by the external action of other extended matter. Thus, for Leibniz the only possible relation between mind and body is the extrinsic parallelism instituted by God's "pre-established harmony". Accordingly, we can roughly say that Leibniz, not far from Descartes, conceived of the soul (*res cogitans*) as a purely immaterial principle of thought lacking any contact or relation whatsoever with extended matter (*res extensa*) and which only manifests in the human mind. But is this the only way we can conceive of the soul?

In his *Historisches Wörterbuch der Biologie*, Georg Toepfer makes a convincing case that, if we go back to the Greek term *psyché*, the Aristotelian (and, even more, the Galenic) notion of "soul" can be understood as a synonym for the "organization" of a body for several reasons:

> (1) Like the soul, the organization of a body is not itself a body, but belongs only to it; (2) soul and organization equally do not exist independently of a particular body; (3) soul and organization both designate the principle of unity and identity of a system; and (4) soul and organization can be understood as the principle of the movement of a body (Toepfer 2011, II: p.755).

In this sense, with the term 'soul' Stahl referred to can be understood as nothing other than a "supervening order in the object, its principle of organization" (Zammito 2018, p. 33).

Seen from this point of view, the debate between Leibniz and Stahl appears as a veritable *querelle des anciens et des modernes* concerning the notion of organism and the nature of life, with Leibniz defending a fundamentally modern understanding of the organism as a machine of nature designed by God and Stahl adhering to the ancient approach that sees living bodies as intrinsically purposive entities animated by a "naturalistically" conceived soul. If we interpret

Stahl's "soul" as an elaboration of this Aristotelian-Galenic tradition, the centuries-old image of Stahl as a mystical endorser of immaterial principles in medical science gets turned upside down. Indeed, if at first glance the *Negotium Otiosum* seems to position Stahl as a "conservative", metaphysical thinker in relation to the more "progressive", naturalist-minded Leibniz, this is largely because of the "dull, tedious, unnecessarily high-flown and abusively slow" (Huneman and Rey 2007, p. 226) style of Stahl's responses. Yet if we situate Stahl's animism in relation to the Aristotelian-Galenic understanding of soul as a "principle of organization", he instead emerges in a much different light, with a notion of soul that is much more "naturalized" than Leibniz's (largely Cartesian) immaterial *res cogitans*.

The considerations I have just laid out concern precisely what Jonas refers to as the "artificial sundering of *res cogitans* and *res extensa* in the heritage of dualism, with the extrusion between them of 'life'" (Jonas 1966, p. 21). This sundering "depended on the condition that 'soul' was banished from the context of nature, and this again was possible only insofar as in the explanation of nature in general soul was not needed as a *cause of motion*. In fact, in a nature without mind, 'soul' as a dynamic principle would be a source of irrationality and promote disorder rather than law" (Jonas 1966, p. 71). Although we find no detailed analysis of Stahl's work in Jonas' discussion of early modern natural philosophy, the role Stahl ascribed to the soul as a principle of motion in living organisms pairs strikingly well with Jonas' co-extensive understanding of life and mind.

As I hope this rough outline suggests, Jonas' thought offers an alternative perspective on the genealogy of the dominant metaphysics of our era. Contemporary naturalism (at least as understood in the strict physicalist sense criticized by De Caro and Macarthur) is largely a legacy of Cartesian dualism. As Jonas acutely points out, the only possible way for any metaphysics after early-modern natural philosophy to get beyond the Cartesian juxtaposition of mind and matter was a choice between them: while a minority of thinkers (among whom Jonas counts Berkeley on idealism and Leibniz on monadology) took the side of the former, most of early-modern natural science and philosophy strongly opted for the latter. Against this choice, materialism emerged as "the real ontology of our world since the renaissance, the real heir to dualism" (Jonas 1966, p. 20).

From this point of view, in which matter is the only "real" thing out there in the world, the question of whether the bacterium swimming in a glucose gradient could have something like a "soul" sounds purely metaphysical. On the other hand (provided that we understand the bacterium as an animal in a very broad sense) if we consider the "soul" the dynamic unity of a living body striving towards its own self-preservation, we can provide a thoroughly naturalized answer to the question: yes, animals, qua living beings, have souls. This implies however that we ascribe even the simplest form of biological organization a minimal degree of *cognition* and *agency* – or, as Jonas would put it: *freedom*.

## 3 A View From Within

Given the historical background provided in the previous section, I will now address the question of theoretical methodology. Or rather, *how do we understand and conceptually formalize the peculiar forms of adaptivity and agency that characterizes biological systems?* I will focus on Jonas' criticism of cybernetics and system theory as he develops a solution to this dilemma, which takes the form of a phenomenological-style "view from within".

In the mid-twentieth century, cybernetics offered the most coherent and influential framework for accounting for goal-directed behavior in organisms and artifacts alike – presenting itself as the science of "control and communication in the animal and in the machine" (Wiener 1948). As the title of Wiener's seminal work suggests, cybernetics largely developed as a machine theory of the organism that reserved the term "teleology" for a specific form of behavior regulated by a mechanism of negative feedback.

This "machine theory" is evident in the examples laid out by one of the seminal papers of the discipline, *Behavior, Purpose and Teleology* (Rosenblueth, Wiener, Bigelow 1943). Providing examples of non-feedback behavior, the authors refer to both a machine insensible to light programmed to interfere with a luminous object (and thus unable to modify its own behavior according to it) and a snake striking a frog with no visual report from the prey after its initial movement (Rosenblueth, Wiener, Bigelow 1943, p. 20). In contrast, "the behavior of some machines and some reactions of living beings involve a continuous feed-back from the goal that modifies and guides the behaving object" (Rosenblueth, Wiener, Bigelow 1943, p. 20).

Before turning to Jonas' analysis of cybernetics, I would like to return briefly to the theoretical analysis of the early-modern view of mechanism that, following Jonas' lead, I developed in the previous section. What is the relationship between this cybernetic conception of the machine and the one elaborated by early-modern natural philosophy? In the previous section, I identified two fundamental theses and a corollary: (1) an "ontological thesis", (2) an "epistemological thesis" and (3) the extrinsic teleology thesis. In what follows, I will sketch out how cybernetics adheres to each.

(1) The *ontological* thesis considered organisms as "machines of nature". In this respect, I emphasized some variation between a strict "Cartesian" model, according to which organisms are considered identical to man-made machines

except that they are designed by God, and a more "deflationary", chiefly Leibnizian, model that emphasized some differences between artifacts as "human machines" and organisms as "divine machines". I highlighted that the former constitute a simple assembly of parts while the latter are perfectly organized even in their "most minute parts". Cybernetics can read as a "Leibnizian" machine-theory of the organism, since its main defendants argue for a strong behavioral analogy between machines and living organisms while insisting on their structural differences in material composition.[7]

(2) The *epistemological* thesis holds that the physiology and behavior of living organisms should be explained solely in terms of efficient causations among component parts, without any reference to final causes. Although structurally involved in an investigation of the notion of purpose, cybernetics seems to subscribe to this thesis as well, since one of its main concerns is to detach the analysis of purposive behavior form any concept of final cause. Within cybernetics, purposive behavior is addressed by studying "servomechanisms" that work according to negative feedback-loops and do not imply an opposition to causal determinism.[8]

(3) If organisms are analyzed as analogous to machines whose behavior is regulated by servomechanisms, then what about the *extrinsic teleology* thesis? The answer to this question relies largely on how we understand natural purpo-

---

[7] "A further comparison of living organisms and machines leads to the following inferences. The methods of study for the two groups are at present similar. Whether they should always be the same may depend on whether or not there are one or more qualitatively distinct, unique characteristics present in one group and absent in the other. Such qualitative differences have not appeared so far. [...] While the behavioristic analysis of machines and living organisms is largely uniform, their functional study reveals deep differences. Structurally, organisms are mainly colloidal, and include prominently protein molecules, large, complex and anisotropic; machines are chiefly metallic and include mainly simple molecules" (Rosenblueth, Wiener, Bigelow 1943, pp. 22–23).

[8] "In classifying behavior, the term 'teleology' was used as a synonymous with 'purpose controlled by feed-back.' Teleology has been interpreted in the past to imply purpose and the vague concept of 'final cause' has been added. The concept of final causes has led to the opposition of teleology to determinism. A discussion of causality, determinism and final causes is beyond the scope of this essay. It may be pointed out, however, that purposefulness as defined here, is quite independent of causality, initial or final. Teleology has been discredited chiefly because it was defined to imply a cause subsequent in time to a given effect. When this aspect of teleology was dismissed, however, the associated recognition of the importance of purpose was also unfortunately discarded. Since we consider purposefulness a concept necessary for the understanding of certain modes of behavior we suggest that a teleological study is useful if it avoids problems of causality and concerns itself merely with the investigation of purpose" (Rosenblueth, Wiener, Bigelow 1943, pp. 23–24).

siveness to work. In the case of a typical cybernetic self-correcting device such as a thermostat, the notion of "purpose" refers to the "end state" of the device itself; in the case of the thermostat, the end state, programmed by an engineer, is the maintenance of a certain temperature, and it is obtained through an error-sensing mechanism.[9] If we apply the same line of reasoning to living organisms, we reach the conclusion that cybernetics, as the science of "control and communication in the animal and in the machine," surreptitiously implies the extrinsic teleology thesis, in the sense that teleological behavior necessarily implies previous intelligent design in the form of "pre-established structural arrangements" (Bertalanffy 1951, p. 353). Of course, the identity of the engineer is not so easily answered in this case as in early-modern natural philosophy, but it remains true that, from a philosophical point of view, the cybernetic theory of the organism still moves within the conceptual space of post-Cartesianism.

Masterfully describing the theoretical situation I just laid out, Jonas summarized this point by saying that "Darwinists, behaviorists, cyberneticians, adopt in effect the Cartesian position without its metaphysical cargo" (Jonas 1966, p. 91). Against this theoretical backdrop, his philosophical biology thus attempted to overcome the machine theory (outlined by the three mechanistic theses above) he saw defining our understanding of life since the scientific revolution. General system theory seemed to provide Jonas with a theoretical springboard towards that end: in 1951, he participated in a special issue of the journal *Human Biology* dedicated to general system theory as a new approach to the unity of science. The issue included three essays (introduction, conclusion, and an original article entitled "towards a physical theory of organic teleology: feedback and dynamics") by Ludwig von Bertalanffy, as well as comments by Carl Hempel, Robert Bass, and Jonas himself. This is an important document for assessing Jonas' position towards system theory as a possible response to cybernetics: a position characterized by significant alignment but also by the critique that system theory did not go far enough in counteracting materialism.

Although Jonas indubitably considered Bertalanffy's system theory a valuable asset for overcoming the post-Cartesian machine theory of the organism underlying the cybernetic program, in his view Bertalanffy was still guilty of taking a 'view from without' – considering the organism from the point of view of the *res extensa*. This is where Jonas' phenomenological approach led him to the ar-

---

[9] "We have restricted the connotation of teleological behavior by applying this designation only to purposeful reactions which are controlled by the error of the reaction – i.e., by the difference between the state of the behaving object at any time and the final state interpreted as the purpose." "Teleological behavior thus becomes synonymous with behavior controlled by negative feedback" (Rosenblueth, Wiener, Bigelow 1943, pp. 23–24).

gument that the only way to truly understand organic purposiveness is by taking a "view from within", or considering the first-person experience of our own lived bodies.

Both Bertalanffy's and Jonas' theoretical endeavors were conceived as a means of overcoming cybernetic machine theory, but their difference lies in how they proposed to achieve this goal. For Bertalanffy, general dynamic principles should provide the theoretical foundations for an adequate understanding of organic goal-directedness and organization – of which cybernetic regulation represents only a derivative subcategory. Yet how this should happen, or in what exactly these dynamic principles should consist, Bertalanffy does not say. Indeed, one would expect Jonas to address precisely this critical point, but his comments go in a different direction, criticizing system theory as still too materialistic and focusing on the necessity of taking a "view from within". As mentioned above, I believe this is where his contribution to a non-reductionist but still naturalist-minded theory of the organism drifts toward non-naturalist shores.

In Jonas' framework, system theory becomes a method of description that can be applied to certain biological phenomena, making use of differential equations to model the internal dynamics of biological processes such as embryonic development or metabolic self-regulation. It describes how a system behaves if certain formal conditions are fulfilled, but it does not say anything about the origin of that behavior itself.

It is not clear whether Jonas is arguing that an explanation of metabolism is simply out of reach for objective natural science or if he considers it something we are not yet able to explain with the scientific tools currently at our disposal. Certainly, however, he maintains that, no matter how far we go in "complexifying" our scientific toolkit, as long as its philosophical underpinnings remain bound to a materialist conception of living matter as first and foremost *matter*, we are still moving within the boundaries of Cartesianism, where living beings are automata explicable by the tools of physical science. For Jonas, system theory effectively refines the theoretical tools by which we try to make sense of organic goal-directedness. Still, even if its mathematical models are sufficient to overcome the explanatory swamps of reductionism, from a philosophical point of view they will never be able to give us a fundamental sense of the intrinsic purposiveness that defines living beings as such.[10]

---

[10] In Jonas' words, the theoretical weakness of system theory thus consists in trying "to make the concept of wholeness somehow available for structural and functional description, and to that extent to escape the analytical monopoly of physics, without infringing upon the ontology of physics itself which implies a ban of teleology. This would, if the concept of wholeness were

Jonas argues, in classic phenomenological style, that the only way we can access that intrinsic purposiveness is to observe from within ourselves the peculiar drives and desires that are manifest in us and to apply them through analogy to all other living beings. To return to the examples laid out in the introduction, when I take a "look from without" on the agency of a cat or a bacterium, I can claim that they are actively striving for food only because I have a direct experience "from within" of my very own strivings as a living being. In this sense, Jonas' answer to biological machine theory, from Descartes through cybernetics, is this "inwardness" of living organisms, which we can access through our first-person experience of "what it's like to be an organism". This "inwardness" ultimately stands as the core of Jonas' philosophical biology – adequately synthesized in the motto that "life can be known only by life" (Jonas 1966, p. 91).

## 4 Jonas' Legacy and What to Do with It

In this chapter, I have taken up the potential legacy of Jonas' philosophical biology for a theory of the organism, following Varela's and Weber's suggestion that he represents one of the "giants" on whose philosophical shoulders we should be able to ground a new understanding of organisms as "natural purposes".

I began by questioning what role Jonas' theory of the organism might play within current debates on philosophical naturalism. These debates address questions like: How we can ground specifically human phenomena from the point of view of natural science? Is it possible to retrieve the naturalistic foundations of intentionality and value? Is it possible to ground subjectivity naturalistically? I believe these questions index a theoretical endeavor largely overlooked by contemporary naturalism, especially its mainstream physicalist and reductionist branches, and that in this sense Jonas' philosophical biology points toward a potential way forward for contemporary philosophical discussions. What we can retain from Jonas' account of the relation between life and mind is a relevant conceptual indicator of how a naturalized account of normative agency might overcome the dualisms still characterizing our understanding of the relationship between human and non-human nature.

Jonas' essay on the philosophical aspects of Darwinsim makes this case quite convincingly, as it argues that, by establishing the radical continuity of descent between humans and non-human animals, Darwinian evolution abolished

---

taken seriously, amount to the attempt to make an omelette without breaking the eggs" (Jonas 1951, p. 51).

the special position of the human that had characterized fundamental ontology since early modern natural philosophy and Descartes' declaration of substance dualism. In this respect, Jonas' call for an "ontological revolution" pairs strikingly well with recent calls for a "normative naturalism" capable of overcoming the century-long gap between the "realm of laws" and "space of reasons".[11]

This challenge can be exemplified by the ostensibly metaphysical question of whether animals have "souls". As Jonas' analysis suggests, this question provides important insights on the theoretical genealogy linking a certain kind of reductive physicalism to the metaphysical background of Cartesianism. Indeed, what we take to be the least "metaphysical" position (hard naturalism) turns out historically to imply rather significant metaphysical baggage, including a "disembodied" conception of cognition as immaterial thinking. On the other hand, such re-examination of Cartesian dualism and early-modern natural philosophy can perhaps provide the theoretical starting point for reconsidering apparently metaphysical positions, such as Stahl's "animism", by situating them within the framework of a new kind of non-reductive "vital materialism" in which the "soul" is considered a dynamic principle of organization of a living body – thus providing a naturalistic grounding for sense-making and normative agency in biological systems.[12]

---

[11] To see this point, it is sufficient to compare the following quotes: (1) "Either to take the presence of purposive inwardness in one part of the physical order, viz., in man, as a valid testimony to the nature of that wider reality that lets it emerge, and to accept what it reveals in itself as part of the general evidence; or to extend the prerogatives of mechanical matter to the very heart of the seemingly heterogenous class of phenomena and oust teleology even from the 'nature of man,' whence it had tainted 'the nature of the universe' – that is to alienate man from himself and deny genuineness to the self-experience of life" (Jonas 1966, 37). (2) "There are several ways to go here. One would be to deny that subrational behavior is truly normative action. Another would be to say that not all action is truly normative, but that a sort of 'subnormative' action also exists. Yet another would be to bite the bullet and admit that our original distinction was misguided, and that the higher animals (at least) are fully capable of action in the normative sense." "From the foregoing considerations, we may conclude that all of the elementary normative concepts, as well as the concept of agency, are properly ascribable to organisms as such – i.e., organisms are properly regarded as agents in the full normative sense of the term. In other words, the proper scope of application of our concept of normative agency is living systems as such" (Barham 2012, pp. 94–96).

[12] In recent years, this approach has been productively taken by Ezequiel Di Paolo (among others), who shows how the concept of 'adaptivity' can be used to provide a naturalistic account of sense-making. Di Paolo defines the notion of adaptivity as "the capacity of an organism to regulate itself with respect to the boundaries of its own viability" (Di Paolo 2005, p. 430), or the ability to normatively 'evaluate' its encounters with the surrounding environment as contributing to or endangering the maintenance of its own autopoiesis – and to make behavioral choices

This kind of account of sense-making would effectively naturalize the threads developed in Jonas' philosophical biology without having to resort to notions of inwardness or the transcendental lived body. Within our endeavor to construct a theory of the organism, the phenomenological "view from within" championed by Jonas at best plays the role of "regulative principle" in the Kantian sense of the term, i.e., a hermeneutic tool providing an alternative philosophical outlook (that of "inwardness") on organic nature with the goal merely of providing "objective" explanatory frameworks and theoretical models. If mistaken for a genuine explanatory ground, the reference to inwardness is ultimately the part of Jonas' legacy that we *do not* need if we are to develop a scientifically viable theory of the organism.

# References

Barham, James (2012): "Normativity, agency and life". In: *Studies in History and Philosophy of Biological and Biomedical Sciences*, 43(1), pp. 92–103.

Bertalanffy, Ludwig (1951): "Towards a physical theory of organic teleology: Feedback and dynamics". In: *Human Biology*, 24(4), pp. 346–361.

---

accordingly. In this sense, the notion of adaptivity provides the theoretical foundations for a naturalized concept of agency that conceives of organisms as 'living subjects.' Charles Wolfe has qualified this introduction of subjectivity in biological discourse as "biochauvinism", understood as the modern-day version of "vitalism" in the derogatory sense (Wolfe 2014). I believe this statement to be true if we understand "subjectivity" as a transcendental form of disembodied cognition not accessible to naturalistic inquiry (which seems to be the case for Jonas' reference to inwardness). On the other hand, this criticism misses its target if "subjectivity" is not understood in this transcendental/phenomenological sense but rather as a specific form of dialectical relation between an autopoietic system and its *Umwelt* that establishes a meaningful perspective on a world. To return to our starting example, "bacteria possessing this capability will be able to generate a normativity *within* their current set of viability conditions and *for themselves*. They will be capable of appreciating not just sugar as nutritive, but the direction where the concentration grows as useful, and swimming in that direction as the right thing to do in some circumstances" (Di Paolo 2005, p. 437). In this sense we can legitimately argue that "the organization of biological systems is inherently teleological, which means that its own activity is, in a fundamental sense, first and foremost oriented toward an end… Biological organization determines itself in the sense that the effects of its activity contribute to establish and maintain its own conditions of existence: in slogan form, biological systems are what they do. Self-determination implies therefore a circular relation between causes and effects: the organization produces effects (e.g., the rhythmic contractions of the heart) which, in turn, contribute to maintain the organization (e.g., the cardiac contractions enable blood circulation and, thereby, the maintenance of the organization)" (Mossio and Bich 2017, p. 1090).

Breitenbach, Angela (2009): "Teleology in biology: A Kantian approach". In: *Kant Yearbook*, 1, pp. 31–56.
Butterfield, Herbert (1959): *The Origins of Modern Science, 1300–1800*, New York: Macmillan.
Canguilhem, Georges (2008): *The Knowledge of Life*, Fordham: Fordham University Press.
Carvallo, Sarah (2004): *La controverse entre Stahl et Leibniz sur la vie, l'organisme et le mixte*, Paris: Vrin.
Dawkins, Richard (1986): *The Blind Watchmaker*, New York: Norton and Company.
De Caro, Mario and Macarthur, David (2004): *Naturalism in Question*, Cambridge MA, Harvard University Press.
De Caro, Mario and Macarthur, David (2010): *Naturalism and Normativity*, New York: Columbia University Press.
Di Paolo, Ezequiel (2005): "Autopoiesis, adaptivity, teleology, agency". In: *Phenomenology and the Cognitive Sciences*, 4(4), pp. 429–452.
Di Paolo, Ezequiel (2009): "Extended life". In: *Topoi*, 28, pp. 9–21.
Duchesneau, François and Smith, Justin E. H. (2016): *The Leibniz-Stahl Controversy*, Yale University Press.
Franzini Tibaldeo, Roberto (2009), *La rivoluzione ontologica di Hans Jonas: Uno studio sulla genesi e il significato di "Organismo e Libertà"*, Milano: Mimesis.
Franzini Tibaldeo, Roberto (2015), "The meaning of life: Can Hans Jonas's 'philosophical biology' effectively act against reductionism in the contemporary life sciences?". In: *Humaniora*, 1(9), pp. 13–24.
Ginsborg, Hannah (2004): "Two kinds of mechanical inexplicability in Kant and Aristotle". In: *Journal of the History of Philosophy*, 42(1), pp. 33–65.
Ginsborg, Hannah (2006): "Kant's biological teleology and its philosophical significance". In: Bird., Graham (ed.): *A Companion to Kant*, Oxford: Blackwell, 455–469.
Gould, Stephen J. and Lewontin, Richard (1979): "The spandrels of San Marco and the panglossian Paradigm: A critique of the adaptationist programme". In: *Proceedings of the Royal Society*, 205, pp. 581–598.
Goy, Ina and Watkins, Eric (eds.) (2014): *Kant's Theory of Biology*, Berlin: De Gruyter.
Huneman, Philieppe (ed.) (2007): *Understanding Purpose. Kant and the Philosophy of Biology*, North American Kant Society Studies in Philosophy, 8, Rochester: University of Rochester Press.
Huneman, Philippe (2017): "Kant's concept of organism revisited: A framework for a sossible Synthesis between developmentalism and adaptationism?". In: *The Monist*, 100(3), pp. 373–390.
Huneman, Philippe and Rey, Anne-Lise (2007): "La controverse Leibniz-Stahl dite *Negotium Otiosum*". In: *Bulletin d'Histoire et d'épistémologie des sciences de la vie*, 14(2), pp. 213–238
Huneman, Philippe and Walsh, Denis (2017): *Challenging the Modern Synthesis*, Oxford: Oxford University Press.
Huneman, Philippe and Wolfe, Charles (2010): "The concept of organism: Historical, philosophical, scientific perspectives". In: *History and Philosophy of the Life Sciences*, 32(2–3), pp. 147–154.
Jonas, Hans (1951): "Comment on general system theory". In: *Human Biology*, 24(4), pp. 328–335.

Jonas, Hans (1966): *The Phenomenon of Life: Toward a Philosophical Biology*, New York: Harper and Row.
Kauffman, Stuart (2000): *Investigations*, Oxford: Oxford University Press.
Koyré, Alexandre (1939): *Études galiléennes*, Paris: Hermann.
Kreines, James (2005): "The inexplicability of Kant's *Naturzweck:* Kant on teleology, explanation and biology". In: *Archiv für Geschichte der Philosophie*, 87(3), pp. 270–311.
Laland, Kevin N., Uller, Tobias, Feldman, Marc, Sterelny, Kim, Müller, Gerd B., Moczek, Armin, Jablonka, Eva, Odling-Smee, John (2014): "Does Evolutionary Theory Need a Rethink?". In: *Nature*, 514, pp. 161–164.
Lories, Danielle and Depré, Olivier (2003): *Vie et liberté: Phénoménologie, nature et éthique chez Hans Jonas*, Paris: Vrin.
Michelini, Francesca, Wunsch, Matthias, Stederoth, Dirk (2018): "Philosophy of nature and organism's autonomy: On Hegel, Plessner and Jonas' theories of living beings". In: *History and Philosophy of the Life Scineces*, 40(3), pp. 40(56), pp. 1–27.
Mayr, Ernst (1974): "Teleological and teleonomic: A new analysis". In: Cohen, Robert and Wartofsky Marx (eds.): *Methodological and Historical Essays in the Natural and Social Sciences*, Dordrecht: Reidel, pp. 91–118.
McLaughlin, Peter, (1990): *Kant's Critique of Teleology in Biological Explanation: Antinomy and Teleology*, New York: Mellem Lewinston.
Moss, Lenny and Nicholson, Daniel J. (2012): "On nature and normativity: Normativity, teleology and mechanism in Biological Explanation". In: *Studies in History and Philosophy of Biological and Biomedical Sciences*, 43(1), pp. 88–91.
Mossio, Matteo and Bich, Leonardo (2017): "What makes biological organisation teleological?". In: *Synthese*, 194(4), pp. 1089–1124.
Nicholson, Daniel (2014): "The return of the organism as a fundamental explanatory concept in Biology". In: *Philosophy Compass*, 44(4), pp. 669–678.
Nunziante, Antonio Maria (2011), *G.W. Leibniz, Obiezioni contro la Teoria medica di Gerog Ernst Stahl. Sui concetti di anima, vita, organismo*, Macerata: Quodlibet.
Pigliucci, Massimo and Müller, Gerd (2010): *Evolution, The Extended Synthesis*, Cambridge MA: MIT Press.
Richards, Robert J. (2000): "Kant and Blumenbach on the *Bildungstrieb*: A historical misunderstanding". In: *Studies in History and Philosophy of Biological and Biomedical Sciences*, 31, 1: 11–32.
Rosenblueth, Arturo, Wiener, Norbert, and Bigelow Julian (1943): "Behavior, purpose and teleology". In: *Philosophy of Science*, 10(1), pp. 18–24.
Stahl, Ernst Georg (1859). *On the Difference Between Mechanism and Organism (Disquisitio de mecanismi et organismi diversitate)*. In: Œuvres médico-philosophiques et pratiques, translated by Théodore Blondin, ed. Louis Boyer, vol. 2. Paris: J.-B. Baillière.
Steigerwald, Joan (2006): "Kant's concept of natural purpose and the reflective power of judgment". In: *Studies in History and Philosophy of the Biological and Biomedical Sciences*, 37(4), pp. 712–34.
Stroud, Barry (2004): "The charm of naturalism". In: De Caro and Macarthur 2004, pp. 21–35.
Toepfer, Georg (2011): *Historisches Wörterbuch der Biologie*, Stuttgart-Weimar: Metzler.
Van den Berg, Hein (2013): "The wolffian roots of Kant's teleology". In: *Studies in History and Philosophy of Biological and Biomedical Sciences*, 44(4), pp. 724–734.

Varela, Francisco (1991): "Organism: A meshwork of selfless selves". In Tauber, Alfred (ed.): *Organism and the Origin of Self*, Dordrecht: Kluwer, pp. 79–107.
Varela, Francisco (1997): "Patterns of life: Intertwining identity and cognition". In: *Brain and Cognition* 34, pp. 72–87.
Walsh, Denis (2015): *Organisms, Agency and Evolution*, Cambridge: Cambridge University Press.
Weber, Andreas and Varela, Francisco (2002): "Life after Kant: Natural purposes and the autopoietic foundations of biological individuality". In: *Phenomenology and the Congnitive Sciences*, 1(2), pp. 97–125.
Wiener, Norbert (1948): *Cybernetics: Or, Control and Communication in the Animal and the Machine*, Cambridge MA: MIT Press.
Wolfe, Charles (2010): "Why there was no controversy over life in the Scientific Revolution". In: Boanza, Victor and Dascal, Marcelo (eds.): *Controversies over the Scientific Revolution*, Amsterdam: John Benjamins, pp. 187–220.
Wolfe, Charles (2014): "Holism, organicism and the risk of biochauvinism". In: *Verifiche*, 43(1–4), pp. 39–57.
Zammito, John (2006): "Teleology then and now: The question of Kant's relevance for contemporary controversies over functions in biology". In: *Studies in History of Biological and Biomedical Sciences*, 37(4), pp. 748–770.
Zammito, John (2012): "The Lonoir thesis revisited: Blumenbach and Kant", *Studies in History of Biological and Biomedical Sciences*, 43(1), pp. 120–132.
Zammito, John (2018): *The Gestation of German Biology: Philosophy and Physiology from Stahl to Schelling*, Chicago: University of Chicago Press.
Zuckert, Rachel (2007): *Kant on Beauty and Biology: An Interpretation of the Critique of Judgement*, Cambridge: Cambridge University Press.
Zumbach, Clarck (1984): *The Transcendent Science: Kant's Conception of Biological Methodology*, Dordrecht: Springer.

Pierfrancesco Biasetti
# Dialectical Thinking and Science: The Case of Richard Lewontin, Dialectical Biologist

**Abstract:** Richard Lewontin's dialectical approach to biology emphasizes the relationship between the organism, its development, and the environment, providing an alternative view to the one provided by "mechanistic" and "reductionist" paradigms. This alternative view can be seen as the most lucid attempt made in recent times to apply to a particular science the dialectical tradition flowing from Engels' Antinti-Dühring and the unfinished Dialectics of Nature. By analysing Lewontin's critique of mechanistic biology and his constructivism, a general assessment of the pretension of the dialectical approach in science will be attempted

## Introduction

In this chapter I will focus on Richard Lewontin's dialectical approach to biology[1]. Lewontin's views are interesting for the theme of this book not only for their emphasis on the relation between organisms and environment: they also form the most lucid attempt made in recent times to apply to biology the peculiar tradition of dialectical thinking originating from Engels' *Anti-Dühring* and the unfinished *Dialectics of Nature*.

I will start from the roots of this tradition in order to grasp its main points (Part 1). I will then sketch Lewontin's peculiar position in the tradition by focusing on his critique of mechanistic biology (Part 2) and his constructivism

---

[1] Lewontin's dialectical leanings are explicitly formulated in a series of papers collected in the book *The Dialectical Biologist*, co-authored with fellow Harvard biologist Richard Levin (1985). Another collection of dialectical essays –largely less technical – is *Biology Under the Influence*, again with Levins (2007). Lewontin's impressive career produced a rather long list of publications. Beside the aforementioned collections of essays, the best volume to pick in order to grasp his application of dialectical thinking to biology is, perhaps, *The Triple Helix* (2000a). The first three chapters were first published in Italian (1998), and for the most part they expand the seminal paper *The Organism as the Subject and Object of Evolution* (1983). More militant books are *Not in Our Genes*, co-authored with neuroscientist Steven Rose and psychologist Leon Kamin (1984), and *Biology as Ideology* (1991).

**Pierfrancesco Biasetti,** Leibniz Institute for Zoo and Wildlife Research

(Part 3). In the last part of the paper I will try to assess the pretension of the dialectical approach in science through an analysis of Lewontin's case (Part 4).

# 1 Dialectics and Metaphysics

In the introduction to his *Anti-Dühring*, Friedrich Engels stated that "nature works dialectically and not metaphysically"[2]. The consequences are that scientists should abandon the *metaphysical* frame of theorizing, and embrace dialectical thinking (henceforth: DT). But what is the meaning here of these two terms, "dialectical" and "metaphysical"?

Let us start with the latter. A conventional distinction between metaphysics and ontology[3] is that ontology *catalogs existing entities*, while metaphysics *explains what they are*. According to Engels, we could say, natural scientists must engage in both tasks in order to understand natural phenomena. However, they usually fail at metaphysics because they ignore philosophy, and, as such, fall prey to *bad* philosophy. Engels, thus, is not claiming that scientists should stop dealing with the questions posed by what we could call *metaphysics₁* – *what natural entities are*. He claims instead that they should stop framing their answers according to what we could call *metaphysics₂*: the *metaphysical₁* view opposed to *dialectics*.

## 1.1 Dialectics of Nature: Metaphysics₁, Epistemology and Sociology of Science

Metaphysics₂ can be resumed in the habit of

> observing natural objects and processes in isolation, apart from their connection with the vast whole; of observing them in repose, not in motion; as constants, not as essentially variables, in their death, not in their life.[4]

Metaphysics₂ is grounded in dichotomy, analysis, and a static view of identity. As it is based on commonsense, it supplies us with a good heuristic for grasping

---

[2] "Nature is the proof of dialectics, and it must be said for modern science that it has furnished this proof with very rich materials increasing daily, and thus has shown that, in the last resort, nature works dialectically and not metaphysically" (Marx and Engels 1987a, 23–4).
[3] See Varzi (2005). The distinction can be traced back at least to Quine (1948).
[4] Marx and Engels (1987a), 20.

simple and ordinary pictures of things. But its use in natural science is limited, because it defies two properties of the world: complexity and change. According to metaphysics$_2$, everything *has* its place and everything *is* in its place. As such, the best science it can produce is Newton's model of celestial bodies and Linneo's classification of species.

At the opposite spectrum of metaphysics$_2$, DT stresses change and relation. While for the *metaphysician*$_2$ all fundamental properties of things are intrinsic and all relations external, for the dialectician everything is connected and interacting. As such, no property can be understood by abstracting the object from its context, and all relations are internal and not accidental. This picture of the world is dynamic, such that states of equilibrium are not the norm, but the exception. Change is ubiquitous, and the difference between constants and variables is only contextual – due to the different scales we employ. In fact, the whole reality can be seen "as a complex of processes"[5], with no ultimate bottom line of analysis – no *fundamental level*, no *atoms* whose ultimate properties reverberate bottom-up. For the dialectician *there is no such a thing as a thing*.

Engels states that during the eighteenth century the metaphysical$_2$ view of nature began to be challenged by more dialectically inclined theories. The first of these was Kant's nebula hypothesis, followed by Lyell's geology, and Dalton's and Lavoiser's chemistry. The latest (by Engels' time) was Charles Darwin's theory of the evolution of living beings[6]. The view of the world shared by Kant, Lyell, Dalton, Lavoisier, and Darwin differed from the one given by metaphysics$_2$ by taking change and relations seriously. However, all these scientists did not formulate and employ DT *consciously:* they stumbled upon it. Like Moliere's *Bourgeois Gentilhomme*, who spoke in prose without knowing it, they embraced dialectics without formulating its principles or even being aware of them. This, according to Engels, explained the scientific "mess" of his times:

---

[5] "The great basic thought that the world is not to be comprehended as a complex of ready-made things, but as a complex of *processes*, in which the apparently stable *things*, no less than their mental images in our heads, the concepts, go through uninterrupted change of coming into being and passing away, in which, for all apparent accidentality and despite all temporary retrogression, a progressive development asserts itself in the end – this great fundamental thought has, especially since the time of Hegel, so thoroughly permeated ordinary consciousness that in this generality it is now scarcely ever contradicted." (Marx and Engels 1990, 384).
[6] See the introduction to the *Dialectics of Nature*, Marx and Engels (1987a), 323–7. In *Ludwig Feuerbach and the End of Classical German Philosophy*, Engels names three discoveries that made metaphysics$_2$ obsolete: the cell, transformation of energy, and Darwin's theory of evolution (Marx and Engels 1990, 385–386).

> Naturalists who have learned to think dialectically are few and far between, and this conflict of the results of discovery with preconceived modes of thinking explains the endless confusion now reigning in theoretical natural science.[7]

Engels' aim, in his *Dialectics of Nature*, was to clear this confusion, by giving DT a coherent frame and a materialist backbone, in order to make it useful for scientists, and urge them to abandon all residue of *metaphysics$_2$* and theology that still polluted their work. As is well known, he never completed this work. But his project highlights the essential connection made in DT between *methodology* and *metaphysics$_1$*, given the fundamental assumption of an isomorphism between reality and the way we apprehend it. DT advances not only a *metaphysical$_1$ claim* – that nature has a dialectic structure – but also an *epistemological claim* – that scientists need to adopt a dialectic methodology in order to comprehend nature better. Moreover, besides these two, it also advances a *sociological claim*. To the dialectician, reality is a process, and DT itself is also the result of a progressive enterprise during which different strategies of apprehension are attempted and then discarded. Science is not a neutral, detached enterprise, governed by its own rules and tribunals: in fact, scientists tend to transpose values, prejudices, and thinking structures from their social context into their work. A classic example of this analysis is Marx and Engels' claim that the concept of *struggle for life* was a transposition into nature of Hobbes' *bellum omnium contra omnes*, capitalism as regime of competition, and Malthus' theory of population[8]. In this way,

---

[7] Marx and Engels 1987a, 24. See also 486.
[8] See, for instance, Marx and Engels 1987a, 583. Engels's and Marx's attitudes towards Darwin differ. Engels was one of the few hundred persons to obtain a copy of the first edition of The Origin. He read it almost immediately, and gave to Marx an enthusiastic review ("absolutely splendid" – Marx and Engels 1983, 551). Marx replied on the subject only one year later, acknowledging the importance of Darwin's work, and how it provided, "in the field of natural history", the basis for their views (Marx and Engels 1985, 232). However, he soon grew less enthusiastic of Darwinism, believing he had found a better alternative in French ethnologist and archaeologist Pierre Trémaux, who contested the concept of struggle for life, and thought that environmental causes were a better source of explanation for evolution than natural selection. Marx recommended Trémaux's work to Engels ("it represents a very significant advance over Darwin"), who, in turn, was not lured by the Lamarckian arguments of the Frenchman, and denounced him as a charlatan (Marx and Engels 1987b, 304–305; 320). However, as further exchanges show, Marx's persisted in his opinion. That he wanted to dedicate *Das Kapital* to Darwin is a story that has been proven a fake (see Feuer 1975; Fay 1978), and Darwin's name recurs in Marx's masterpiece only a few times, without any particular enthusiasm. Marx's Lamarckian leanings and disaffection with Darwin had probably political roots, as it is evident in an 1862 letter to Engels, where he judged the core arguments of The Origin as a projection of Malthus, Hobbes and English society into nature (Marx and Engels 1985, 381). As several quotations

the scientific image of the world tends to progress not only thanks to new and more detailed observation, but also due to societal transformation.

## 1.2 Specifying the Metaphysical₁ Claim

The metaphysical₁ claim advanced by DT is compatible with various forms of process metaphysics₁. However, it assumes specific and unique traits when placed within the frame of DT. On the one hand, the metaphysical₁, epistemological and sociological claims form an interdependent structure, and cannot be isolated one from each other except in abstraction. On the other hand, a further set of assumption of DT specifies the kind of processes that compose reality, describes the modes of transformation, and defines the nature of relations subsisting within the whole, specifying the *why* and *how* of the metaphysical₁ claim.

With regard to this, Engels famously gave three laws of dialectics[9]: a) conversion of quantitative changes into qualitative changes (and vice versa); b) interpenetration of opposites; c) negation of negation. These laws, in turn, presuppose a longer series of assumptions: the ontological reality of contradiction – that is, that contradiction is a real feature of the world, and not something exclusively made up by minds; that reality is not ruled by harmony, but by conflict – as contradiction gives rise to opposition which, in turn, gives rise to change; that the result of this change is further negated by a successive opposition; that the result of this double process produces a stage similar but somehow superior to its starting point; that there is no bottom nor uppermost level in this progression; that processes are both continuous and made of discrete leaps. All these assumptions were developed by Hegel, as "mere laws of thought"[10], while Engels' point was to show that they could be extrapolated from both nature and history. They form the crucial core of DT – and perhaps, in their application to nature, its

---

show (Marx and Engels 1987a, 583; Marx and Engels 1991, 106–10), Engels in the end accepted this reading. However, he also tried to reconcile it with his admiration for Darwin's work. He did so by separating humankind from natural history, and crafting the myth of Darwin and Marx as complementary thinkers – the first being the discoverer of "the law of development or organic nature", the second the discoverer of "the law of development of human history" (quoted from Engels' speech at Marx's funeral, Marx and Engels 1989, 467).

**9** See Marx and Engels (1987a), 356. In the "Outline of the general plan" for the work Engels cites what is possibly a fourth law, or, more probably, a general consequence of all three: "spiral form of development" (317).

**10** Marx and Engels (1987a), 356.

most controversial part, even among Marxists[11]. I will say something more on this subject in the last part of the paper. Let us now depart from Engels and the general analysis of dialectics of nature, and start reviewing how DT is applied in Lewontin's case.

## 2 Dialectic Vs Un-dialectic Approaches to Biology

"Virtually the entire body of modern science", Lewontin writes, "is an attempt to explain phenomena that cannot be experienced directly by human beings"[12]. In order to build such explanations, we must refer to other phenomena that we may not perceive directly: because they are too small (like molecules), too big (like the universe), totally defy our senses (like electromagnetism), or are born from too complex interactions (like the development of an organism). In this way, scientific explanations, when not expressed in an invented technical language, must resort to *metaphors* to be comprehensible.

It is the need for metaphors that makes science vulnerable, for three reasons: a) the choice of a particular metaphor is *not neutral*; our image of nature is modeled on the society we live in, and the metaphors we use are inspired by societal structure and values; b) metaphors quickly exceed their explanatory role, and become *models* for research programs; c) in the end, metaphors cease to be considered *just metaphors*, and become a commitment to how things are.

### 2.1 The Machine Metaphor

According to Lewontin, the chief metaphor of pre-modern science was *the body*, the indissoluble whole that, paraphrasing Pope, *we murder in the moment we dissect*. Modern science replaced it with new metaphor: *the machine*[13].

---

[11] Starting from Lukacs, many Marxists believed that Marxism – as a *philosophy of praxis* – and its dialectical analysis could not be applied outside the social and historical dimensions. From different perspectives, the belief was also shared by Gramsci and Sartre. Moreover, after Stalin's (1938) vulgarization, Engels' *Dialectic of Nature* has been traditionally seen as crude and deterministic.

[12] Lewontin (2000a), 3.

[13] The turning point was Descartes' *Discours de la méthode*, where the metaphor was applied to every physical thing in the universe – both inorganic and organic (see Descartes 1902 [1637],

The machine metaphor is a paradigm of metaphysics$_2$ and it is grounded in the fundamental premise of modern capitalism[14]: ontological priority of individuals. As societies are believed to be born and shaped by the collision of self-owning and autonomous individuals, machines (unlike *bodies*, which are, by definition, indivisible) are made of autonomous and pre-existing pieces, from which they derive their properties and functions. In this sense, the machine metaphor reflects the way into which capitalist society is organized.

Once formulated, the machine metaphor ceased almost immediately to be seen just as a metaphor, and acquired a distinct *epistemic* value, becoming a model for research. Nowadays no one believes that organisms literally resemble clocks or Jacques de Vaucanson's *canard digérateur*, yet, the machine model, in refined forms, is still present in analogies between brains (or minds) and computers, in adaptationist claims concerning optimization, and in characterization of biology as a peculiar branch of engineering[15].

In the end, the machine metaphor became the standard metaphysics$_1$ view. In fact, while Descartes formulated it in what he conceived as a *discour de la méthode*, the metaphor was taken progressively as depicting things as they are:

> Cartesian reduction as a method has had an enormous success in physics, in chemistry, and in biology, especially molecular biology, and this has been taken to mean that the world is like the method.[16]

However, the machine metaphor, given its epistemic and metaphysical$_1$ interpretations, introduced a series of arbitrary – and *wrong* – assumptions in the scientific enterprise.

The first of these is *alienation*. Mechanistic science is based on a never-ending series of arbitrary separations. Things are believed to be separable into discrete, unrelated, and interchangeable pieces, and, as such, their properties are studied independent from the whole of which they are part. In biology, alienation affects the organism, which is separated, on the one hand, from its development, and, on the other hand, from its environment. This originates several sub-

---

part V). Lewontin's judgment on Descartes is an interesting point of departure from Engels, who saw the French philosopher as one of the few dialectical thinkers in an era of un-dialectical philosophers and scientists. However, Engels' opinion was not born from Descartes' biology, but from his formulation of conservation of momentum, interpreted as the original expression of the law of conservation – a "dialectical law" that presupposes *transformation* of energy.

14 Levins and Lewontin (1985), 270.
15 See, for instance, Dennett (1995). For a critical survey on the use of engineering metaphors in contemporary biology see Boudry and Pigliucci (2013).
16 Levins and Lewontin (1985), 2.

metaphors: the "blueprint" sub-metaphor concerning genes, the "program" sub-metaphor concerning development, and the "optimization" and "adaptation" sub-metaphors concerning evolution.

The second assumption is *reductionism*. Reductionism can be defined as the attempt "to explain the properties of complex wholes – molecules, say, or societies – in terms of the units of which those molecules or societies are composed"[17]. According to the machine metaphor, the whole and its properties are caused by the pieces: these, then, are the only things we need to study in order to build an explanation. A reductionist, in this sense, is committed to an ontological hierarchy which, in turn, is reflected in an epistemological hierarchy of different sciences ranked according to their "level" of explanation – each one more or less fundamental than the other. This ontological and epistemological hierarchy is clearly reflected in biology, whereas molecular biology has obtained a position of hegemony[18]. Yet, this hegemony is grounded – in Lewontin's view – in *faith*, as it merely derives from the peculiar kind of reductionism ingrained in the machine metaphor.

The third assumption is *determinism*. This is the idea that the behavior of every complex whole can be inferred and foretold by collecting enough information on its basic components. This is, on the one hand, a consequence of reductionism – as it presupposes that the causal vector goes straightforwardly from pieces to wholes – and, on the other hand, a consequence of alienation, as it *a priori* rules out any intervention of chance or of other external condition. According to Lewontin, in biology determinism shapes mainstream views on development, restricting the domain of causal factors to the genes.

The machine metaphor does not only slow down possible progress in science due to the dogmatic commitment to rigid and, in the end, false assumptions[19]. It also produces negative societal effects. The alienated view of organisms and environments, for instance, is the cause for the unrealistic tasks assigned to conservation biology[20]. Reductionism produces false expectations on much publicized

---

[17] Rose, Lewontin and Kamin (1984), 5.
[18] The preeminence of molecular biology is discussed especially in Lewontin (1991). See Kitcher (1999) for an analysis and a critique.
[19] Lewontin compares the progresses of mechanistic science to the strategy of medieval armies: "Cities are laid siege to, and most surrender, but a few hold out indefinitely. The army sweeps around these, leaving behind some of its troops, who settle down to a long and frustrating encirclement. This has certainly been the case in biology, where the extraordinary progress made in molecular studies has been the consequence of a straightforward reductionist program, while the understanding of embryonic development and of the functioning of the central nervous system have remained in a rudimentary state" (Levins and Lewontin 1985, 269).
[20] Lewontin (2000a), 66–67.

(and subsidized) research programs, like the Genome Project [21]. Determinism, under the guise of biological determinism, reinforces the prejudices ingrained in society, and become a justification for past and present injustices[22].

For all these reasons, the role of dialectic in science should be two-fold: on the one hand, by replacing the machine metaphor, it should be able to foster a better scientific view of the natural world. And, as better science replaces the old, alienating one, it should help the formulation of better social policies.

## 2.2 Dialectic of Biology

Returning to the holistic mysticism of the body metaphor would be an impossible journey back to the pre-scientific world. Hence, in order to overcome the failings of mechanistic biology, the solution is a *sublation* surpassing both views while recovering the partial truths they contain.

Lewontin, alongside Levins, individuates four principles of DT applied on biology. The first is that "a whole is a relation of heterogeneous parts that have no prior independent existence as parts"[23]. This is a clear rejection of the fundamental idea at the core of the machine metaphor, and has the important function of undermining reductionism, as it establishes that there is no absolute correct way (in a methodological *and* metaphysical$_1$ sense) to divide a whole: division into pieces depends every time on the particular level of explanation we are pursuing.

The second is that "in general, the properties of parts have no prior alienated existence but are acquired by being parts of a particular whole"[24]. In other words, parts have properties that exist only insofar we consider the whole as their cause. Defining a gene $X$, for instance, as the cause of a phenotypic trait $Y$, is wrong, because the gene per se has no power of producing trait $Y$ outside the whole developmental system of which it is part. The property of gene $X$ to provide for the trait $Y$ is not an intrinsic property of gene $X$.

The third is that "the interpenetration of parts and wholes is a consequence of the interchangeability of subject and object, of cause and effect"[25]. This prin-

---

21 Lewontin (2000b).
22 In this case, the main target of Lewontin's social critique has been biological determinism and its alleged incarnations in sociobiology (Wilson 1975) and selfish gene theory (Dawkins 1976).
23 Levins and Lewontin (1985), 273.
24 Levins and Lewontin (1985), 273.
25 Levins and Lewontin (1985), 274.

ciple follows from the previous two: as there is no fundamental "level" to look for, and, as causal explanations are born from the interaction of whole and parts, then both poles of a relation, depending on the circumstances, can be seen as subject and object, cause and effect. This principle grounds Lewontin's thesis that organisms are both the subjects and the objects of the evolutionary process – while the adaptationist paradigm tends to see them as passive objects.

These three principles are meant to refine Engels' law of the compenetration of the opposites. A fourth principle sums up instead Engels' first and third laws. This principle states simply that "change is a characteristic of all systems and all aspects of all systems"[26]. This kind of change is different from the one theorized by mechanistic science. The latter is a kind of change without *real change:* it is an *unfolding* of something already there. According to the gene's eye point of view on evolution, for instance, development is simply an implementation of a program already contained in the genes. This happens, again, because mechanistic explanations sever the interpenetration of things, and lose in this way the dynamic aspect, reconstructing it as a one-way causal vector. On the contrary:

> What characterizes the dialectical world, in all its aspects, (…) is that it is constantly in motion. Constants become variable, causes become effects, and system develop, destroying the conditions that gave rise to them. Even elements that appear to be stable are at a dynamic equilibrium of forces that can suddenly become unbalanced (…). Yet the motion is not unconstrained and uniform. Organisms develop and differentiate, then die and disintegrate. Species arise but inevitably become extinct. (…) The development of systems through time, then, seems to be the consequence of opposing force and opposing motions.[27]

Equilibrium is not the norm: *self-negation* is the norm, and this leads to a discontinuous process of change – the ultimate image of nature from the standpoint of dialectical biology.

## 3 Development, Evolution, and Adaptation

In the "alienated" world painted by the machine metaphor the organism is separated from its developmental process, its historical evolution, and even from its concrete existence as a whole. The organism is seen as just a more or less complex collection of particular functional traits, entirely shaped by internal (the genes) and external (the environment) causal forces and, as such, without

---

[26] Levins and Lewontin (1985), 275.
[27] Levins and Lewontin (1985), 279.

any relevant ontological and explanatory value. Against this picture, Lewontin stresses the role of the organism both as the object *and* as the subject of biological processes like development and evolution.

## 3.1 Development

According to Lewontin, standard development theories are modeled upon the concept of *unfolding*. Historical changes in an individual organism are seen as products of internal forces in charge of the process. This view can be called *internalism*[28], as (individual) organisms are passively modeled in the course of their life by self-sufficient inner programs. In the standard versions of internalism, the information necessary to implement the programs is carried out by the genotype. In its more sophisticated versions, the model incorporates environmental interactions triggering or affecting the advancement of certain parts of the program.

Internalism is defined by Lewontin as a kind of *transformational* theory of change[29]. Transformational theories describe changes occurring to members of a set dominated by the *same historical laws*. These laws describe an ideal vector of progressive stages that the members of the set normally traverse during their history. Change is then the passage from a stage to another, according to a fixed model that describe the ideal, "Platonic" trajectory of development. In this sense, this trajectory is normative, and protracted stop or regression are considered anomalies.

From the standpoint of DT, internalism is a peculiar form of change without change, because everything that is expressed in the final stage is already contained in the beginning of the process, and real change – *individual differences* – are explained as anomalies caused by external factors. Moreover, a transformational view of development is reductionist (and deterministic), since it reduces causal factors to a single set of simple, atomic and self-sufficient entities: *genes*.

Put in another way, there is an implicit *preformationist* assumption in these models of development:

> It is usually said that the epigenetic view decisively defeated preformationism. [...] Yet it is really preformationism that has triumphed, for there is no essential difference, but only one

---

[28] See for instance Godfrey-Smith (2001). This internalist view on development must not be mistaken for the *externalist* view of adaptation inherent in mechanistic biology: see the next sub-section.
[29] See Lewontin (2000a), 8, and Levins and Lewontin (1985), 86.

of mechanical details, between the view that the organism is already formed in the fertilized egg and the view that the complete blueprint of the organism and all the information necessary to specify it is contained there.[30]

The result is that organisms are "alienated", and as such, irrelevant:

> The view of development that sees genes as determinative, or even a view that admits interaction between gene and environment as determining the organism, places the organism as the end point, the object, of forces. The arrow of causation point from gene and environment to organism.[31]

The problem with this transformational view of development is that there is no way to know how an organism will come to be just by collecting all information available on its "program" or "blueprint". The reason cannot be imputed to the impossibility of predicting environmental factors capable of altering the movement of the organism along the progression of stages, as *there is no real progression of stages*. What we have is instead a ramified pathway along which environment and developmental noise – random events happening at molecular or cellular level – play a role alongside genetic causes in forming the unique trajectory pursued by the organism. Genetic causes are very important when it comes to examining the difference between organisms of different species – why humans, for instance, can talk, while sheep cannot, or, on the other hand, why sheep have a woolly fur, while humans don't – but when we come to explain why two members of the same species, or even two sides of a same organism, differ, environmental and casual factors become crucial.

It could be argued that even in this view the organism remains, in the end, a mere product of the various elements constituting the process, and as such, alienated. However, Lewontin contrasts this argument with two objections.

The first is that at any specific time of the developmental process, change depends upon the present status of the organism and not on how it came to be how it currently is (an application of the principle of the existence of irreducible level):

> Small seeds give rise, in general, to small seedlings, which grow slowly because they are shaded by competitors. It does not matter whether the seed was small because of the maternal plant's genotype or because it set seed in an unfavorable habitat.[32]

---

**30** Lewontin (2000a), 5–6.
**31** Levins and Lewontin (1985), 96.
**32** Levins and Lewontin (1985), 96.

In this example it is *size*[33] that determines possible slowing down in growth: and it does not matter if this characteristic derives from the genotype or from environmental factors. In this sense, "the organism, irrespective of the internal and external forces that influenced it, enters directly in the determination of its own future"[34].

The second argument is that organisms *build* their environment, and, as such, they influence the environmental conditions that may affect their development: a view that we will now explore with more details.

## 3.2 Evolution

Darwinian and neo-Darwinian theories of evolution are modeled upon the concept of *adaptation*. Historical changes in species are seen as products of *external* forces that control the process. This view of evolution – *externalism* – predicts that organisms are passively modeled through generations by external conditions – that is, by the environment. In the most extreme versions of externalism[35], adaptation is considered both as the unique (or quasi-unique) force at work in shaping organisms, and as an all-powerful mechanism[36]. More moderate views take a pluralistic stance concerning the forces leading to evolution, and concede that optimization is only a working hypothesis for explaining the origin of particular traits. However, they still rely on the idea that populations are mainly shaped by the necessity to adapt to the external environment.

The kind of change implied by adaptation is variational, not transformative[37]. Variational change happens when the composition of a group of similar

---

[33] Lewontin insists in various occasions on the "size factor". The size of an organisms, for instance, determines the forces that rule its existence: bacteria are affected by gravity and Brownian motion in a very different way than human beings (see, for instance, Lewontin 1991, 76 – 7). This shows that, while genes do not straightforwardly determine the characteristics of an organism, neither the environment has an "autonomous" existence from its relation to organisms, as these can "select" the crucial external aspects that will affect their existence. It is worth noting that this insistence on size may come from influence from another Marxist biologist, J.B.S. Haldane, who wrote an essay *On Being the Right Size* (collected in Haldane 1927).
[34] Levins and Lewontin (1985), 96.
[35] Like those criticized in the famous "spandrels" paper: see Gould and Lewontin (1979).
[36] Alfred Russell Wallace is usually considered the forefather of this view. For an historical exposition and a defense of these "ultra" adaptionist positions see Cronin (1991).
[37] Lamarckian and vitalistic explanations of adaptations are instead transformative, as they postulate an inner tension in the organism to change in order to better adapt to the environment. According to Lewontin, Darwin's major insight is not to have claimed that evolution occurs –

(yet not identical) individuals varies through time as some properties shared by some members of the group vary in frequency. Concerning organisms, this happens every time a group satisfies three conditions[38]: a) *Variation* in morphological, physiological and behavioral traits among individuals; b) *Heredity:* variation is in part heritable, so that offspring resemble more their parents than other unrelated individuals; c) *Differential fitness:* different variant of individuals leave different numbers of offspring.

These conditions do not necessarily imply adaptation. In fact, adaptation is introduced as a possible way to explain differential fitness, and, as a sub-metaphor of the machine metaphor, it relies on several assumptions of mechanistic biology. The first assumption is the divisibility of the organism into *discrete traits* – morphological parts, physiological systems, behavioral outputs. The second assumption is that traits absolve one or several *functions*. The third assumption is that adaptation is an *optimization process*, or, in other words, that natural selection works as an optimizing "agent". The fourth assumption is that explanation of adaptations can be achieved by *reverse engineering* traits[39].

However, in order to be fully defended, adaptation also needs a fifth assumption – from a paradigm predating modern science: *natural theology.* Machines are man-made artifacts that serve a purpose: they solve *problems.* When organisms are interpreted in analogy with machines, they too are seen as artifacts to solve problems. However, from a logical and temporal standpoint, *problems must precede solutions:* as adaptation is a process, it cannot happen "without the ideal model according to which the adaptation is taking place"[40]. In a pre-Darwinian world, this did not constitute a problem: organisms were God-made, and they were equipped with functional features capable of solving problems posed by the equally God-made environment:

> The divine artificer created both the physical world and the organisms that populated it, so the problems to be solved and the solutions were products of the same schema. God posed the problems and gave the answers[41].

---

that had already been claimed – but to have understood that changes in the occurrence of traits in a population are variational and not transformative. See Levins and Lewontin (1985), 31.
**38** These are sometimes called "Lewontin's three conditions for evolution", as they were stated in Lewontin (1970), and later revised in Lewontin (1977a) and Lewontin (1977b). For an analysis and a critique see Godfrey-Smith (2007).
**39** See for instance Dennett (1995) or Pinker (1997).
**40** Levins and Lewontin (1985), 67.
**41** Levins and Lewontin (1985), 67.

The post-Darwinian world dispensed with God as an explanatory cause, but left intact the assumption that the environment is a landscape of problems to which organisms must fit in order to survive and reproduce. This gave rise to the concept of *ecological niches:* the idea of the environment as a collection of pre-existing holes that species try to fill by becoming more adapted, via natural selection, to the holes' shape[42].

As development understood as mere unfolding is a process that determines the organism from within, adaptation determines the organism from the outside: the environment "poses" problems, natural selection "picks" the variations best suited to solve them. Again, the organism is a mere by-product of forces outside its reach: it is a passive object of biological process. Again, Lewontin's response is to show that organism and environment form a dialectic unity in which every pole is at the same time object and subject, cause and effect[43]. His basic idea is that adaptation has to be substituted with the metaphor of *construction:*

> The environment is not a structure imposed on living beings from the outside, but is in fact a creation of those beings. The environment is not an autonomous process but a reflection of the biology of the species. Just as there is no organism without an environment, so there is no environment without an organism. The construction of environments by species has a number of well-known aspects that need to be incorporated into evolutionary theory.[44]

Organisms employ different strategies in constructing their environment[45]. They choose what is relevant among the different things and properties of the environ-

---

[42] In order to explain the fact that species change through time, we need to assume that even holes change: evolution is then a perpetual chase between organisms and their niches – as per the Leigh Van Valen's "Red Queen Hypothesis" (see Van Valen 1973). However, this still does not explain the phenomenon of *diversification* of species. On an adaptationist view, this is explicable only if we postulate that niches preexist species – that is, that the environment shapes evolution (on this latter issue see Lewontin 1978).

[43] Similar ideas were formulated by the XIX century Anglo-American Hegelians: see Pearce (2014). The striking resemblances may be due to similarity in the starting conditions: Anglo-American Hegelians needed to navigate between anti-evolutionism and Spencerian reductionism in the same way Lewontin is committed to avoid both post-modernist anti-scientific stance and neo-Darwinian reductionism. However, it is probably the use of the same *dialectical* methodology that draws both theories very close.

[44] Levins and Lewontin (1985), 99. The root of this idea of *construction* may be perhaps be traced in an extension (from humankind to every living beings) of Marx's third thesis on Feuerbach, as it was edited by Engels: "The materialist doctrine that men are products of circumstances and upbringing, and that, therefore, changed men are products of other circumstances and changed upbringing, forgets that it is men who change circumstances and that the educator must himself be educated" (Marx and Engels 1976, 7).

[45] Levins and Lewontin (1985), 99–104; Lewontin (2000a), 51–66.

ments they encounter. They alter the composition and the structure of their surroundings. They determine *via* their biology the actual physical nature of signals from the outside. They transform and modulate the statistical pattern of environmental variation in the external world. For all these reasons, organisms cannot be reduced to mere chunks of clay shaped by the environment. They are just as much determined by natural selection as they are capable of imposing their own conditions to the selection process.

## 3.3 Parts and Whole

In the alienated world of mechanistic biology, the organism is an epiphenomenon. Genes and the environment are the only true constitutive elements: environment talks, genes provide, all under the inflexible gaze of natural selection. By again establishing the organism as an irreducible causal factor, constructionism operates a delicate synthesis between different elements. Development is no longer a matter of genes expression mediated by the environment. And, at the same time, evolution is no longer a matter of environmental shaping of the gene pool. Both processes instead see the organism as the acted and the actor.

However, it remains a last issue to be addressed. According to mechanistic biology, the organism can be divided in *autonomous* and *discrete* parts each possessing particular properties capable of solving specific problems: that is, organisms can be divided in *functional traits*, just like a clock can be disassembled into gearwheels, springs and levers. However, there is no simple way to partition an organism:

> In what natural sense is a fin, leg, or wing an individual trait whose evolution can be understood in terms of the particular problem it solves? If the leg is a trait, is each part of the leg also a trait? At what level of subdivision do the boundaries no longer correspond to "natural" division?[46]

Adaptationists reply by claiming that traits can be defined in terms of the cluster of genes expressing them: these latter are, after all, definite and discrete entities (at least more discrete than the anatomical features of an organism). From a di-

---

[46] Levins and Lewontin (1985), 72. See also Lewontin (1978), 217. This insistence on the inadequacy of the whole/parts categories when discussing organism echoes Engels (see, for instance, Marx and Engels 1987a, 494), who, in turn, echoes Hegel.

alectical perspective this is untenable, as genes are only one factor contributing to development[47].

Moreover, traits cannot be really autonomous, as their functional properties usually depend on interaction with other parts. Accepting this latter fact does not necessarily involve a retreat to the body metaphor and a negation of evolution. Organisms, while they cannot be understood as mere sums of discrete traits, have the capability (through the generations) of changing small features of their phenotype without destabilizing their general structure. According to Lewontin, the whole/parts dialectics of organism has to be understood through two features. The first is *continuity:* "very small changes in morphology, physiology, and behavior usually have only a small effect on the ecological relations of the organism"[48]. The second is *quasi-independence* of characters: "there must exist a large number of possible phenotypic correlations between a given character change and other aspects of the phenotype"[49], otherwise, no particular feature of an organism could develop without altering the rest in a nonadaptive way.

It is not even clear why we should accept the idea that every trait absolves one or more functions – that is, that all traits of an organism are adaptive – or that adaptation always strives for optimality. These are disproved assumptions originating from natural theology (many traits, like fingerprints, hardly have any functions, while other, like the beginning of the optic nerve in the back of the retina, are clearly sub-optimal solutions), and are dangerous even as working hypothesis, because they can lead to the postulation of *ad hoc* hypotheses – the so-called *just so stories* – just in order to fulfill their requirements.

# 4 Assessing DT: The Case of Lewontin's Dialectical Biology

As we have seen, DT is a complex cluster of interrelated claims providing, at the same time, a description of nature's fundamental characteristics, a frame for scientific research, and an explanation of the development of our understanding of the natural world. However, while its advocates claim that DT is explanatorily decisive in so many ways, it has been always hard to explicate, at least in a clear

---

[47] Further critique on the "gene's eye point of view" of evolution are contained in Sober and Lewontin (1982).
[48] Levins and Lewontin (1985), 63.
[49] Levins and Lewontin (1985), 64.

and simple way, its specific contribution to the scientific enterprise – *its epistemic value*. Even Engels, for instance, admitted that many important scientific contributions were made by scientists who did not operate within an explicit dialectical frame of thinking. Is DT then useful for science, or do its pretensions need to be scaled down, or worse, abandoned?

## 4.1 Some Standard Arguments Against DT

Attempts to apply DT in biology and other natural sciences have often been criticized on several grounds. One has been, for instance, *obscurity*. On many occasions, DT has been labeled as mumbo jumbo, a complicated cant used to dignify emptiness of content with lavish words. The commonest target of this accusation has been the principle of *negation of the negation*[50], which relies on Hegel's notion of contradiction, so very different from its meanings in ordinary discourse and standard logic. A discussion on this technical point will lead us far astray, but two observations can be made. Scrupulous faith in this principle has led many – starting from Engels – to overuse it[51]. Moreover, its applications in biology have produced rather naive and pointless results, from Engels' barley seed negating itself into the plant and then in new seeds[52], to Haldane's mutation that negates heredity and it is (usually) negated, in turn, by natural selection[53].

In fact, DT has often been accused of being just a *rigid scheme superimposed on facts* – one that does not add nothing to standard scientific explanations – and would amount to nothing more than a pigeon hole into which to push, in a more or less arbitrary fashion, scientific discoveries made independent from any

---

[50] John Maynard Smith, himself a Marxist for a period of his life, confessed that he could never come to understand the meaning of this "law" (see Maynard Smith 1989, 37).

[51] For instance, in mathematics, where Engels, like Hegel, was very eager to find "contradictions". He believed he had found one particularly blatant in complex numbers ("a real absurdity", Marx and Engels 1987a, 112), and he compared strict belief in imaginary units and non-euclidean geometries with belief in spiritualism (Marx and Engels 1987a, 354). Calculus, in his opinion, was another nest of contradictions, that separated the field of "higher" mathematics from standard – "lower" – mathematics. This need to find "contradictions" in mathematics led Engels to some rather crude mistake on the subject, as when he incorrectly claims that "in every system with an odd base, the difference between even and odd numbers disappears" (Marx and Engels 1987a, 538).

[52] Marx and Engels 1987a, 126.

[53] Haldane (1941b).

dialectical considerations⁵⁴. Behind this and the previous criticism lays the idea that DT is, in the end, useless. On the one hand, it cannot predict or demonstrate its main claims: all it can do is to connect *a posteriori* different elements through a sloppy and obscure concept of contradiction. Clearly, this is not a fruitful framework for scientific research: it is, instead, a parasitic enterprise, a particularly byzantine way of organizing and describing scientific production.

Moreover, while dialecticians accuse mechanistic scientists of being narrowed down in their work by *a priori* assumptions, the accusation can also be reversed. Engels, for instance, had a tendency to refuse or to consider problematic those scientific results that clashed with his ideas⁵⁵. And some historical examples of "dialectical science" have been terrible – Michurinism⁵⁶ being the extreme case.

---

54 In the context of biology, this accusation was made for instance by the Russian born (and Soviet expat) economist Abram Lerner in a commentary piece (1938) on Haldane's paper *A Dialectical Account of Evolution* (1937). Here, Haldane proposed three dialectical "triads" as examples of DT at work in biology: (a) Heredity/Mutation/Variation, (b) Variation/Selection/Evolution, and (c) Selection of the fittest/Consequent loss of fitness/Survival of the non-competitive species. Lerner claimed that the methodology adopted to assemble the triads was inconsistent. In (a), mutation negates *strict heredity*, and the synthesis is then just a rephrasing of the antithesis; in (b) selection *does not negate variation*, it is just something different, and the synthesis is hence reached by addition of thesis and antithesis; in (c) the antithesis *is entailed by the thesis*, and the synthesis proceed from a further inference from the antithesis. Moreover, Lerner claimed, the particular proposed grouping of these concepts added nothing to their significance in biological theory.

55 An example is his aversion to the hypothesis of a "heat death" of the universe (Marx and Engels 1987a, 331–5), born from ideological reasons: in order for materialism to be true, Engels believed, the universe must be infinite and eternal. As he explains to Marx: "I am simply waiting for the moment when the clerics seize upon this theory as the last word in materialism. It is impossible to imagine anything more stupid. Since, according to this theory, in the existing world, more heat must always be converted into other energy than can be obtained by converting other energy into heat, so the original *hot state*, out of which things have cooled, is obviously inexplicable, even contradictory, and thus presumes a god" (Marx and Engels 1988, 246). It is interesting to note that, while some western Marxists coped with the issue by abandoning Engels' dogmatic views (see, for instance, Haldane 1941a, 266–268), the prejudice conditioned Soviet and Chinese cosmology for much of the XX century, even at a time when favorable empirical observations for the big bang theory – like galactic redshifts – were available (see Kragh 2012).

56 Between 1948 and 1953, Soviet biology banned genetics and adopted Michurinism, a neo-Lamarckian doctrine elaborated during the 1930s by Trofim Lysenko from the works of Ivan Vladimirovich Michurin. Official support of Michurinism by Soviet authorities continued up till 1965, despite its obvious and predictable failures. Levins and Lewontin (1985, Ch. 7) argue that Michurinism had nothing to do with DT – it was a poison ivy born from the rotten soil of Stalin's DIAMAT (see also Mayr 1997). However, this is a very controversial point: John Maynard Smith (1989), for instance, argued that Michurinism's soil was DT, as its main adversary,

Related to this latter point, DT can be accused of *arbitrarily separating humankind from nature*. Engels explicitly stated that dialectics knows *no hard and fast lines*, as proved for instance by Darwinism[57]. However, this was not taken to mean that *there were no lines at all*. Engels, for instance, thought that scientific laws and concepts were not easily transferable from the natural world to human societies:

> The most that the animal can achieve is to *collect*; man *produces*, he prepares the means of subsistence, in the widest sense of the words, which without him nature would not have produced. This makes impossible any unqualified transference of the laws of life in animal societies to human society.[58]

Engels' dialectic of nature has often been accused by western Marxists of being "reductionist" and "scientistic". However, the opposite charge can be raised: that DT serves the purpose of arbitrarily negating the possibility of advancing claims, from the perspective of natural science, on human nature[59].

## 4.2 The Case of Lewontin's Dialectical Biology

There is no doubt that Lewontin cannot be accused of obscurity and triviality, as he is a serious scientist and usually a clear writer, with no trace, in his scientific works, both of the sort of hocus pocus and of the *a posteriori* exercises of pigeonholing sometimes attributed to dialectical thinkers. Moreover, he cannot be accused of dogmatism, nor of having promoted bad science or even quackery, and, while he is certainly no friend of the idea of a "human nature"[60], he has repeatedly shown his willingness to defend this position on the scientific ground, without retreating to dogmatic trenches. Overall, his dialectical biology can then be seen as a particular promising test case to assess the epistemic value of DT.

---

Mendelian genetic, was clearly a non-dialectical theory: for this reason scientists influenced by Marxism were already inclined to look on it with suspicion.

57 Marx and Engels 1987a, 493.
58 Marx and Engels 1987a, 584. See also 330 – 1.
59 Steven Pinker, for instance, claims that "commitment to the 'dialectical' approach" permits one to "deny human nature and also deny [to] deny it" (2002, 127).
60 As is explicitly stated in Rose, Kamin and Lewontin: "this is why about the only sensible thing to say about human nature is that it is 'in' that nature to construct its own history" (1984, 14).

As seen before, Lewontin's approach reformulates the general principles of DT, on the one hand adjusting them to the specific of biological discourse, on the other hand by focusing on the dialectics of wholes and levels. Moreover, it claims that these principles do not uphold "law" status: instead, *they supply us with the correct epistemological and ontological frame for understanding the world.* Is this last point true? More specifically: could it be possible to criticize the mechanistic edge of modern synthesis biology, in order to embrace a new, more system-centered, understanding of development and evolution, without resorting to the full arsenal of DT? Could we obtain similar – or even better – results in this way? Or is DT the best approach?

The concept of *construction*, as opposed to *adaptation* is, probably, Lewontin's major contribution outside genetics – his main field of research – and a very fertile one. We find it, for instance, at one of the roots of eco-evo feedback theory[61]. Similarly, it has sparked the whole new field of research of niche-construction theory (NCT)[62]. This, in turn, is part and parcel of the extended evolutionary synthesis (EES)[63], an approach to biology born from a dissatisfaction with externalism, gene-centrism, "blueprint" metaphors, and similar concepts of mechanistic biology[64]. In fact, many (but certainly not all) features of EES are compatible with Lewontin's views. However, all the main proponents of NCT and EES, while they have some objectives and insights in common with Lewontin, do not share his commitment to DT. For some of them, this may be for "strategic" reasons[65] – that is, in order to avoid accusations of mixing "political" concepts into evolutionary theory. Nevertheless, this explanation cannot be generalized.

Is it possible then to be "anti-reductionist" without being dialectic? In Engels – but also in Lewontin – metaphysics$_2$ and dialectics are *contradictory* positions: only one of them is true, but also one of them *has* to be true. For this reason, Engels thought that successful, non metaphysical$_2$ scientists "unconsciously" adopted DT: *tertium non datur*. The same explanation could be given for contemporary theories that supply an alternative to standard evolutionary theory (given that this latter is a form of metaphysical$_2$ theory): they are dialectical, even if they do not explicitly adopt the dialectical frame. However, this explanation is problematic for (at least) two reasons. The first is that it does not take as a possibility that *there could be a third way:* a non "metaphysical$_2$", yet non "dialectic" approach. The second is that it indirectly questions the epistemic

---

61 Post and Palkovacs (2009).
62 Odling-Smee et al. (2003).
63 Laland et al (2015).
64 Laland et al (2015), Pigliucci and Müller (2010).
65 Svensson (2018).

value of DT. What is the point in working a precise dialectical methodology, if scientists can arrive at the same results with or without it?

A possible answer could be to circumscribe the epistemic role of DT to a cautionary principle. As Lewontin and Levins claim, "dialectical materialism is not, and has never been, a programmatic method for solving particular physical problems"[66]. It has to be considered as a set of warnings that should alert us every time our theories turn towards metaphysics$_2$. This would explain, on the one hand, why it is possible to do good science even ignoring DT, on the other, why we still need to reaffirm and clarify DT: to use an analogy, cautious drivers can do without speed limits, but these are still needed for all those who tend to be heavy on the gas.

This is of course a major scaling down of DT. And, even in this way, the first objection (there could be a *third way*) remains unresolved. The issue could be settled only by showing that all *real* alternatives to reductionistic approaches are dialectics – albeit "unconsciously" – or that they are not really alternatives. In order to sort this out it is necessary first to determine what could count as a *true* dialectic explanation – what is the *original element* of dialectic explanations apart from the particular vocabulary it implies. And this is not an easy task.

One particular feature of DT, as it applies to organisms, seems to be its stress on *integration*. However, integration cannot be *absolute*, as for the holistic – and pre-scientific – paradigm of the body metaphor. Integration has instead to be a variable with a finite (and non-zero) value determined by circumstances. And this is something easily accepted even by "reductionists". In fact, the extreme position that gives zero value to integration is a straw man. This is demonstrated by the overlapping of metaphors happening sometimes between "reductionists" and "dialecticians". For instance, in order to explain how "wholes are composed of units whose properties may be described, but the interaction of these units in the construction of the wholes generates complexities that result in products qualitatively different from the component parts", Lewontin and his associates adopt in one case the analogy of baking a cake – all the ingredients and procedures are susceptible to being analyzed separately, but the final product is not analyzable by taking the elements separately[67]. Richard Dawkins, to many the prototype of the arch-reductionist, noted how he had used the same analogy a few years before, and sarcastically asked if he too maybe was to count as a dialectical biologist[68].

---

66 Levins and Lewontin (1985), 191.
67 Rose, Lewontin and Kamin (1984), 11.
68 Dawkins (1985). The cake example is in Dawkins (1981).

This could mean that the difference between what we have labeled as a dialectical and a metaphysical$_2$ approach is more a matter of grade than of substance. After all, Lewontin's introduction of concepts like *continuity* and *quasi-independence* in order to explain otherwise inexplicable phenomena like evolutionary convergence is a major concession to "reductionism", even if it excludes the possibility that every kind of restructuring of organisms is possible. In fact, it means that we can take a standard adaptationist program as a good approximation of what is going on – in many cases, probably, even a *sufficient approximation* from an operative standpoint[69].

In order to avoid such reduction of DT into a particular tone of a palette that also contains reductionist approaches at its extreme, Lewontin painstakingly tries to differentiate dialectics from *interactionism* – the idea that complex phenomena are the resultant of *interactions* between simpler concepts – for instance, human psychology as a resultant of interaction between nature and nurture, or development as a resultant of interaction between genes and environment. In his views, dialectical explanations are distinguished from interactionist explanations because in these latter the causal pathway proceed straightforwardly in one direction. For instance, an interactionist explanation of development would see the organism as the passive effect of the interaction of outer and inner forces, while a dialectic explanation would also take in account the organism as a cause of this interaction – as organism intervene in their own development, by constructing and selecting their environment. But is this really a decisive difference? Could this kind of explanation be brought back to the general schema of interactionism by taking it as a peculiarly complex form of interaction?

Suppose that, during its development, at time $t_1$ an organism $o_1$ showing a certain phenotype $p_1$, given the adverse environmental conditions $e_1$, is capable of changing these latter into the favorable conditions $e_2$, which, in turn, redirect, together with chance and other external factors, the developmental process into producing the phenotype $p_2$. At time $t_2$ – that is, after $o_1$ has remodeled its environment – the organism can be said to stand as both the subject and the object of its development. However, the whole process can be easily rephrased in an interactionist discourse just by resorting to the vocabulary of *feedback*. For the first interaction (the one producing the state of affairs at $t_1$), triggers both the ret-

---

[69] Dennett (1992) generalizes this point: "It is true, as Lewontin has often pointed out, that the chemical composition of the atmosphere, for instance, is as much a product of the activity of living organisms as a precondition of their life, but it is also true that it can be safely treated as a constant, since its changes in response to local organismic activity are usually insignificant as variables in interaction with the variables under scrutiny" (35).

roaction on the environment, and the subsequent second interaction that produces the state of affairs at $t_2$.

Put in other words, the constructive relationship between the organism and its environment – their *co-development* – can be seen as an effect of the original interaction between the inner and the outer forces at play at time $t_1$. And while the whole process is more complicated than this original interaction, it may nevertheless be analyzed in terms of a chain of *multiple* separate interactions. In real cases, this analysis may be hard to bring forward due to various epistemic limits: interaction can be a mess. Yet, this does not mean that we should consider the whole process as on principle intractable from the standpoint of causality. On any other interpretation, the concept of *construction* risks being obscure – if not *mystical*.

In fact, the whole dialectical vocabulary is not employed by proponents of NCT, who prefer to speak of *reciprocal causation*[70] or of *cyclical causation*[71]. They are basically the same concept, and, as it has been noted[72], they have been part of evolutionary theory long before their explicit formulations in EES and NCT. It may even be that they are the principal models of causation at work in many fields investigated by biology – more than "standard" linear models of causation – and that the main ground of distinction between "reductionist" and "non-reductionist" is, in fact, the correct distribution of reciprocal/cyclical and linear phenomena of causation in evolutionary theory[73]. More importantly, reciprocal/cyclical causation can be understood without any reference to any of the assumptions of DT: the only conceptual piece that is needed is that of *feedback*[74].

However, Lewontin and Levins clearly state that the dialectical conception of cause/effect cannot be conflated with feedback circuits, as in these latter "there is no ambiguity about which is causing subject and which is cause object"[75]. The main issue here seems to be that feedback still maintains a *discrete analysis* of

---

[70] Laland et al. (2011).
[71] Laland (2004).
[72] For instance, in Brodie (2005) and in Svensson (2018).
[73] This is indeed the main difference between NCT and Dawkin's *extended phenotype* theory (Dawkins 1982) concerning the relation between organisms and environment: see Dawkins (2004) and Sultan (2015).
[74] Svensson (2018) seems to disagree when he claims that "reciprocal causation is nevertheless a good example of how Marxist philosophy and dialectical thinking have had a positive influence on the development of our field" (11). However, he employs the concept of *feedback* in order to explain reciprocal causation, and temporally distinguishes in an analytical manner cause from effects: two feature that contradict DT's approach to cause and effect as defined by Lewontin and Levins.
[75] See Levins and Lewontin 1985, 269.

cause and effect: while the action may be reciprocal, it can still be separately analyzable in principle. According to DT, instead, there is an element of ambiguity in reciprocal causality. And this ambiguity seems to be both inescapable and unintelligible, as Lewontin tries nowhere to clarify it. On the one hand, it is the only ingredient that avoids any reduction of its dialectical view of causation to some form of feedback theory; on the other hand, it is clearly a "magical" ingredient, because no one knows what it is and how it works. And worse, it is also a *useless* ingredient, since from an operative standpoint it does not produce a superior explanatory situation to reciprocal/cyclical causation.

In fact, the situation produced is probably explanatory *inferior*. According to an example provided by Lewontin, *construction* happens when a) an organism changes some of its features in order to cope with the environment, filtering away some of its elements and making others relevant; or when b) an organism changes a feature of the environment. As Godfrey-Smith has noted[76], only the second case can account for niche-construction. This happens because the first case entails a change in some relational property, but only the second involves a real change of intrinsic properties of the environment. And if we allow for every change in relational properties to count as niche-construction, then the concept becomes trivial – as nearly every activity of an organism, from its metabolism to locomotion causes a change in the relational properties between it and the environment. However, this latter distinction between relational and intrinsic properties is something negated by DT. According to DT, all properties are in the end relational, as nothing stands isolated from the overall network of processes.

These views on properties and causation are both essential elements of DT: they are part of the set of assumptions that specify its metaphysical$_1$ claim, and permit us to characterize the dialectical approach in a distinctive way. However, they also seem to provide DT with a weak spot. On the one hand, it is not necessary to assume them in order to criticize mechanistic biology and build a more system-centered theory. On the other hand, they provide us with a weaker understanding of the complex interactions between organisms and environments. DT then, at least in the case we have examined, stands as a sharp and fertile approach, yet probably not as the *best* approach.

---

76 Godfrey-Smith (1996) and Godfrey-Smith (2001).

# References

Boudry, Maarten & Pigliucci, Massimo. (2013). *The Mismeasure of Machine: Synthetic Biology and the Trouble with Engineering Metaphors*. "Studies in History and Philosophy of Biological and Biomedical Sciences", 44, 660–668.

Brodie, Edmund D. (2005). *Caution: Niche Construction Ahead*. "Evolution" 59–1, 249–251.

Cronin, Helena. (1991). *The Ant and the Peacock. Altruism and Sexual Selection from Darwin to Today*. New York: Cambridge University Press.

Dawkins, Richard. (1976). *The Selfish Gene*. New York: Oxford University Press.

Dawkins, Richard. (1981). *In Defence of Selfish Gene*. "Philosophy" 56–218, 556–573.

Dawkins, Richard. (1982). *The Extended Phenotype: The Long Reach of the Gene*. New York: Oxford University Press.

Dawkins, Richard. (1985). *Sociobiology: The Debate Continues (Review of Lewontin, Rose, & Kamin's "Not in Our Genes")*. "The New Scientist", 24, 59–60.

Dawkins, Richard. (2004). *Extended Phenotype – but not Too Extended. A Reply to Laland, Turner and Jablonka*. "Biology and Philosophy" 19, 377–396.

Dennett, Daniel. (1992). *Hitting the Nail on the Head. Commentary on Thompson, Palacios and Varela*. "Behavioral and Brain Sciences" 15–1, 35.

Dennett, Daniel (1995). *Darwin's Dangerous Idea. Evolution and the Meaning of Life*. New York: Simon and Schuster.

Descartes, René. (1902). *Ouvres*. Vol. 6. Paris: Léopold Cerf Imprimeur Éditeur.

Fay, Margaret A. (1978). *Did Marx Offer to Dedicate* Capital *to Darwin?*. "Journal of the History of Ideas" 39, 133–146.

Feuer, Lewis S. (1975). *Is the "Darwin-Marx Correspondence" Authentic?*. "Annals of Science" 32, 1–12.

Godfrey-Smith, Peter. (1996). *Complexity and the Function of Mind in Nature*. Cambridge: Cambridge University Press.

Godfrey-Smith, Peter (2001). *Organisms, Environment, and Dialectics*. In Rama et al. (2001).

Godfrey-Smith, Peter (2007). *Conditions for Evolution by Natural Selection*. "Journal of Philosophy" 104, 489–516.

Gould, Stephen Jay & Lewontin, Richard (1979). *The Spandrels of San Marco and the Panglossian Paradigm: A Critique of the Adaptionist Program*. "Proceedings of the Royal Society of London", Series B, 205–1161, 581–598.

Haldane, John Burdon Sanderson. (1927). *Possible Worlds and Other Essays*. London: Chatto & Windus.

Haldane, John Burdon Sanderson. (1937). *A Dialectical Account of Evolution*. "Science and Society" 1–4

Haldane, John Burdon Sanderson. (1941a). *Dialectical Materialism and Modern Science. I. Everything has a History*. "Labour Monthly" June, 266–268.

Haldane, John Burdon Sanderson. (1941b). *Dialectical Materialism and Modern Science. IV. Negation of the negation*. "Labour Monthly" October, 430–2.

Kitcher, Paul. (1999). *The Hegemony of Molecular Biology*. "Biology and Philosophy" 14, 195–210.

Kragh, Helge. (2012). *The Universe, the Cold War, and Dialectical Materialism*. arXiv:1204.1625v2 [physics.hist-ph]. Last retrieved: February 27–2019.

Laland, Kevin N. (2004). *Extending the Extended Phenotype.* "Biology and Philosophy" 19, 313–325.

Laland, Kevin N., Sterelny, Kim, Odling-Smee, John, Hoppitt, William, & Uller, Tobias. (2011). *Cause and Effect in Biology Revisited: Is Mayr's Proximate-Ultimate Dichotomy Still Useful?.* "Science" 334 (6062), 1512–1516.

Laland, Kevin N., Uller, Tobias, Feldman, Marcus W., Sterelny, Kim, Müller, Gerd B., Moczek, Armin, Jablonka, E., & Odling-Smee, John. (2015). *The Extended Evolutionary Synthesis: Its Structure, Assumptions and Predictions.* "Proceedings of the Royal Society Publishing" 282.

Lerner, Abram. (1938). *Is Professor Haldane's Account of Evolution Dialectical?.* "Science and Society" 2–2, 232–242.

Levins, Richard & Lewontin, Richard. (1985). *The Dialectical Biologist.* Cambridge (Mass.) & London: Harvard University Press.

Levins, Richard & Lewontin, Richard. (2007). *Biology Under the Influence. Dialectical Essays on Ecology, Agriculture, and Health.* New York: Monthly Review Press.

Lewontin, Richard, Rose, Steven & Kamin, Leon. (1984). *Not in Our Genes. Biology, Ideology, and Human Nature.* New York: Pantheon Books.

Lewontin, Richard. (1970). *The Units of Selection.* "Annual Review of Ecology and Systematics", 1–1.

Lewontin, Richard. (1977a). *Evoluzione.* In "Enciclopedia Einaudi", vol. 1. Torino: Einaudi. English translation in Levins & Lewontin (1985).

Lewontin, Richard. (1977b). *Adattamento.* In "Enciclopedia Einaudi", vol. 1. Torino: Einaudi. English translation in Levins & Lewontin (1985).

Lewontin, Richard. (1978). *Adaptation.* "Scientific American", 293–3, 156–169.

Lewontin, Richard. (1983). *The Organism as the Subject and Object of Evolution.* "Scientia", 188, 65–82.

Lewontin, Richard. (1991). *Biology as Ideology. The Doctrine of DNA.* Toronto: House of Anansi Press.

Lewontin, Richard. (1998). *Gene, organismo e ambiente.* Bari/Roma: Laterza.

Lewontin, Richard. (2000a). *The Triple Helix. Gene, Organism, and Environment.* Cambridge (Mass.) & London: Harvard University Press.

Lewontin, Richard. (2000b). *It Ain't Necessarily So. The Dream of Human Genome and Other Illusions.* New York: New York Review of Books.

Marx, Karl & Engels, Friedrich. (1976). *Collected Works.* Vol. 5. New York: International Publishers.

Marx, Karl & Engels, Friedrich. (1983). *Collected Works.* Vol. 40. New York: International Publishers.

Marx, Karl & Engels, Friedrich. (1985). *Collected Works.* Vol. 41. New York: International Publishers.

Marx, Karl & Engels, Friedrich. (1987a). *Collected Works.* Vol. 25. New York: International Publishers.

Marx, Karl & Engels, Friedrich. (1987b). *Collected Works.* Vol. 42. New York: International Publishers.

Marx, Karl & Engels, Friedrich . (1988). *Collected Works.* Vol. 43. New York: International Publishers.

Marx, Karl & Engels, Friedrich. (1989). *Collected Works.* Vol. 24. New York: International Publishers.

Marx, Karl & Engels, Friedrich. (1990). *Collected Works*. Vol. 26. New York: International Publishers.
Marx, Karl & Engels, Friedrich. (1991). *Collected Works*. Vol. 45. New York: International Publishers.
Maynard Smith, John. (1989). *Did Darwin Get It Right? Essays on Games, Sex, and Evolution*. New York: Chapman and Hall.
Mayr, Ernst. (1997). *Roots of Dialectical Materialism*. In Edouard I. Kolchinsky (ed.), "Sovetskaia Biologiaa v 20–30 kh Godakh", 12–17.
Odling-Smee, F. John, Laland, Kevin N., Feldman, Marcus W. (2003). *Niche Construction. The Neglected Process in Evolution*. Princeton and Oxford: Princeton University Press.
Pearce, Trevor. (2014). *The Dialectical Biologist Circa 1890: John Dewey and the Oxford Hegelians*. "Journal of the History of Philosophy", 52–4, 747–778.
Pigliucci, Massimo & Müller, Gerd (eds.). (2010). *Evolution. The Extended Synthesis*. Cambridge (Mass.): The Mit Press.
Pinker, Steven. (1997). *How the Mind Works*. New York: Norton.
Pinker, Steven. (2002). *The Blank State. The Modern Denial of Human Nature*. New York: Penguin Books.
Post, David M. & Palkovacs, Erik P. (2009). *Eco-evolutionary Feedbacks in Community and Ecosystem Ecology: Interactions Between the Ecological Theatre and the Evolutionary Play*. "Philosophical Transaction of the Royal Society" B 364, 1620–1640.
Quine, William Van Orman. (1948). *On What There Is*. "Review of Metaphysics", 2.
Singh, Rama S., Krimbas, Costas B., Paul, Diane B. & Beatty, John (eds.). (2001). *Thinking About Evolution: Historical, Philosophical, and Political Perspectives*. Cambridge: Cambridge University Press.
Sober, Elliott & Lewontin, Richard. (1982). *Artifact, Cause and Genic Selection*. "Philosophy of Science", 49, 157–180.
Sultan, Sonia E. (2015). *Organism and the Environment. Ecological Development, Niche Construction, and Adaption*. Oxford: Oxford University Press.
Svensson, Erik I. (2018). *On Reciprocal Causation in the Evolutionary Process*. "Evolutionary Biology" 45, 1–14.
Van Valen, Leigh. (1973). *A New Evolutionary Law*. "Evolutionary Theory" 1, 1–30.
Varzi, Achille. (2005). *Ontologia*. Roma-Bari: Laterza.
Wilson, Edward O. (1975). *Sociobiology. The New Synthesis*. Cambridge (Mass.): Harvard University Press.

Lenny Moss
# Can Normativity be the Force of Nature that Solves the Problem of *Partes Extra Partes?* Episode IV – A New Hope – Natural Detachment and the Case of the Hybrid Hominin

**Abstract:** From a subjective point of view, we take the existence of integrated entities, i.e., ourselves as the most unproblematic given, and blithely project such integrity onto untold many "entities" far and wide. However, from a naturalistic perspective, accounting for anything more integral than the attachments and attractions that are explicable in terms of the four fundamental forces of physics has been anything but straightforward. If we take it that the universe begins as an integral unity (the singularity referred to as "the cosmic egg") and explodes into progressive stages of internal detachment, then we can also fathom the idea that eddies of relative detachment becoming increasingly integral. Helmuth Plessner made a powerful case for the onset of "positionality" constituting one of the major transitions in nature. Surely, the emergence of "entities" (i.e., life-forms) that position themselves in relation to their surround marks a decisive transition in relative levels of detachment and some would say "autonomy." It would follow, with no less force, that where and when entities can be seen to be normatively integrated, and indeed to be the agents of their own normativity, that another threshold of detachment has been crossed. The paper introduces and explores the idea that normativity, embedded in a wide-ranging theory of natural detachment, can be considered an emergent force of nature that is requisite to accounting for levels of integration beyond that which is explicable in terms of the four fundamental forces of physics. Following this line of enquiry, we argue that the first expression of a fully, normatively-integrated life-form is neither a spoken language user nor for that matter an individual but rather the neoteny-based, Homo erectus Group. In so doing we claim to have made an inroad into embedding the force of normativity into a wide-ranging naturalist framework, to have provided philosophical anthropology with a new (post-individualist) point of departure, and at least playfully, to have given some naturalistic grist to Hegel's proclamation that spirit (Geist) is the truth of nature.

**Lenny Moss,** University of Exeter

# I Natural Integration and Natural Normativity

The problem of natural integration, or perhaps unification – the constitution of a unity – is a truly hard problem that has seldom if ever been addressed as such. Indeed, it is perhaps *the* hard problem of the philosophy of life and mind. Granted, how a natural entity can have interiority, i.e., subjectivity, is a hard problem, but if the question of integration and unification is not identical to the "Hard Problem of Consciousness" (Chalmers 1995) it is also inseparable from it and surely a presupposition of the very possibility of interiority. Nor would even an understanding of how subjectivity could be resident to a single cell tell us how consciousness could become an integrated unity across many cells. If we assume consciousness is a physically based phenomenon, and that it draws upon the activity of various parts of the brain, let alone constituent cells, then we must face our deficits in understanding how the experience of a unified consciousness is realized at the level of an integration of some cells but not others albeit in the absence of evident, non-arbitrary, physical boundaries.

Both more fundamental than the Hard Problem of Consciousness, and more expansive in scope, the problem of integration/unification is also central to the problem of the origin(s) of life. If merely conventionally mechanistic associations of parts outside of parts (*partes extra partes*)[1] were sufficient for creating a living unit, it would have been achieved long since. The fact that a natural bacterial chromosome could be successfully replaced by an artifactual chromosome only further testifies to the irreducible importance of a naturally integrated unit, the bacterial cell, as the prior condition of possibility for any molecular configuration, natural or artificial, to *be* a chromosome.[2] Perennial claims by biologists to have solved the problem of the origins of life have either been projections based upon the largely uncritical metaphysics of informational idealism (masquerading as consensual science), or based on the largesse of liberal promissory notes fueled by commercial marketing interests.

---

[1] I first heard the wonderfully mellifluous phrase "partes extra partes" when uttered by Charles Taylor in lectures, possibly his lectures on Hegel at Berkeley in 1981. Taylor himself took the phrase from Merleau-Ponty.

[2] When Craig Venter announced that he had created artificial life in 2010 it resulted in splash headlines in media outlets as respectable as The Guardian and convinced lay readers that Venter had succeeded in creating life *de novo*. https://www.theguardian.com/science/2010/may/20/craig-venter-synthetic-life-form. The project cost 40 million dollars and was addressed by the responsible investigators themselves as representing an advance in the technology of synthetic biology, not in terms of relevance to questions of the origins of life (Gibson et al. 2010).

Lest there be any doubt, I do not aspire to overcome the problem of integration *tout court* in this paper. Beyond thematizing it as a problem (a worthy endeavor in itself, I claim) I will want to explore the role of "normativity" in hominin/human integration and toy with the idea of normativity being *a force of nature*. Clearly doing so, if feasible at all, would require a radical expansion, along with a capacity for scaling, of our current concept of normativity. I hope to at least provide some adumbration of what that might look like.

For Leibniz, Kant and Hegel (among others) normativity plays a central role in the understanding of life and mind, albeit in different ways. Hegel, for our present *naturalistically* oriented purposes, is of most interest because he comes the closest to offering an account of the transition between a pre (or perhaps proto?) normative account of Nature and a fully normative account of Spirit. As Brandom tells us: "*How one understands the relation between these, both conceptually and historically, is evidently of the first importance in understanding what Hegel has to teach us about the normative realm he calls 'Geist'*" (Brandom 2019).[3]

Hegel, as most clearly and emphatically presented by Brandom, provides a fundamental account of how the very glue that enables the human realm of culture to exist, that integrates and unifies it, is the glue of normativity. Normativity on this level is about a force which constitutes and unifies human life, albeit not a force that can be reckoned in terms of the vocabulary of classical mechanics. Human life entails the existence of self-conscious subjects who are constitutively enmeshed in the force field of reciprocal accountabilities and entitlements. To be a self-conscious subject, is to have a self-identity, to be not just an entity in-itself but an entity for-itself. An entity for-itself doesn't exist in a vacuum. To be an entity for-itself is an ongoing social achievement dynamically realized within the normative fabric of human life. As George Herbert Mead argued, a 'self' is a reflective objectification that we discover from a certain social vantage point. There is the self we are as son or daughter, as husband or wife, as father or mother, as teacher or carpenter, and so on. We are neither the passive residue of forces out of our control nor the sovereign dictators of the self we are, we are active in taking stands in each social location about who we are, but our ability to find ourselves to be that self is mediated by the constructive recognition of others. All the possibilities of being a certain self are normatively defined (they are located with

---

[3] Brandom's much anticipated tome on Hegel had not yet be released as of this writing. The quotations are derived from his 2014 draft of the text made available on his home page https://www.pitt.edu/~brandom/spirit_of_trust.html. The quotation above is from A Spirit of Trust: A Semantic Reading of Hegel's Phenomenology, Part Three Self-Consciousness and Recognition, Chapter Seven: The Structure of Desire and Recognition: Self-Consciousness and Self-Constitution.

the framework of entitlements and accountabilities). To be a being-for-itself is thus an on-going accomplishment achieved necessarily within the fabric of social-normative space.[4]

Within the fabric of normatively structured reciprocal accountabilities and entitlements we take a stand on whom we claim we are entitled to be recognized as. And we do so in the context of reciprocally recognizing those whose recognition we require. To speak of entitlements, obligations and accountabilities is to speak of forces (or *power* in the Foucaultian sense). Hegel famously refers to the *I that is a We and the We that is an I*.[5] It can only be a We that enables me to be an I, and thus reciprocally a We can only be a We if it is situated by an I. The obligate relation of I and We describes a special kind of palpable integration and unity. Hegel has done original and foundational work with respect to characterizing the kind of normatively powered unity and integration that constitute the phenomenon of Geist (only in the context of which selves can be selves). But that is not to say that Hegel offers a full-fletched, naturalistic answer to the problem of integration as we've defined it.

For both Leibniz and Kant, normatively resonant entities must necessarily be presupposed and cannot be theoretically derived from pre-normative precursors. As fundamental substance Leibniz's monads are as basic as it gets and are always already normatively-infused fully integrated units. Only God could provide the recipe for how to make one. Likewise, the Critical Philosophy holds in principle that neither the origins of the ostensibly purposive functional and organizational unity of any living being, nor certainly that of the normativity of pure practical reason that infuses "Man's moral vocation", could ever be accounted for by (human) theoretical means.

Hegel, by contrast, appears to provide a developmental account in which animal desire becomes an entryway into Geist when desire becomes a desire for recognition. But even the assumption of animal desire is already assuming a unity of purpose that takes the fundaments of what I am calling the really hard problem as an assumed given. Hegel has at least offered the prototype of a dialectical account of movement from a lower to a higher level of unity. Inasmuch as Hegel is, as his famous title suggests, performing a phenomenology of mind, some manner of mindedness is always already assumed. Hegel is thus not accountable to the kind of really hard problem that a contemporary naturalism would pose as it would not be immanent to his phenomenological standpoint.

---

[4] For an excellent discussion of Mead's concepts of I and self and of the intersubjective origins of self-hood see Habermas 1992.

[5] Hegel 1977, §177, 110.

The Absolute, and the Concept, which are ontological primitives for Hegel, supersede the problem of integration and unification in advance, albeit by phenomenological fiat.

Neither a phenomenological nor a transcendental philosophy alone can offer a fully naturalistic account of the basis of integration and unity, given the constitutively methodological assumptions and constraints of each approach. And this is no less the case for Heidegger's existential phenomenology that takes its point of departure from *Dasein's* practical involvement in her normatively integrated world. But nor is a reductive empiricism the answer to going beyond the limits of a *partes extra partes* vision of nature. Expressed in very simple terms, neither a science that limits itself entirely to a bottom-up approach, nor a science that limits itself to only a top-down approach, will be able to solve the problem of integration and unification at all levels of nature. What we require is an approach that can bring both of these together in some way. In recent work Arran Gare has made a compelling case for how this involves a renewed acceptance and embrace of speculative thinking (Gare 2018). For physicist Lee Smolin and philosopher Roberto Mangabeira Unger (2015), whose collaborative work over eight years has resulted in a fascinating book entitled *The Singular Universe and the Reality of Time*, bringing together the big picture of the universe as a whole with the micro-level advances of physics, entails no less than a return to the practice of Natural Philosophy proper. In referring to "normativity as a force of nature", I have thus far only used as a teaser, or perhaps appetizer, the idea that normativity might be able to span the chasm between the bottom up and the top down, and thus serve as a conceptual lynchpin of a renewed speculative natural philosophy. There will be more to come. But it should also be acknowledged that attempting to bring together a top-down with a bottom-up approach, even to also address questions of integration and unification is not entirely new.

Just shy of 100 years ago, Helmuth Plessner, who had trained in both biology and philosophy, put forward the theme of "positionality" as the key concept in an attempt to weave together empirical and phenomenological, bottom-up and top-down, big picture and micro-detail, perspectives into a coherent natural philosophy and philosophical anthropology. Plessner included fundaments about biological structure, organization, physiology, ecology, and ethology in his attempt at elaborating a non-dualist, and systematically coherent framework for all forms of life (Plessner 2019).

Using positionalilty as a point of departure for elaborating a logic of life, including human life, is in fact tantamount to simultaneously treating the question of integration, or unification, as a point of departure. As Plessner makes clear, it is only with the living state that boundaries become something other than arbitrary and contingent. To the extent that life is defined by anything, it is defined by the dynamic enactment of a distinction between inner and outer, Plessner's so-called "double aspectivity." "Positionality" then is the categorical universal and *sine qua non* of the living state and conceptually defines the boundary between physical phenomena that, for all intents and purposes, are *partes extra partes*, from those for which some non-trivial level of system unity and integrity has been achieved. Plessner does not specifically take up the language of *normativity* and yet one can find it to be implicit in his account. The realization of a life, even at the most rudimentary, let's say the simple, single-celled level, already entails a form of active mediation between reaching outward and enforcing a boundary, that suggests the regulatory enactment of an implicit norm. For Plessner, there is a logic of dialectical building of positional levels upon levels of reflection that culminate in the human level of "excentric positionality" whereby the "shared" (and invariably normative) perspective of the *universal Other* is always reflexively embodied and reflectively available. Which is to say that Plessner has long since offered a body-mind neutral account in which human-level normativity is located on a natural continuum in which questions of dynamic system integration are at least implicitly fundamental. Where Plessner fell short, I suggest, is in (only) deriving a largely monological account of the emergence of human-level normativity.[6] The following may be viewed as in part an attempt to offer the complementary perspective, albeit with the full reconciliatory and synthetic engagement to appear in subsequent work.

## II Natural Detachment and the Hybrid Hominin

### A The Idea of Detachment

The idea of "detachment" may seem on first pass as a very odd way to begin a discussion about the basis of integration. However, if we begin with some notion of the universe as a whole in a state of a kind of primordial integration then perhaps we can fathom how detachments from a simple and primordial integration would be a pre-condition for new and perhaps more interesting and intricate forms of integration. When we take the human self (however "natural" doing so may feel) as an unproblematic given we quickly obscure the issue. No matter

---

[6] Cf. my preliminary account of this criticism in my Notre Dame Philosophical Reviews discussion of the recent English translation (Plessner 2019) https://ndpr.nd.edu/news/levels-of-organic-life-and-the-human-an-introduction-to-philosophical-anthropology/

how you cut it, from any naturalistic point of view, the self is highly derived state of affairs. We can't afford to forget this lest we risk inadvertently tumbling back into metaphysical dualism. There will be a fairly compelling case to be made for the relevance of the idea of detachment to the transition (or transitions) constitutive of hominin evolution and indeed this idea goes back to the late eighteenth century. The more speculative mood deigns to propose that detachment can be scaled and tracked *all the way down*. For present purposes the objective is not to aspire for the idea of detachment to assume the status of a new master theory but rather to lobby for its value in a perspectival sense and in so doing simultaneously affirm the wisdom of some measure of perspectivism in our thinking.

The intuition I wish to arouse is that an entity that claims some measure of detachment is to that extent standing on its own feet, and the greater the level of detachment the more self-standing it is. Detachment then would also suggest internal unity and thus integration, and so must be a kindred concept, and yet detachment is not identical to integration. It can't be identical with integration because, as we will see, there is also a special kind of detachment which is "downward" and disintegrative. More on this later.

Detachment, we will have to assume, begins at the beginning, i.e., with the Big Bang. Prior to the Big Bang, all existence (whatever that means) was confined to an infinitely dense, infinitesimally small singularity in which there was no space or time and all four basic forces of nature (as we now understand them) were united into one. The universe was born, space-time emerges, in an explosion of detachment. Without getting too bogged down in the technical details of high-energy physics and cosmology, the take-home lesson is that a logic of detachments-built-upon-detachments is set into motion from... the beginning. It has been theorized that cosmic detachment began with the detachment of gravity from the unity of fundamental forces, resulting in the formation of elementary particles and anti-particles followed by an *inflation* into space-time triggered by the further detachment of the strong nuclear force. As yet inexplicable asymmetries in the appearance of baryons (matter) versus antibaryons (anti-matter) were a *sine qua non* for the early persistence of our universe. It is now believed that the possibility of mass was predicated upon the detachment of the particle called the Higgs boson and with that the associated Higgs field. The detachment of the Higgs boson, and thus of mass, then constitutes a horizon for all subsequent material detachments in our universe.

Physicists characterize the possibility *state space* of a simple system in terms of its "degrees of freedom." For example, a simple atom like hydrogen can respond to a perturbation (such as being hit by a photon) by moving in space along three axes (3), rotating (4), or elevating the energy level of an electron (5). It is thus accorded five degrees of freedom. A simple diatomic molecule, like $H_2$, can also vibrate along its common axis so adds an additional degree of freedom. *Detachment is always about the emergence of higher degrees of relative independence.*

We can already see a reciprocal relationship between detachment and integration. Two atoms each with five degrees of freedom, shed their independent degrees of freedom through integrating, by way of a covalent bond, to form a new entity, the diatomic molecule, with six degrees of freedom. The more degrees of freedom the greater the detachment. As our universe has evolved it has given rise to constituent entities with greater and greater abilities to buffer themselves against "ambient winds" (be that bombardment by radiation or predation by voracious carnivores). Following the same logic, a particle with rest mass that creates a well in space-time is more detached than a particle (like a photon) with no rest mass. A macromolecule, like a protein-based enzyme, whose folding history affects its future actions is more detached than a simpler molecule whose structure is solely determined by thermodynamic necessity (and thus has no history). A major transition in detachment occurred when a system emerged that actively constituted its own boundary (Plessner's *positionality*) and actively sustained its ability to do so. We typically associate this level of detachment with what we recognize as "life".

All states of detachment are relative, none are absolute. Levels of detachment exist in nested hierarchies. When a new level of detachment emerges, such as the boundary constituting, self-sustaining system (a simple cell), it also creates a space for *downward detachments* that may be viewed as parasitic on the higher level of detachment upon which it depends. Viruses emerged as expressions of downward (parasitic) detachment. Parasites and their hosts, lower and higher levels of detachment, dialectically interact resulting in the transformations of each and the appearance of new capacities that neither side of the equation alone could have produced. What begins as an "arms race" between parasites and hosts has resulted in *de novo* resources that enable parasite and host to re-integrate and make the jump to a new higher level of detachment. Internal compartmentalization and thereby an increase of organizational complexity has been a response strategy of "eukaryotic cells" to parasitic challenge (Koonin 2016) but through symbiotic re-integration has provided the basis for eukaryotes to engage in exploratory processes that lead the way toward both *de novo* ontogenetic and phylogenetic adaptations. The capacity for active exploratory processes ratchets up the degrees of freedom and level of detachment of a eukaryotic entity by exponential measures. When "variation" (in the Darwinian sense) is no longer principally about the stochastic roll-of-the-dice of enol-keto tautomerism in DNA replication, but rather is facilitated by the active processes of a detached entity and instigated by sensitivity to contingent ambient conditions, when might we be warranted to say that normative-choice making has come into play?[7]

The stark inadequacy of a narrowly-survivalist, Neo-Darwinism was amply revealed by the genome-theoretic debunking of the "junk DNA" thesis, which itself had been a leading inspiration for the far more popularly influential doctrine of the "selfish gene." It has been long-since empirically well-established that the vast majority of human genomic DNA not only doesn't code for unique protein sequences, nor even for non-coding regulatory elements, but rather is highly repetitive and virus-like in its structure. But rather than being merely a genomic "free-rider" as the proponents of the "junk DNA" thesis proclaimed, the endogenous, and potentially transposable and self-replicating sequences, have been shown to be instrumental in processes of genetic segmental duplication that result in species specific, perhaps even species defining, gene families (Moss 2006). The dialectics of detachment suggest that transitions to higher order levels of detachment will attract new rapprochements between competing entities from lower and higher levels of detachment, and thus that downward detachments, seen dialectically, are means toward a variety of different possible sequelae.

When Hegel announced that Spirit is the truth of Nature, he didn't imagine a story being told at a cell and molecular level and yet he offered a perspective that we may find insight-inspiring to generalize upon. The preponderance of philosophical normativists have kept naturalism, one way or another, at arm's length. The very idea, however, of an expansive movement of detachment, indeed a dialectics of detachment, anchors an impetus towards an inevitability of normativity as a force of integration, well beyond the imagination of our speculation-allergic normativists. If the legacy of propositionally-delimited reflection on the nature of normativity inhibits our ability to fathom normativity outside of modern human practice, then perhaps nature is calling upon us to start thinking beyond the constraints of this legacy. The case of the hybrid hominin, I suggest, will

---

[7] Kirschner and Gerhart (2005), two leading contemporary cell and developmental biologists, updated our understanding of the cellular processes involved in generating novelty, including that of "exploratory processes" and referred to these as the basis of "facilitated variation." That these concepts, based upon unimpeachable empirical evidence, and which radically change the perception of the status of cells and organisms as agents in evolutionary change, haven't filtered into general understanding should well raise questions about the warrants for privileging the continued promulgation of outdated versions of Neo-Darwinism as being more rational than myth, ideology or creed. The proper response to the potential dangers of Creationist irrationalism (if such there is) is not the defensive (or offensive) petrification of "classic ideas" that have been made refractory to new findings, new insights and new theories.

provide an enabling pathway toward this end (Moss 2016a). That is also to say, in the spirit of a renewed philosophical anthropology, that when we have met the true *missing link*, she will be us.

## B The Hybrid Hominin

Remarkably, the great majority of both popular and disciplinary accounts of human nature, human evolution, cultural evolution and the like stand oblivious to two elephants in the room, the biological juvenilizing of the human organism, and the uniquely human orientation toward cooperativity and "we-ness." Whereas the former insight has its origins in common-sensical observations of the late eighteenth century which progressively gained more recondite confirmation over the centuries,[8] the latter is arguably the leading psychological discovery of only twenty-first century human sciences[9]. Despite this temporal disparity, I will argue that these insights reflect mutually implicative aspects of the hominin we became and continue to be. Hominin neotenous juvenilizing and the hominin orientation toward 'we-ness' will be part of a story about a transition to a new level of detachment that will offer an exemplar for normative integration at a pre-conceptual level with implications ranging both back into pre-hominin existence and forward into the symbolically structured human world as well.

Our two elephants are typically ignored because they fail to coincide with deeply ingrained assumptions about the human individual; the individual human body and the individual human mind. These assumptions have been secured by various traditions and practices not limited to methodological individualism, Neo-Darwinist reductionism, neo-classical economics, rational choice theory, liberal political theory, and so forth. More recent attempts at elaborating

---

[8] This insight is best thought of as beginning with Gottfried Herder's 1772 award winning "Essay on the Origin of Language" (1986 [1772]) taking theoretical shape with late nineteenth-century theories of neoteny, heterochrony and juvenilization and reaching its philosophical pinnacle with Arnold Gehlen's masterwork *Der Mensch* (1940) translated into English as *Man: His Nature and Place in the World* (1988). For details of the late nineteenth-century biological theories see Gould 1977, a recent exposition of the human neoteny view can be found in Bromhall 2003.
[9] The work of Michael Tomasello and co-workers at the Max Planck Institute for Evolutionary Anthropology has been a game-changer insofar as establishing the inherent orientation of human infants toward cooperative interactions and the centrality of the human capacity for social understanding and the capacity for "we-mode" as the unique cognitive differentia that distinguishes humans from other higher primates. An easily approachable introduction to these findings can be found in Tomasello 2009 and further explored in Tomasello 2019.

cultural evolution theory from a Neo-Darwinist perspective simple follow suit in treating norms as bits of information that competing, self-interested individuals choose to adopt or not for reasons of individual instrumental benefit (Moss 2016b). At its most basic level, human sociality as such, has typically been side-stepped as if it were a non-question. Detachment theory offers a very different account (Moss 2014).

For cognitive psychologist Merlin Donald, the "riddle" of *Homo erectus* was the provocation that led to a breakthrough in thinking about human sociality (Donald 1991). How was it possible for *Homo erectus*, a species that endured for over a million years, that domesticated fire and lived in permanent encampments, that produced tools including the Acheulean hand-axe that would have required training to produce, that in greatest likelihood engaged in organized big mammal hunts that would have required a division of labor and who managed to leave Africa and colonize all of the contiguous Euro-Asian land masses (adapting to highly disparate environments and biomes), to do all of this without the benefit of spoken language?

*Homo erectus*, evidence compels us to conclude, had to have established a fully normatively integrated form of life, and for all intents and purposes was the first form of life to have done so. The thesis being proposed is that the evolution of *Homo erectus* constituted a radical transition in levels of detachment and we will proceed to draw on the work of various investigators, including Donald, to support this claim. Should this claim be accepted as warranted its implications would include radically undercutting the entire legacy of methodological, ontological and epistemological individualism, making common cause with some like-minded contemporaries in the areas of social ontology and phenomenology and reconfiguring the proper understanding of human freedom and autonomy.

If paleontologists agree on anything, they have agreed that early hominin survival, with the loss of the arboreal cover, required a level and a form of social cohesion unprecedented amongst higher primates. Everything we know about *Homo erectus* supports the view that an entirely new form of expanded sociality was achieved. But how was this possible? As early as the eighteenth century, Enlightenment thinkers turned their sights onto the human organism and thought about its relationship to the human mind. The common observation was that as organisms, humans are under-specialized weaklings compared to our fellow great apes. It was Gottfried Herder however who grasped the significance of this in detachment theoretic terms (Herder 1986 [1772]). For Herder, the loss of physiological specialization constitutes a detachment from the beck and call of nature (or natural stimuli) and thereby the precondition for a new form of directed attention he called *Besonnenheit*. On the basis of this simple insight

Herder was able to become the grandfather of cultural anthropology. In the absence of behavioral determinations governed by instinctive stimulus-response mechanisms, a new level of integration could take place (and indeed had to!). The hominin group could emerge on the basis of unprecedented degrees of freedom to constitute a way of life not by way of inborn-fixed response patterns but by way of the contingent constitution of group norms. Herder referred to a normatively integrated form of life as a Folk (*Volk*). The practices of a folk are its means of expressively constituting its way of being. Herder was thereby the founder of the expressive-constitutive theory of language[10] and of the study of folk practices in general that anticipated the birth of cultural anthropology as a discipline. When Franz Boas founded cultural anthropology in the last decade of the nineteenth century, it was expressly understood as a further manifestation of the Herder-Humboldt Volksgeist tradition (Bunzel 1996).

The transition to a new level of detachment in which system integration has become fully normative must have involved both losses and gains of functions such as to result in a net increase in degrees of freedom albeit at the level of the Group. The widespread observation of the radical dependency of the human neonate and the comparative weaknesses of the human individual speak to the evident loss of function; but from a biological standpoint how could these have come about? Evolutionary theory has increasingly become aware of the role that heterochrony, or changes in developmental timing, plays in evolutionary transitions. In the 1920s, the Dutch anatomist Louis Bolk put forward the fetalization thesis suggesting that major human features came as a package through the evolutionary retention of much of the fetal phenotype of the ancestral ape (Gould 1977). This thesis was revived in the form of a contemporary popular science account by English zoologist, writer and filmmaker Clive Bromall (2003). The ensemble of characteristic human features that resemble those of the fetal but not mature chimpanzee includes the following: an upright head, largely hairless body, massive brain with bulbous skull, flattened face, short lower jaw, small teeth, everted lips, in the female the retention of the outer labia, hymen and frontal position of the vulva, in the male the lack of a fully protective foreskin, lack of a penis bone and lack of spines along the penis. The Belgian anatomist Jos Verhulst also claims that human lungs and heart resemble that of the infant ape (Verhulst 2003). Alongside the fetalization thesis is the observation that the growth pattern of the human neonate's brain

---

[10] Charles Taylor has written on this extensively over his career but most comprehensively in his recent *The Language Animal – The Full Shape of the Linguistic Capacity* (2016).

for its first year follows that, not of a great ape neonate, but rather that of a fetal brain giving rise to the idea of the human (or hominin) *extra-uterine year*.

By the lights of the fetalization thesis, the comparative enlargement of the human brain came as part of a systematic developmental package and was not initially selected for enhanced brain power (as the individualist outlook has traditionally assumed). Consistent failure to correlate differences in human intelligence with differences in human brain size would lend some credence to this view. The transition to a fully normatively regulated form of life, a transition that I refer to as the "First Detachment" (of hominin evolution), as already suggested, would necessarily involve both losses and gains – losses with respect to specialized response mechanisms triggered by particular natural stimuli, and gains with respect to the wherewithal for social integration and co-ordination. The extra-uterine pattern of brain growth can be seen as serving both of these requirements simultaneously. By relocating formative stages of brain development to the post-natal environment, basic structural formations become re-situating in a socio-cultural context. This transition fits hand in glove with the seminal work of behavioral ecologist Sarah Hrdy (2009). Hrdy's work is crucial for understanding the affective basis of the emergence of the Hominin Group (or supergroup). Comparing relative frequencies of parturition, length of developmental dependency and relative energy cost, Hrdy concluded that it would have been impossible for hominin mothers to have raised their offspring alone. The support of extended caregivers, *allo-parents*, would have been obligately required. Subsequent studies on post-natal and child-rearing behavior comparing women from various tribal communities with that of great ape mothers confirmed that great ape mothers are far less willing to allow others to hold their infant, and for over greater period of time, than human mothers, and further that only in the case of human mothers is there ever a rejection of a newborn due to some imperfection. Hrdy refers to the affective transition that is introduced by the sharing of infant caregiving at the earliest of stages as the onset of *emotional modernity* and she likewise concurs that allo-parenting and emotional modernity (i.e., first detachment) begins with Homo erectus. The hominin infant, underdeveloped and under-specialized at birth, and raised by an extended community of caregivers, became the first primate to be affectively well-suited to be a highly integrated member of a social group. The under-specialized neonate not only lacks the innate obstacles to normatively structured, socio-cultural inclusion, but is also in dire need of compensation for what she lacks. The hominin/human infant, as has become well established, has an appetitive drive toward shared attention and cooperative involvement (Tomasello 2009) driven by the need for compensation. The normatively structured world of the Group is the compensation.

Merlin Donald (2001) has led the way in terms of emphasizing that hominin evolution has been about the evolution of sociality. The hominin group coheres on the basis of shared emotions, shared perceptions and shared practices. Brains have evolved in relation to the cultures of cognition of which they are part. Mindedness is not an individual phenomenon but a cultural one.

Minds and cultures, being two reciprocal and interdependent aspects of a single phenomenon, are subject to changes that may be looked from either or both bottom up or top-down directions. But contrary to the common assumptions emanating from the reflective mind of the philosopher, there is no reason to assume that the nature of consciousness has been any more static than the nature of cultures and this will point will be elaborated upon further below. Does the kind of consciousness that allowed for First Detachment and the emergence of a flexibly adaptive normative form of life require or presuppose fully and characteristically (as *we* know it to be) self-conscious individuals? I don't see why it must (and indeed this will become a hallmark of "Second Detachment"). But even if full-on self-consciousness is not necessary for the transition to a fully normative form of life neither is merely the loss of environmentally oriented specialization nor is the affective openness to others sufficient. The lifestyle of *Homo erectus* was sufficiently complex that there had to be some medium for shared understanding at the requisite level of complexity, and a medium with sufficient semantic degrees of freedom to allow for coordinated responsiveness to contingencies beyond that which merely emotional contagion could provide for. Donald approached this problem, taking a cue from Vygotsky, and contemplating what possible game-changing innovation could lie within a plausible *zone of proximate evolution*.[11] Great apes, notably chimps, gorillas and bonobos, are seen to possess fairly sophisticated levels of social cognition, yet only within the context of an in-the-present, social episode. Just on the basis of a fully upright stature, early small-brained hominins (i.e., australopithecines) would have already been able to take full advantage of the upright body as a canvass for expressive gesture within the episode. Donald reasoned that in light of an already highly developed capacity for motor coordination, hominins would be within an evolutionary stone's throw of gaining a *de novo* capacity for enhanced volitional movement outside of the episode, that mimics movements that have recognizable meaning within an episode. He referred to this capacity as that of *autocuing* (Donald 1991*)*. The capacity to autocue sequences of movements

---

[11] Vygotsky analyzed stages and transitions in human development guided by the concept of a zone of proximal development that spoke to what new developmental capacity could plausibly be within reach of what was already present. Donald "exapted" this concept for the evolutionary analysis.

would have allowed *Homo erectus* deliberately to redeploy sequences of movements that already have meaning, with communicative intent. Donald refers to this as mimesis and makes the case for how mimesis would have transformed the culture and the cognition of *Homo erectus*. For present purposes, we can delimit our attention to the way in which mimesis could transform the capacity for normative integration. What does it mean to refer to an entity, or a group of entities, as engaging in a practice? For an activity to be a practice is to ascribe a normative content to it. To engage in a practice is to conform to the right way of doing something. Producing the Acheulean hand-axe was a practice, organizing a division of labor for a hunt would have been a practice, and any ritual, such as a ritual dance, was a practice. Donald suggests that with autocuing *Homo erectus* gained the capacity for "kinematic imagination", by which he meant the ability to imagine one's body acting in social context *as if* from an outside perspective. This would seem to coincide well with Helmuth Plessner's concept of "ex-centric positionality" (Plessner 2019) and suggests that ex-centric positionality would have likely begun with *Homo erectus*. This is not to say that *Homo erectus* would have enjoyed a full sense of selfhood as we know it, but rather that the embodied foundation of selfhood would have been established and provided the basis upon which symbolically mediated structures could eventually be built. With autocuing, hominins could become *implicitly* accountable for their actions. The culture and cognition of mimesis created the fabric, for the first time, of a fully normatively integrated form of life. With the culture of mimesis, Hominins entered the realm of Geist, not because they were engaged in explicit self-to-self recognition but because the medium of social integration was that of relations of normative as opposed to physical causality. *This claim does constitute an explicit, if perhaps subtle, challenge to the orthodox Hegelian view.*

## III Normativity, Pre-Reflective Plural-Self Awareness and Cerebral Asymmetry

In recent literature concerned with collective intentionality there is a growing debate around the idea of "plural pre-reflective self-awareness." While it is somewhat more widely accepted that there is a minimal pre-reflective me-ness that accompanies our on-going experience (which has been understood by Dreyfus and others as Heidegger's view), there are many hesitations about the assertion of a pre-reflective *us-ness*. But what these discussions lack, is exactly the genealogical account that the present theory provides. In order to round out our account of that normatively structured form of life that begins with First Detachment we must draw upon the recent renewal of research on cerebral hemispheric asymmetry inspired by the brilliant work of Scottish neuroscientist Ian McGilchrist (2009). The story he has to offer, and upon which I have drawn some further extrapolations, provides exactly the kind of alternative to the individualistic and cognitivist account of the place of the normative that has impeded naturalistic enterprises from the get-go.

Why have cerebral asymmetry to begin with? The right and left hemispheres perform distinctly different and in principle complementary (and yet also competitive) functions. The right hemisphere is oriented toward the big picture, the holistic context, whereby the left hemisphere is oriented toward a discrete focus. The right hemisphere is responsible for all forms of attention except focused attention, i.e., vigilance, sustained attention, alertness and most of divided attention. Where the left hemisphere is analytic and logical and means-ends oriented, the right hemisphere is both the source of one's emotional style and character and the place where the emotions of another can be interpreted and understood. The right hemisphere is oriented towards end-in-themselves. Where the left hemisphere can judge logical consistency, it can't judge and detect the soundness of premises, even when patently absurd. The right hemisphere can judge the soundness of premises but not necessarily the analytic validity of inferences. As the right hemisphere is context sensitive it can detect and understand a frameshift (a change in context) where the left hemisphere cannot. The right hemisphere provides the location of the body schema and so was presumably instrumental in the transition to autocuing and kinematic imagination. The left hemisphere is parasitic on the right hemisphere to the extent of drawing upon its content for its own form of focal analysis.

That the right hemisphere is heavier in social animals provides some clue as to its history and trajectory. What seems very likely is that the right hemisphere played a role, especially for animals subject to predation, in providing an around-the-clock vigilance of the general surroundings. A moment's reflection will confirm that a) such general ongoing vigilance would make perfect sense for a creature subject to predation, and b) general vigilance and focused attention are two very different lines of work and having them separated into different hemispheres makes perfect sense as well. When and where mammals become social, the vigilance would concern itself also with the social dynamics of its surrounding environment. The key move I want to make at this point is to suggest that the right hemisphere was poised to become a normativity detector with the transition of First Detachment. We can gain a valuable picture onto what this might look like by referring back to the thesis put forward by former Princeton psychologist Julian Jaynes (1976) in his celebrated *The Origin of Consciousness in the Breakdown of the Bicameral Brain.* Jaynes argued that there was no indication of self-consciousness as recent as the Iliad and that individuals acted according to what they experienced as the dictates of the voices of the gods that were heard as auditory hallucination produced by the right hemisphere. Jaynes further supported this thesis by the findings of studies done in Wilder Penfield's laboratory, whereby volunteers were subject to electrical stimulation of their right hemisphere resulting in auditory experiences, some but not all of which were experienced as voices, but all of which were experienced as originating from a source external to the subject. It would follow, and all the more so given what we now know about the right hemisphere, that the right hemisphere was active in interpreting the normative meaning of its environment which it then announced to the left hemisphere. Consciousness, for Jaynes, in the familiar sense, begins when the left hemisphere commences the narration of a story to oneself about oneself which is based on extrapolation. Whether Jaynes is correct in his reading of the Iliad and his dating of the emergence of self-consciousness to as recently as within 3,000 years ago is not crucial for present purposes. What Jaynes depicts with respect to the right hemisphere, pre-self-consciously monitoring the normative indications of its social environment is precisely what would enable *Homo erectus*, in the absence of speech, to achieve and enact a normatively integrated form of life prior to language. Jaynes' conjectures about the right hemisphere are consistent with the conclusions of McGilchrist decades later, which are also further supported by a wealth of intervening data.

The proposed primordial dominance of the right hemisphere and its role in ongoing normative perception, along with the affective and cognitive resources previously discussed, go a long way towards providing a plausible account of a pre-linguistic group existence, but I will suggest that one more piece of this puzzle must be put in place. I suggest that for a fully normatively integrated form of life to have flourished, as *Homo erectus* did for something approaching a million years, that there had to be a primordial norm around which all else paid implicit obeisance to, one and only one universal norm, and this was the norm that held that the Group is the Good. It is this Ur-norm, this magnetic north, around which all the practices of the group cohered and towards which they were inflected. It is this Ur-norm which allows an affectively and cognitively competent form of life to be constituted by practices that can thereby cohere together. The thesis of the hybrid hominin is that while we are no longer only creatures of the First Detachment Group and are now left-hemisphere dominant, self-narrating, self-conscious individuals, that we are *also still* creatures of the Group, hence the hybridity. The path toward Second Detachment individuation can also be approached in terms of the integrative force of normativity. Language, the spoken capacity for which is nested in the left hemisphere, was driven by the integrative benefit of rendering the content of ritual into myth. Likewise, the integrative benefits of individual accountability were initially strengthened by the capacity for reflective self-identification and linguistic mastery of the system of personal pronouns. In detachment-theoretic terms however, human individuation is also a form of downward parasitic detachment, which dialectically speaking has both been instrumental in challenging the limits of the traditional Group in the name of greater, more universal goods, and yet has also served to subvert the normative coherence of any level of human association. A more adequate elaboration of the normative implications of a dialectics of human detachment however must await another day.

There are many implications of the hybrid thesis which pertain to the scope and presence of the "normative force". Hans Bernhard Schmid (2014), who has defended the idea of a pre-reflective *plural self-awareness*, albeit as yet with no thought to a genealogical account, has called to evidence the experience of members of a group cleaving toward a normative desire for agreement as an apparent end-in-itself. Schmid's observation only touches the tip of the iceberg. The legacy of First Detachment continues to orient us toward abstract norms, the diffracted rays of the lost primordial group, above and beyond accountabilities to merely actual people at particular times and places. Children of immigrants, for example, whose linguistic exposure is almost entirely limited to their parents during their critical language acquisition phase still manage to become native speakers of the adopted language with no residual accent. These offspring unconsciously privilege their perception of the ambient norm. Even the most accomplished intellectuals will curiously feel themselves to be at a moral disadvantage if they are outnumbered two to one in an argument, despite fully knowing that the vast majority of the as yet uncommitted remaining 7 billion extant humans may well side with their position. Empirical studies in social psychology have shown that, for example, being informed about the facts of waste disposal, waste pollution and recycling has less impact on the typical recycling practices of an individual than finding out what their neighbors are doing (independent of who their neighbors happen to be) (Kesibir 2012). Every term I teach I will invariably encounter students whose internalized norm about not raising their voice will trump the pleadings of a slightly hearing-impaired professor to speak up. The same student who refuses to ask a question, or express a view, *in propria persona,* will eagerly enter into a protracted peroration just so long as it is understood as the representation of a group of three. One may want to argue that there is "safety in numbers" but I would suggest that this adage is an easy cover for what is at root about the psychology of the Group and not an individualistic expression of a rational choice (about the benefit of numbers). Less anecdotally, I would suggest that the highly influential, if pragmatically ambivalent, Foucauldian concept of Power could be better materialized and made serviceable for human, and even meta-human[12] benefit, if it was reconstructed as an expression of normative force and further contextualized within a dialectics of detachment. Again, a topic for forthcoming work.

# IV Coda

Finally, to return to the original question of the very hard problem. Even if it is granted that the thesis of the hybrid hominin can yield new insights into how the force of normativity does and/or doesn't constitute the glue of higher order unities in hominin and human life, is there any hope for imagining that any of this analysis can be backtracked down into progressive strata of pre-hominin life, let alone even so far as the early stages of the universe? Does a dialectics of detachment hold any greater promise for taking us closer to a fuller vision of nature beyond that of just so many parts outside of parts? Detachment theory suggests that all unities are only relative, that the unity we have taken to be most unequivocal, the unit of the self, is in fact derivative of the prior unity of the group, as well as a narrative construct that may conceal as much as it self-articulates. Detachment theory identifies a drive toward higher degrees of independence but in an even more speculative vein could we not also propose a *pain of detachment* that perhaps even begins with the Big Bang and constitutes a drive toward compensation for the loss of prior unities? Compensation for detachment? In true dialectical fashion, might not every transition in detachment, constitute a compensatory re-integration on the one side and yet the pain-inducing provocation for further compensation on the other? Might it thereby be the case that if we were to at least loosen the ties of our neo-Darwinian and Newtonian vestments, that the hard problem of interiority (consciousness) and the *very* hard problem of integration might become indistinguishable? Could this be at least a glimpse of that truth that spirit reveals about nature?

---

[12] Post-humanism uncritically takes on board an individualistic misconception of the human and proceeds to build its worldview on the basis of an abstract negation of something it misunderstands. Transhumanism begins with the same misconception and moves in another misguided direction (see Moss 2017). Meta-humanism, building on detachment theory, will offer an alternative route for sublating anthropocentrism, not by eschewing the human but by exposing the expansiveness immanent of the hybrid hominin's normatively structured dialectics of detachment and compensation. We have become modern, but we were never *Human*.

# References

Brandom, Robert (2019) *A Spirit of Trust: A Reading of Hegel's Phenomenology* Cambridge MA:Belknap Press.
Bromhall, Clive (2003) *The Eternal Child* London: Ebury Press.
Bunzel, Matti (1996) Franz Boas and the Humboldtian Tradition in Stocking, G.W. Jr. (ed) *Volksgeist as Method and Ethic: Essays on Boasian Ethnography and the German Anthropological Tradition* Madison: University of Wisconsin Press.
Chalmers, David (1995) "Facing up to the problems of consciousness" *Journal of Consciousness Studies* 2 (3), pp. 200–219.
Donald, Merlin (1991) *Origins of the Modern Mind* Cambridge MA: Harvard University Press.
Donald, Merlin (2001) *A Mind So Rare: The Evolution of Human Consciousness* New York: W.W. Norton.
Gare, Arran (2018) "Natural Philosophy and the Sciences: Challenging Science's Tunnel Vision" Philosophies 3(4) 33–62. https://www.mdpi.com/2409-9287/3/4/33
Gehlen, Arnold (1988) Man: His Nature and Place in the World (New York: Columbia University Press.
Gibson, Daniel G., et al. (23 co-authors) (2010) "Creation of a Bacterial Cell Controlled by a Chemically Synthesized Genome" Science 329:(5987), 52–56.
Gould, Stephen Jay (1977) *Ontogeny and Phylogeny* Cambridge MA:The Belkhap Press of Harvard University Press.
Habermas, Jürgen (1992) "Individuation through Socialization: On George Herbert Mead's Theory of Subjectivity" in Habermas, J. *Postmetaphysical Thinking: Philosophical Essays* (trans. Mark Hohengarten) Cambridge: Polity Press.
Hegel, G. W. F. (1977) *Phenomenology of Spirit*, (trans. A.V. Miller) Oxford: Oxford University Press.
Herder, JG (1986 [1772]) "Essay on the Origin of Language" in *On the Origin of Language Two Essays. Jean-Jacques Rousseau Johann Gottfried Herder* (trans J. Moran & A Gode). Chicago: University of Chicago Press.
Hrdy, Sarah (2009) *Mothers and Others* Cambridge: Harvard University Press.
Jaynes, Julian (1976) *The Origins of Consciousness in the Breakdown of the Bicameral Mind* New York: Houghton Mifflin.
Kesibir, Selin (2012) "The Superorganism Account of Human Sociality: How and When Human Groups Are Like Beehives" *Personality and Social Psychology Review* 16: 233–261.
Kirschner, Marc W. & Gerhart, John C. (2005) *The Plausibility of Life: Resolving Darwin's Dilemma* New Haven: Yale University Press.
Koonin, Eugene V. (2016) "Viruses and mobile elements as drivers of evolutionary transitions" *Philosophical Transactions Royal Society B* 371:20150442
McGilchrist, Iain (2009) *The Master and his Emissary: The Divided Brain and the Making of the Western World* New Haven: Yale University Press.
Moss, L (2006) "Redundancy, Plasticity, and Detachment: The Implications of Comparative Genomics for Evolutionary Thinking" *Philosophy of Science* 73: 930–946.
Moss, L (2014) "Detachment and Compensation: Groundwork for a Metaphysics of Biosocial Becoming" *Philosophy and Social Criticism* 40: 91–105.
Moss, L. (2016a) "The Hybrid Hominin: A Renewed Point of Departure for Philosophical Anthropology" in *Naturalism and Philosophical Anthropology: Nature, Life, and the*

*Human between Transcendental and Empirical Perspectives* (Honenberger, P. ed) New York: Palgrave Macmillan.

Moss, L (2016b) Review: Tim Lewens, Cultural Evolution: Conceptual Challenges, Notre Dame Philosophical Reviews. https://ndpr.nd.edu/news/cultural-evolution-conceptual-challenges/

Moss, L (2017) "New Naturalism and Critical Theory" in Gare, A. & Hudson, W. (eds) *For a New Naturalism*. New York: Telos Press

Moss, L (2019) "Plessner" in Allen, A. & Mendieta, E. (eds) *The Cambridge Habermas Lexicon* Cambridge: Cambridge University Press.

Plessner, Helmuth (2019) *Levels of Organic Life and the Human: An Introduction to Philosophical Anthropology* (trans. M Hyatt). New York: Fordham University Press.

Schmid, Hans Bernhard (2014) "Plural Self-awareness" *Phenomenology and Cognitive Science* 13:7–24.

Taylor, Charles (2016) *The Language Animal – The Full Shape of the Linguistic Capacity*. Cambridge: The Belknap Press of Harvard University Press.

Tomasello, Michael (2009) *Why We Cooperate* Cambridge MA: The MIT Press

Tomasello, Michael (2019) *Becoming Human: A Theory of Ontogeny* Cambridge MA: Belknap Press.

Unger, Roberto M. & Smolin, L. (2015) *The Singular Universe and the Reality of Time. A Proposal in Natural Philosophy*. Cambridge: Cambridge University Press.

Verhulst, Jos (2003) *Developmental Dynamics in Humans and Other Primate: Discovering Evolutionary Principles through Comparative Morphology*. Hillsdale (NY): Adonis Press.

Tom Rockmore
# Towards a Constructivist Approach to Human Nature

**Abstract:** This paper will sketch the outlines of a constructivist approach to human nature. Constructivism is an epistemic approach that emerges in ancient geometry, that comes into modern philosophy and that resembles certain scientific practices. The paper begins with brief remarks on epistemic constructivism before turning to three important illustrations: Kant's Copernican revolution, Bas van Fraassen's constructivist empiricism, and Noam Chomsky's generative linguistics. I argue that the human sciences routinely and successfully construct representations of human nature, not as it is, but as it appears and is known on the basis of available empirical data.

## On Representing Human Nature

The theme of human nature is much discussed in religion, natural and social science, philosophy and other fields. My modest aim in turning to this theme is to offer an alternative to other views but not to provide a definitive response to the familiar question: what is human nature?

Representation of human nature falls into the more general category of epistemic representation. There are many religious, philosophical, scientific and other theories of human nature. Western religious views of human nature include the idea that human beings are made in God's image, the traditional Christian view that human beings are basically evil and have sinned by falling away from God, and the liberal theological conviction that human being is basically good. A short, obviously incomplete list of philosophical views of human nature might include the names of Plato, Aristotle, Descartes, Rousseau, Hume and Kant.[1] There are many different views of human nature in biology, psychology,

---

[1] Instances include the Platonic view of the tripartite soul, the Platonic analogy between the structure of the soul and the state, the Aristotelian teleological view, Aristotle's conceptions of human being as social as well as capable of reasoned speech, and so on. In his rationalist approach, Descartes describes a person as a thinking being as distinguished from either an extended or an infinite being. Rousseau thinks that we do not know the limits of human nature. Hume famously wrote *A Treatise of Human Nature* that, he believed, fell stillborn from the press.

**Tom Rockmore,** Peking University

https://doi.org/10.1515/9783110604665-015

anthropology and allied fields. Darwin appears to deny there is a fixed human nature.

Nothing is to be gained by further multiplying examples. There is already an immense literature about human nature, but nothing resembling general agreement. Suffice it to say that human nature presents a continuing field of research in many different fields, especially the human sciences.

## Some Philosophical Views of Human Nature

Philosophical views of human nature are often included in concepts of the subject. Thus, Augustine and other medieval Christian figures formulate views of the subject capable of moral responsibility. They distinguish between morally correct and incorrect actions to account for human responsibility in the fall away from God. In the early modern tradition, the medieval conception of the subject is reformulated as the capacity to differentiate between ideas that either meet or fail to meet the Cartesian cognitive criterion of clear and distinct ideas.

Augustine, Descartes and others share an indirect approach to the subject as possessing capacities apparently required to account for types of experience, such as acting correctly or claiming to know. Others, who take a more skeptical view, deny we can know human nature, or at least deny we can know the subject, who lies beyond the limits of human cognition. Thus, the new mysterianism denies that certain problems associated with human being, such as the hard problem of consciousness, will ever be solved.[2]

Mysterian skepticism is anticipated by Kant. According to Kant, the subject must accompany all representations and be able to formulate universal principles for action. Yet, in denying the inference from a type of activity to the subject, Kant suggests that the human subject is not known, but unknown, and, in fact unknowable to us. Though we know what a subject is capable of doing, the subject itself lies beyond the limits of cognition.

Kant's skeptical view about the subject is based on his familiar distinction between appearance and reality. The view is clearly stated a number of times

---

In his book on anthropology, Kant focuses on human nature as a question (What is man?) that must be studied from either of two points of view: the physiological perspective in order to see what nature has made of man; or as a freely acting being (*frei handelndes Wesen*) to grasp what man can, ought, or in fact does make of himself.

2 See Flanagan (1992).

in the *Prolegomena* in the analysis of psychological ideas.³ In various passages he relies on his distinction between appearance and reality to argue that the subject is unknown. In reference to the account of the amphiboly of reflection in the first *Critique*, he says that if we remove all accidents or predicates, we arrive at the subject itself, which remains "unknown to us" (P § 46, p. 85). In § 47, in implicitly referring to Descartes, he states that the subject of thinking can be called substance, but this is an "empty" concept. In § 48, he takes up the problem of the persistence of the soul with respect to possible experience, such as we may infer representations, but denies we can prove the persistence of the soul beyond the life of an individual. In § 49, he says we are conscious of what he calls the soul as existing in time, hence the subject of appearances, but, he says, the underlying "being as it is in itself, which underlies these appearances, is unknown to me" (P § 49, p. 88).

Plato is apparently the initial thinker of the first rank to insist on the distinction between appearance and reality. The Kantian reformulation of this ancient distinction recurs in *Being and Time* as the distinction between the various human sciences and Heidegger's theory of *Dasein*. In § 10, Heidegger draws attention to his philosophical conception of the subject as distinguished from anthropology, psychology and biology and even ethnology. According to Heidegger, in such classical views of man as a rational animal and as made in the image of God, "the essence of man" or again "his being," which is mistakenly taken as self-evident, is rather forgotten. Heidegger thinks this is the case for anthropology, as well as psychology and biology. His conviction that the ontological foundations cannot be made visible from empirical materials indicates that an ontological analysis cannot be carried out on an empirical basis. According to Heidegger, the philosophical subject is presupposed by, hence cannot be studied within, the human sciences.

There is a crucial difference between Kant's and Heidegger's approach to human nature. Kant claims that since reality cannot be known, the subject cannot be known. Heidegger, who distinguishes philosophy and the human sciences, claims the subject can be known philosophically. In what follows I will sketch a constructivist approach, if not to human nature, at least to human being as it is known to us through empirical data in enfranchising the human sciences on philosophical grounds. The scientific procedure of finding out about human beings is in a way similar to what is sometimes called the hermeneutical circle. This circle is understood in various ways, such as a relation be-

---

3 See Immanuel Kant (2012), §§ 46–49. References are cited in the text as P followed by the paragraph or § and the page number.

tween part and whole, or again between what one presupposes and what one uncovers or discovers on that basis. At a minimum, it provides a way, if not to determine human nature as it is in independence of us, then to study human nature not as it is but as it appears in experience.

## Representation as an Epistemic Problem

We can start by looking at representationalism. Representation of human nature assumes there is something we can call human nature, hence that it is, but not what it is. This theme presents important difficulties. From the epistemic perspective it raises the question of the nature and limits of knowledge of human nature, or what human beings can legitimately claim to know about themselves. From the scientific perspective, it raises the related question of how to study human nature within the parameters of contemporary science.

Representation of human nature is a kind of cognition, hence bound up with, and subject to the limits of, cognition in general. To understand the limits of cognition, it will be useful to begin with the crucial triple distinction between phenomenon, appearance, and representation. I will understand the term "phenomenon" to refer to what, though present to mind, or in mind so to speak, does not refer beyond itself to something else, for instance to what is not in mind. I will understand "appearance" to relate to what is in mind but that refers beyond itself in establishing a relation between itself and the referent. As used here, "reference" is both a quasi-logical as well as a cognitive relation.

Reference is quasi-logical since it establishes a quasi-logical relation between two objects, one of which refers and the other of which is referred to, or the referent. Reference is often understood in causal terms. In order to refer beyond itself, an appearance might be the effect of what appears is the cause. Yet an appearance need not resemble or otherwise be like its cause, hence enabling us to correctly or again faithfully represent reality. It follows that one cannot infer from an appearance, any appearance, to its cause. In other words, if successful representation is the price of cognition, then there is no cognitive relation between an appearance and what appears.

This identity quickly became a cognitive criterion that echoes through the entire later tradition. According to this model, a successful representation correctly picks out or otherwise identifies what it represents in yielding knowledge of what is. For instance, in a causal theory of reference the effect correctly refers, hence correctly represents its cause or referent. Plato's denial of the backward inference from effect to cause is an important part of his turn to the notorious theory of forms or ideas.

This triple distinction between "phenomenon," "appearance" and "representation" provides a basis to understand "representation" of all kinds, hence representation of human nature. The problem of epistemic reference arises early in the Greek tradition. Parmenides influentially suggests, since thought and being are the same, that there is an identity between thought and being. This identity quickly became a cognitive criterion that runs from ancient philosophy until the present.

Epistemology is the daughter of ontology. The Parmenidean claim for the identity of thought and being suggests three possible cognitive approaches. They include: metaphysical realism, or the view that cognition requires a grasp of mind-independent reality beyond mere appearance; the opposite view, or epistemic skepticism following from the failure to grasp mind-independent reality; and, finally, the alternative claim that we cognize only what we in some sense can be said to construct, or epistemic constructivism.

Everyone will have a favorite illustration of the different interpretations of the Parmenidean cognitive criterion. Suffice it to say that the distinction between appearance and representation divides Plato and Kant. Appearance falls short of cognition, which requires correct representation, or the cognitive grasp of mind-independent reality. Plato rejects the very idea of representation in suggesting that if there is knowledge direct epistemic intuition must be possible.

## Some Views of Philosophical Representation

The highest form of representation is imitation (mimesis). In the *Republic*, Plato draws attention to the distinction between an imitation and reality. According to Plato, artists and poets should be beaten and excluded from the city for proposing an imitation of an imitation as the real or true. He further suggests on grounds of nature and nurture that for knowledge to be possible, some gifted individuals must be able directly to "see" or directly intuit the real.

Plato proposes a speculative solution to the cognitive problem. He takes seriously the Parmenidean view that cognition presupposes a cognitive grasp of the mind-independent real, in his case the forms or ideas to which his theory refers. His view includes two main points. He rejects the backward inference from effect to cause, since correct representation is not possible even in the case of artistic mimesis. According to the theory of forms, cognition circumvents the inability to represent the real, hence, to satisfy the Parmenidean criterion, through direct intuition of the real.

This problem later becomes central to the critical philosophy. Kant thought he was misinterpreted by his early readers. He later suggested that an author is easy to understand if we rely on a view of the whole. Yet several centuries later there is no agreement on even the general outlines of his position.

I believe Kant inconsistently features two different and incompatible approaches to cognition. His early representationalist view, which has support in the texts, regularly leads scholars to infer he is a representationalist. In his mature critical period, Kant rejected representationalism in favor of constructivism. Kant's constructivist view, which lies at the heart of his mature view, is perhaps his most original philosophical contribution. Yet several hundred years later Kantian constructivism is still little studied.

Representationalism is popular in modern philosophy. Yet it has never been made out, and Kant was correct to abandon it. A representational approach to cognition is not plausible since we do not and cannot match up or otherwise compare a proposed representation to what it represents. No plausible way has ever been suggested to show that a representation is correct. A successful representation would cognize a mind-independent object, hence cognize the real. Kant accepts this view, since, as he points out, in dealing a fatal blow to cognitive representationalism at least on a Parmenidean model, no progress has ever been made toward grasping an independent object.

Kant, who claims his theory should be understood in relation to Hume's, is perhaps better understood with respect to Plato. Kant's relation to Plato is unclear. Though we do not know if Kant studied or even read his Greek predecessor, he clearly claims to understand Plato better than the latter did himself.[4] We recall that Plato rejects the backward inference from being to thought, or from the world of appearances to mind-independent reality, in opting for intellectual intuition of the real. Kant, like Plato, rejects the backward cognitive inference from appearance to reality as well as intellectual intuition on which Plato relies. In its place, and to avoid cognitive skepticism, he suggests what I will be calling cognitive constructivism.

# The Copernican Revolution as Cognitive Constructivism

Cognitive constructivism is another name for the frequently mentioned but little studied so-called Copernican Revolution. Kant never uses this term to refer to his own position. But it was used by his contemporaries Reinhold and Schelling to describe the critical philosophy. Kant's Copernican turn is rapidly evoked in the

---

4 See Kant (1998), B 370, p. 396.

B preface to the *Critique of Pure Reason* but routinely overlooked in the debate. It is at least surprising that at a time when the Tower of Babel is nearing completion, and when every aspect of Kant's corpus has received extensive attention, that the most detailed study of Kant's relation to Copernicus, a weighty tome of some 600 pages, concludes that there is no relation.[5]

Kant very rapidly presents his discovery in three points. First, according to Kant, as mentioned above, there has never been any progress towards grasping a mind-independent object. This suggests the impossibility of basing cognition on the supposed grasp of mind-independent reality. In other words, Kant, who describes himself as an *a priori* thinker, is inconsistently suggesting on *a posteriori* grounds that there has never been any progress toward metaphysical realism. Second, Kant famously suggests that knowledge may be possible if we invert the relation of subject and object so that the latter depends on the former. This is a version of the central constructivist epistemic claim: we know only what we construct. And, third, there is an analogy between the critical philosophy and Copernican astronomy since in both cases the cognitive object, or what we seek to know, depends on the subject. In other words, since objective claims to know are centrally linked to a subjective component, for Kant and other constructivists the road to objectivity necessarily runs through subjectivity.

Kant relies on experience in suggesting that knowledge is possible if and only if the object depends on the subject. As a result, Kantian apriorism is compromised by, and perhaps is even a species of, pragmatism. Instead of providing a transcendental theory that conclusively demonstrates the possibility of knowledge in general, Kant can plausibly be read as formulating a speculative approach to cognition dependent on past experience. In rejecting cognition of a mind-independent object, Kant abandons the traditional effort to know the real. In turning to constructivism Kant also gives up representationalism in all its forms.

According to Kant, in order to avoid cognitive skepticism, we should, to avoid skepticism, undertake as an experiment an approach based on the idea that we know only what we construct (*herstellen*). To weigh the importance of this suggestion it will be useful to distinguish between *a priori*, apodictic and *a posteriori*, non-apodictic cognitive claims. Kant's normative view of knowledge as *a priori*, hence apodictic, relies on Euclidean geometry in formulating his constructivist approach. According to plane geometry, the construction of a single geometrical figure, for instance a right-angle triangle, proves the existence of the entire class, in this case the class of right-angle triangles. Kant, who favors

---

5 See Blumenberg (1987).

a normative view of cognition as apodictic, thinks that pure mathematics and pure natural science are both *a priori* sources of knowledge since their objects must conform to cognition.

Kant's constructivist approach for perhaps the first time breaks the Parmenidean link between metaphysical realism, thinking and being. According to Kant, cognition, which is limited to what we construct, excludes knowledge of reality, also known as the noumenon or again the thing in itself.

Kant inconsistently defends, but also gives up, metaphysical realism. According to Kant, if, as he says, the sensory objects given in experience are appearances, then there must be things in themselves, or reality, as their cause.[6] This view is doubly problematic. On the one hand, we do not know and cannot determine that the objects of experience are appearances rather than mere phenomena. On the other hand, if there were appearances, this would lead to the inference that cognition of objects of experience depends on the necessary assumption of an unknown and unknowable real, or metaphysical reality. In this case cognition presupposes a cognitive relation between the knower and the knowable or reality that appears. It is often assumed that cognition requires a grasp of the mind-independent real. Yet no-one has ever shown how this is possible, nor demonstrated that reality appears. If we abandon the view of the object of experience as an appearance and consider it as a mere phenomenon, we can give up talk about its relation to reality in considering its cognitive role for us.

For thought to grasp being, we must either be able to know the real or abandon the in-principle hopeless task in constructing a theory about what we know. Both approaches yield the Parmenidean identity between thought and being. But only the latter, which abandons the theoretical insistence on grasping the real, which in practice cannot be demonstrated, still seems plausible.

In this respect, Kant's critical philosophy is a defining moment in the epistemic debate. With respect to Kant there is a before and after. On the one hand, there was a time before the critical philosophy when, at least since Parmenides, it seemed necessary or at least plausible to insist on a grasp of reality as necessary for cognition. On the other hand, there is a time after the critical philosophy when it no longer seemed plausible to insist on a cognitive criterion that remained equally popular after as well as before the critical philosophy.

Kant's view that there has never been progress toward grasping independent objects has never been answered. There are many different forms of realism. Metaphysical realism, which claims to cognize the real, and constructivism, which

---

6 See P §32.

suggests that though we cannot cognize the real we can at least know what we construct, are incompatible. In Kant's wake, this observation points away from metaphysical realism and toward constructivism.

These rapid remarks suggest a view of epistemic constructivism that has the following properties: first, it maintains the Parmenidean stress on the identity of thought and being; second, it accepts a version of the Kantian constructivist insight that we know only what we in some sense "construct" in leaving that term as undefined; and, third, it gives up reference to metaphysical or mind-independent reality. The Kantian constructivist approach is revised by Hegel as an ongoing process of constructing a theory on the basis of the contents of experience that is, if necessary, later "adjusted" to account for further experience.

We assume, but never know that our theories about the contents of experience apply to the world as given in experience. Such theories are never certain and do not yield truth. At best they yield, as Dewey would say, warranted assertible views on the basis of what we think is correct at a given point in time. Theories arise on the basis of experience, and are tested in further experience, in which they are either accepted, or rejected and modified. I shall take this comparatively slimmer, quasi-Hegelian empirical view of constructivism as a model for research about and the representation of human nature.

## Scientific and Empirical Constructivism

The constructivist view presented in the preceding section is perfectly general. If it is valid, it is valid for science of all kinds, including the human sciences.

Science is in general empirical. Even the most theoretical form of science must take into account the available empirical information. Philosophy of science turns on correctly understanding the nature and limits of an empirical approach to cognition. A constructivist form of science, or scientific empiricism, which is broadly compatible with the view sketched here, is formulated by Bas van Fraassen. Van Fraassen distinguishes broadly between realism, or metaphysical realism – two terms he uses as synonyms – and empiricism as mutually exclusive alternatives. He rejects the former in favor of the latter in defending a moderate form of empiricism suitably cleansed of some of the most extreme aspects of Vienna Circle positivism.

In the very short preface to *The Scientific Image*, he writes:

> I shall present three theories, which need each other for mutual support. The first concerns the relation of a theory to the world, and especially what may be called its empirical import. The second is a theory of scientific explanation, in which the explanatory power of

a theory is held to be a feature which does indeed go beyond its empirical import, but which is radically context-dependent. And the third is an explication of probability as it occurs within physical theory (as opposed to: in the evaluation of its evidential support).[7]

Since science needs to take empirical information into account, it is context-dependent. Van Fraassen sees himself as an analytic philosopher concerned with the relationship of scientific theories to the world. But he does not seem to detect the underlying pragmatic tone linking his view of science with, for instance, Hegel's.[8] Analytic philosophy has often inclined toward neo-Cartesianism, which is committed to variations on the foundationalist theme. In the Vienna Circle movement, foundationalism takes the form of a claim described in the *Manifesto of the Vienna Circle* (1929) in two main points: knowledge comes only from experience; and the scientific conception of the world is linked to logical analysis. To grasp the significance of Van Fraassen's rehabilitation of anti-realism, it will be useful briefly to reconstruct the Vienna Circle view he refutes.

## Empirical Constructivism After Logical Positivism

The logical approach to classical empiricism perhaps reaches a peak in the early Vienna Circle's turn to protocols. In general, the members of the Vienna Circle were committed to scientism, or the ideological view of natural science extending beyond its usual limits. This positivist approach understands science as the main or even the only important source of knowledge. The positivists rejected any form of synthetic *a priori* while insisting on strong, quasi-Kantian claims for knowledge in basing knowledge claims on supposedly indefeasible items of experience. This effort led to an important, complex debate between Carnap, Neurath, Schlick and others about "scientific" empiricism, intended to provide a seamless link between observation and scientific law.

Like Wittgenstein, Carnap understands protocol propositions as "propositions that do not require proof but rather serve as the basis for all the other propositions of science," or as the simplest propositions of the protocol language. In anticipating Sellars, the Vienna Circle positivists think protocol propositions are directly related to the given, or the "immediate contents of experience or phenomena, hence to the simplest cognizable states". According to Carnap, "the propositions of the protocol language, for instance the (basic) protocol sentences

---

[7] Bas van Frassen (1980). Further references to this work as given in the text as F followed by the page number.
[8] See Giladi (2016).

are translatable into the physicalist language" of science.⁹ Sellars assumes there is no more than a single correct analysis of the relation between the sign and what it represents.

Wittgenstein, on whom Carnap relies in his view of protocol sentences, felt he was completely misunderstood by his positivist enthusiasts. Neurath, who opposed the foundationalism implicit in Wittgenstein and explicit in Carnap and Schlick, objected to the very idea of a protocol as little more than a fiction presupposing an ideal language. According to Neurath, "The fiction of an *ideal language* constructed of clean atomic sentences is just as metaphysical as the fiction of Laplace's famous spirit".¹⁰ In this context, he made the important remark that Quine cites in the exergue in *Word and Object* about the similarity between a clean protocol sentence and repairing a ship on the open sea. Long before Quine, he sees that the problem of direct reference has no formal solution.

The controversy between Carnap the empirical foundationalist and Neurath who criticized the positivist project is obviously related to semantics. Neurath's criticism of Carnap suggests that the semantic project running through the entire analytic debate since Frege, and perhaps most closely associated now with Brandom, cannot be carried out. A successful reconstruction of the epistemological ship would presuppose an ideal language allowing neither more nor less than a single correct analysis. Neurath, on the contrary, prefers a real language that permits an unlimited number of possible analyses, the same point Quine later works out in his theory of the indeterminacy of translation.

Dreams tend to persist even if the aim for which they arise cannot be fulfilled. Carnap, who acknowledged the importance of Neurath's critique, immediately abandoned the idea of basing science on incorrigible empirical constatations. In giving up empirical foundations as anything more than an unrealizable ideal he fell back on an ideal language. Neurath's attack on Carnap was later continued by Quine who, in proposing the indeterminacy of translation and in refuting the distinction between a natural and an ideal language, decisively undercut Carnap's project, and perhaps the entire Vienna Circle program as well.

In retrospect, Vienna Circle protocol theory belongs to the effort including metaphysical realism, Platonic intuitionism, Cartesian epistemic foundationalism, post-Hegelian phenomenology, Fregean semantics, and so on. Though they differ widely, these and other approaches all aim at cognizing the real. More than two and half millennia after Parmenides called for the identity be-

---

**9** Carnap (1931), p. 453.
**10** Neurath (1932/1933), p. 204.

tween thought and being, they intend finally to make the transition in linking together, as McDowell has recently suggested, *Mind and World*.

## Van Fraassen's Constructivist Empiricism

Van Fraassen's constructivist empiricism presupposes, but does not present or otherwise work out, the critique of Vienna Circle positivism in rejecting metaphysical realism. The association between metaphysical realism and truth is as frequent as it is indemonstrable. There are, like ice cream, numerous kinds of realism. Van Fraassen's particular form of anti-realism seems squarely directed against the idea that at long last, there is real hope, through some hitherto undreamed-of clever stratagem, of finally getting it right about the real. In different ways, this general claim goes all the way back in the tradition to Plato, perhaps even to Parmenides. Not surprisingly it remains popular since metaphysical realism remains popular. A metaphysical realist claim is still implicit at present in the semantic approach to reference that now seems to be ingredient in recent work in inferentialist semantics. Though it may be helpful to self-esteem to think we are finally, after several thousand years of effort, on the verge of cognizing the real, nothing other than the self-delusion of workers in the philosophical vineyard shows this is either correct or indeed necessary.

Van Fraassen rejects "realism" in embracing empiricism that he regards as anti-realist. He bases his understanding of science on the quasi-Kantian view that, since we cannot grasp the real, we must accept anti-realism. He thinks our best and perhaps only recourse lies in constructing a theory about what we indirectly know on the basis of incomplete experience. In *The Scientific Image*, he claims: "Science aims to give us theories which are empirically adequate; and acceptance of a theory involves as belief only that it is empirically adequate" (F, p. 12).

Many observers believe there is an exclusive alternative between cognition and skepticism. The strength of constructivist empiricism lies in the quasi-pragmatic acceptance of a third view, or empirical adequacy. Empirical adequacy is obviously stronger than epistemic skepticism but weaker than truth claims. This intermediate cognitive claim that is neither cognitive fish nor cognitive foul enables van Fraassen to reject claims for scientific truth that cannot be made out in favor of weaker cognitive claims that can be defended, if not metaphysically, at least empirically. According to van Fraassen, "Science aims to give us, in its theories, a literally true story of what the world is like; and acceptance of a scientific theory involves the belief that it is true" (F, p. 8).

The difficulty lies in an ambiguity with respect to "truth." Van Fraassen depicts constructive empiricism as both maintaining as well as abandoning a traditional epistemic cognitive claim. Instead of turning away from truth and toward empirical adequacy, van Fraassen sometimes seems to seek to have it both ways. He writes: "[A] theory is empirically adequate exactly if what it says about the observable things and events in the world is true – exactly if it 'saves the phenomena'" (F, p. 112). This view seems to conflate two points, roughly getting it right about the world, on the one hand, and any claim for truth, on the other.

What is "empirical adequacy"? According to van Fraassen, empirical adequacy is always potential, never fully realized, since further measurements can always be taken (F, p. 69). He apparently thinks that in the absence of countervailing evidence we can hold that theories are adequate, but we cannot hold that they are true. When the chips are down, van Fraassen believes constructive empiricism is comparatively better than the realist alternative, that is, that it "makes better sense of science, and of scientific activity, than realism does" (F, p. 73).

According to van Fraassen, explanation yields no more than description. He writes that "a success of explanation is a success of adequate and informative description" (F, pp. 156–157). In his view science contributes nothing to explanation over and above the descriptive and informative content of the scientific theory. Hence one must agree with his conclusion that, as he puts it, "the assertion of empirical adequacy is a great deal weaker than the assertion of truth" (F, p. 69). A full-fledged account of constructive empiricism would need to consider such further factors as causality, reasoning to the best explanation, unobservables, the hermeneutical circle, and so on.

## Constructivist Empiricism as a Science of Man

Van Fraassen's constructive empiricism is intended to sketch an interpretation of science, but not to substitute itself for science. Human nature is one among a large number of possible themes for scientific investigation. A constructivist form of empiricism is relevant to the more specific theme of human nature. The speculative approach of constructive empiricism is further exemplified in the wide range of the human sciences, for instance to the general linguistic approach to constructing models to understand the implied capacities that explain ordinary language.

Van Fraassen's constructive empiricism and Noam Chomsky's generative linguistic theory are related as two important illustrations of a constructivist approach to cognition. Language is often described as a unique or at least a specific human capacity, but observers divide about how to understand it. Chomsky is a widely known pioneer in so-called generative grammar. "Generative grammar" refers to a system of rules intended to generate grammatical sentences in a particular natural language.

Chomsky has developed this theory over more than half a century, in a series of different but related forms. At the heart of this theory he postulates unlearned but tacitly known "universals of human linguistic structure" that according to his theory make it possible for children to learn natural languages. In short, the success of the theory turns on the ability to construct a model of human language on the basis of universals of human linguistic structure that are postulated to explain linguistic syntax.

Over the years, Chomsky's program has taken a number of forms, including the standard theory, the extended standard theory, the revised extended standard theory, relational grammar, government and binding principles or parameters theory, and then the minimalist program from 1990 to present. Since this is not a study of Chomskyan linguistics, we do not have to follow its development in any detail. We are concerned here with the representation of human nature, for instance in contemporary linguistics, specifically in Chomsky's generative theory of grammar, as an illustration of constructive empiricism, but unconcerned with the specific differences in its formulation.

Chomsky has described his linguistic theory in different ways in a large number of publications. In a recent study, he addresses the ancient question of the kind of creatures that we are in remarks on language. Chomsky's account is often complex, even daunting, for instance, in his offhand remark about the apparent linguistic preference for minimal structural distance as opposed to minimal linear distance.[11] Chomsky believes that minimal distance is employed in what he calls minimal computation (C, p. 11).

Chomskyan universal grammar is broadly Kantian. Kant's critical philosophy is based on the hypothesis of a small number of categories situated in the mind that can be applied to sensory data that are worked up into cognizable objects. In his theory of universal grammar Chomsky replaces Kantian categories by genetically determined universal grammatical structures (C, p. 11).

Obviously, the status of linguistic universals is key to Chomsky's generative grammar. His specific claim about linguistic universals is unclear. One possibility is that linguistic universals that, like Kantian categories, are hard-wired into the mind. Another possibility might be that Chomskyan linguistic universals are

---

[11] Chomsky (2016), p. 10. References to this book will be given as C followed by the page number in the text.

merely a heuristic linguistic device to represent the linguistic capacity exhibited by normal human beings all over the world.

According to Chomsky, language design ignores order since what he calls the Basic Property is determined hierarchically. If one could show that linguistic universals do not exist, that the learning of language does not depend on linguistic universals, or that comparatively better linguistic explanations are available, then so-called universal grammar, including linguistic universals, would lose their explanatory purpose.

At the time of this writing Chomsky, who is still active, detects no plausible threats to universal grammar on the horizon. He describes his recent interest in what he calls the minimalist program, specifically the so-called strong minimalist thesis designed "to ask how far it can be sustained in the face of the observed complexities and varieties of the languages of the world." (C, p. 24) The most recent instantiation of his theory is empirical, constructivist and speculative. It is empirical in that is constructed on the basis of what we know empirically about language. It is speculative with respect to one or more hypotheses introduced with the aim of bringing available empirical materials and postulated but never directly observed or observable linguistic structures together in a single theory.

More generally, Chomsky's universal grammar joins together empirical and conceptual materials. His theory relates unobservable but postulated linguistic structures to empirical data in a so-called universal grammar. The very idea of universal grammar suggests that Chomsky, like Kant, is aiming at unrevisable knowledge, in Chomsky's case on the basis of empirical data.

But, as the adage that all roads lead to Rome reminds us, there are infinitely many different ways to get from here to there. This suggests there may be scientifically reasonable alternatives to Chomsky's generative grammar. New empirical data or deeper analysis of already available data may later identify the need for new iterations of the theory or, as the case may be, successor theories that account for available data in new and better ways or even new data.

We cannot exclude the need for a basically new theory that at long last gives up the idea of linguistic universals. A better theory can be said to explain everything the prior theory explains plus at least one thing it ought to but fails to describe. An example might be the orbital peculiarity of Mercury that should be, but is not, explained in Newtonian mechanics, but is explained in relativity theory. Hence, relativity theory can be said to be explanatorily stronger than celestial mechanics.

A further difficulty lies in the relation of Chomsky's theoretical edifice to available linguistic data. In his recent book as in many of his other linguistic publications, Chomsky is concerned to construct a general theory without, as in this case, a single reference to a human natural language, perhaps because the relation of the empirical component seems not to interest him more than indirectly.

## Conclusion: Towards a Constructivist Theory of Human Nature

This paper suggests that we can and do make progress toward grasping human nature through research in the human sciences. I have argued that what we take to be human nature in the human sciences is, like all science, perhaps cognition of all kinds, the result of the construction of theories on the basis of the available empirical data in grasping indirectly, by inference, what we cannot otherwise grasp directly. In other words, the construction, testing and reconstruction of such theories slowly yields an empirically-based representation of human nature, not as it is but rather as it appears.

Progress towards knowing who we are in practice contrasts with the theoretical arguments marshaled by philosophers. According to Kant, we cannot know who we are since we cannot go beyond mere appearances to reality. According to Heidegger, we cannot learn about human nature through the human sciences that presuppose a conception of the subject that is however available in his phenomenological ontology.

Both Kant and Heidegger deny we can know the human subject since they set the cognitive bar too high, Kant in order to call attention to any concern to go beyond the limits of experience and Heidegger in order to deflect attention toward his favored *Dasein* or human being and being in general. Kant's negative conclusions follow only if we accept the view that knowledge requires a cognitive grasp of the real, in this case human beings as they are, something that in his position lies beyond cognition, as distinguished from how they merely appear. Yet there is no reason to infer that if we cannot know the human subject as it is that we cannot know it as it appears in various forms of experience. In the same way, Heidegger's negative conclusion derives from a misunderstanding of the hermeneutical circle, more precisely a failure to grasp that we come to know human nature in and through what human beings do.

If we invoke a weaker cognitive standard, then we overcome the theoretical problem about knowledge of human nature that emerges in the human sciences. Peirce pointed out roughly a century ago that what we mean by the real is, or at least should be, what our views converge on in the long run. In the same spirit, I would like to suggest that what we take to be human nature is what a constructivist approach to the human sciences tells us in the long run about human beings.

## References

Blumenberg, Hans (1987): *The Genesis of the Copernican World*, trans. by Robert M. Wallace, Cambridge: MIT Press.
Carnap, Rudolf (1931): "Die physicalische Sprache als Universalsprache der Wissenschaft". In: *Erkenntnis* 2.
Chomsky, Noam (2016): *What Kind of Creatures Are We?* New York: Columbia University Press.
Flanagan, O. J. (1992): *Consciousness Reconsidered.* Cambridge: MIT Press.
Giladi, Paul (2016): "Pragmatist Themes in Van Fraassen's Stances and Hegel's Forms of Consciousness. In: *Journal of Philosophical Studies* 24. pp. 95–111.
Kant, Immanuel (1998): "Critique of Pure Reason". In: *The Cambridge Edition of the Works of Immanuel Kant – Critique of Pure Reason.* Translated by Paul Guyer and Allen Wood. New York: Cambridge University Press.
Kant, Immanuel (2012): "Prolegomena to Any Future Metaphysics". In: *Immanuel Kant: Prolegomena to Any Future Metaphysics That Will Be Able to Come Forward as Science: With Selections From the Critique of Pure Reason.* Edited by Gary Hatfield. New York: Cambridge University Press.
Neurath, Otto (1932/1933): "Protokolsätze". In *Erkenntnis* 3.
Van Fraassen, Bas (1980): *The Scientific Image.* New York: Oxford University Press.

# Index of names

Adorno, Theodore 148–150
Akimoto, Yusuke 5, 137
Allison, Henry 54, 56
Alznauer, Mark 114
Angleraux, Caroline 5, 153
Ariew, André 49
Aristotle 3, 11, 13–15, 18, 23–26, 28–30, 33, 35f., 45f., 80, 114, 186, 215, 219, 249, 251–253, 315
Atran, Scott 36
Augustine 316
Auletta, Gennaro 40
Ayala, Francisco J. 20

Bach, Thomas 76
Barandiaran, Xavier 41
Barbagallo, Ettore 67
Barham, James 246, 259
Barrett, Justin 36
Bataille, Georges 237f.
Bateson, Patrick 34
Baum, Manfred 69
Beisbart, Claus 66
Beiser, Frederick C. 162
Bergson, Henry 223f.
Berlin, Sir Isaiah 190f.
Bernstein, Jay 128
Bertalanffy, Ludwig 256f.
Biasetti, Pierfrancesco 1, 7, 265
Bich, Leonardo 260
Bigelow, Julian 254–256
Birch, Jonathan 42
Bloch, Ernst 142
Bloomfield, Leonard 179
Blumenberg, Hans 321
Bolk, Louis 304
Boltzmann, Ludwig 48
Bonnet, Charles 153
Bordignon, Michela 3, 65, 84
Bosanquet, Bernard 6, 177, 190
Boucher, David 177f., 184–187, 195
Boudry, Maarten 271
Bourguet, Louis 153

Bradshaw, Anthony 39, 50
Brandom, Robert 295, 325
Breidbach, Olaf 77
Breitenbach, Angela 59, 66, 244
Brinkmann, Klaus 79
Brodie, Edmund 288
Bromhall, Clive 302
Brook, Andrew 191, 193
Bunzel, Matti 304
Burge, Tyler 41, 51, 55, 124
Butterfield, Herbert 249

Calvo Garzón, Paco 39
Canguilhem, Georges 19, 250
Caporael, Linnda 39
Carnap, Rudolf 324f.
Carvallo, Sarah 251
Chalmers, David 294
Chambers, Jack J. 179
Cheung, Tobias 14f.
Chiereghin, Franco 66, 69
Chomsky, Noam 315, 327–329
Clericuzio, Antonio 13
Cohen, Alix A. 66
Collingwood, Robin George 178, 184, 191
Coopersmith, Jennifer 47
Corti, Luca 102, 117f., 124, 243
Craver, Carl 34, 44
Cronin, Helena 277
Cruysberghs, Paul 74
Cudworth, Ralph 14
Cummins, Robert 43

Dahlstrom, Daniel 66, 69
Dalton, John 267
Dangel, Tobias 114
Darwin, Charles 2, 4–6, 12f., 20, 30, 36, 54, 154f., 169, 172–174, 177, 178–182, 184–186, 188–190, 192, 194f., 214, 256, 258, 267–269, 277–279, 284, 300–303, 311, 316
Darwin, Erasmus 187
Dasgupta, Shamik 124

Davidson, Donald 96 f.
Dawkins, Richard 37, 250, 273, 286, 288
De Caro, Mario 30, 246 f., 253
De la Mettrie, Julien Offray 249
de Lavoisier, Antoine-Laurent 267
de Maistre, Joseph 189
de Vaucanson, Jacques 271
De Volder, Burchard 157, 161
Deleuze, Gilles 16
Dennett, Daniel 271, 278, 287
Depré, Olivier 249
Des Maizeaux, Pierre 157
Descartes, René 3, 11, 13, 15–17, 19 f., 24, 160, 217, 245, 248–254, 256–259, 270 f., 315–317, 324
Desmond, Hugh 3, 8, 33, 39, 46
DeVries, Hugo 178 f.
DeVries, Willem 69
Dewey, John 323
D'Hondt, Jacques 74
Di Paolo, Ezequiel 41, 248, 259 f.
Dilthey, Wilhelm 245
Dobzhansky, Theodosius 36
Donald, Merlin 303, 306 f.
Driesch, Hans 211
Du Bois-Reymond, Emil 168
Duchesneau, François 13 f., 17 f., 251
Duijn, Marc 39
Dunayevskaya, Raya 143
Durkheim, Emil 223
Düsing, Klaus 66, 69 f., 74, 78

Engels, Friedrich 265–271, 274, 279 f., 282–285
England, Jeremy 48

Fay, Margaret A. 268
Ferrarin, Alfredo 114
Ferrini, Cinzia 79
Feuer, Lewis A. 268
Feuerbach, Ludwig 139 f., 143, 267, 279
Fichant, Michel 164
Findlay, John N. 69, 74, 81
Fischbach, Franck 74
Fisher, Mark 66
Fisher, Ronald 35 f.
Flanagan, Owen J. 316

Foster, John Bellamy 137 f., 146–149
Franken, Daan 39
Franzini Tibaldeo, Roberto 249
Frege, Gottlob 96, 325
Friston, Karl 48 f.
Fritzman, J.M 131

Gambarotto, Andrea 7, 243
Gare, Arran 297
Garelli, Gianluca 67, 69
Gehlen, Arnold 302
Gerhart, John C. 301
Giacché, Vladimiro 69
Gibson, Daniel G. 294
Gibson, James J. 45
Gibson, Molly 131
Gigerenzer, Gerd 44
Giladi, Paul 324
Ginsborg, Hannah 66, 244
Glennan, Stuart 45
Godfrey-Smith, Peter 39, 275, 278, 289
Goto-Jones, Christopher 236
Gould, Stephen Jay 186, 250, 277, 302, 304
Goy, Ina 66, 244
Grafen, Alan 35, 43, 45
Gramsci, Antonio 270
Greene, Murray 69
Grier, Michelle 56, 58
Griesemer, James 39
Grmek, Mirko Drazen 13

Habermas, Jürgen 142, 296
Haeckel, Ernst 5, 153 f., 167–174
Halbig, Christoph 113
Haldane, John Burdon Sanderson 6, 36, 247, 277, 282 f.
Haldane, John Scott 6, 199–204, 206, 209–211
Hamilton, Matthew 38, 47
Hampe, Michael 14
Hanczyc, Martin 41
Harvey, William 18
Hegel, Georg Wilhelm Friedrich 2–9, 65 f., 68–84, 89 f., 92–109, 111–133, 137–139, 142–145, 148 f., 177, 185, 187–190, 192, 194, 217, 220, 235 f., 267, 269,

Index of names — **335**

279 f., 282, 293–297, 301, 307, 323–325
Heidegger, Martin   12, 90–92, 215, 223, 245, 249, 297, 307, 317, 330
Heisig, James   217 f., 228, 230, 234
Hempel, Carl   42, 256
Herder, Johann Gottfried   159, 302–304
Hobbes, Thomas   268
Hooker, Cliff   41, 181
Horibe, Naoto   41
Horkheimer, Max   149
Hösle, Vittorio   75 f.
Houlgate, Stephen   131
Hrdy, Sarah   305
Hume, David   183, 190, 315, 320
Huneman, Philippe   3, 8, 33, 43, 46, 51–53, 55, 66, 244, 247, 251, 253
Husserl, Edmund   14, 20
Huxley, Julian   36
Huxley, Thomas Henry   178, 181–183, 185

Ikaheimo, Heikki   113
Ikegami, Takashi   41
Illetterati, Luca   4, 66, 71, 75, 79 f., 89, 94, 106
Itabashi, Yūjin   6, 199 f., 208, 210

Jacob, François   37
Jaeschke, Walter   83
Jaspers, Karl   223
Jaynes, Julian   308 f.
Jonas, Hans   7, 18, 243, 245, 247–251, 253 f., 256–260

Kahneman, Daniel   43
Kamin, Leon   265, 272, 284, 286
Kant, Immanuel   2–9, 15, 33, 35 f., 47, 50–58, 65–71, 73, 80, 104, 108, 153 f., 177, 184, 187, 189–196, 215, 243–247, 260, 267, 295 f., 315–317, 319–324, 326, 328–330
Keijzer, Fred   39
Kern, Andrea   114
Kesibir, Selin   310
Kirschner, Marc W.   301
Kitcher, Paul   272
Kitcher, Philip   20

Klotz, Christian   79
Knappik, Franz   116
Kocis, Robert   5, 177
Koonin, Eugene V.   300
Koyré, Alexandre   249
Kragh, Helge   283
Kreines, James   66, 244
Kuhn, Thomas   184, 195

Labarierre, Pierre-Jean   95
Laland, Kevin N.   247, 285, 288
Lamarck, Jean-Baptiste   187, 268, 277, 283
Leibniz, Gottfried Wilhelm   2–5, 8 f., 11 f., 14–16, 19–30, 47, 153–166, 168, 170, 172–174, 248 f., 251–253, 255, 295 f.
Lerner, Abram   283
Levey, Samuel   158
Levins, Richard   7, 265, 271–281, 283, 286, 288
Lewens, Tim   34, 37, 49, 56
Lewontin, Richard   7, 250, 265 f., 270, 281, 283–289
Lories, Danielle   249
Lovibond, Sabina   128
Lugarini, Leo   69
Lukács, Georg   150, 270
Lyell, Charles   181, 267
Lyon, Pamela   39 f.
Lysenko, Trofim   283

MacArthur, David   30, 246 f., 253
Machamer, Peter   45
Malthus, Thomas Robert   180, 268
Manheim, Karl   149
Manser, Anthony   69
Marmasse, Gilles   81
Marques, Victor   80
Marx, Karl   5, 7, 137–150, 266–270, 277, 279 f., 282–284, 288
Masham, Damaris   157
Matthen, Mohan   49
Maynard Smith, John   43, 282 f.
Mayr, Ernst   34, 37, 178 f., 247, 283
McDowell, John   4, 12, 111 f., 114, 116, 122, 127–132, 326
McFarland, John D.   66
McGilchrist, Iain   308 f.

McLaughlin, Peter   52, 66, 244
Mead, George Herbert   295 f.
Menegoni, Francesca   66, 69
Michelini, Francesca   249
Michurin, Ivan Vladimirovich   283
Mill, John Stuart   193, 195
Millikan, Ruth Garret   34, 43
Miolli, Giovanna   113
Molière   267
Monod, Jacques   37
More, Henry   14
Moreno, Alvaro   35, 40 f., 43
Morisato, Takeshi   6, 213
Moss, Lenny   8, 112, 127, 246 f., 293, 301–303, 311
Mossio, Matteo   35, 40 f., 43, 260
Müller, Gerd   34, 247, 285
Müller, Otto Friedrich   153 f., 166–168
Mure, Geoffrey   78

Nachtomy, Ohad   155
Neander, Karen   34, 43
Neill, Edmund   186
Neurath, Otto   324 f.
Ng, Karen   74
Nicholson, Daniel J.   246 f.
Nicholson, Peter   188, 191
Nicoglou, Antonine   39
Nishida, Kitarō   6, 199–211, 223
Noé, Keiichi   200
Nunziante, Antonio M.   3, 11, 13, 15, 25, 28, 156, 251

Odling-Smee, F. John   285
Okasha, Samir   44
Oken, Lorenz   167
O'Shea, James R.   29

Palkovacs, Erik   285
Paltridge, Garth   48
Parmenides   319 f., 323 f., 325 f.
Pasini, Enrico   14 f., 158
Passmore, John   186
Pearce, Trevor   279
Pérez, Juana   40
Perini-Santos, Ernesto   128
Peters, Julia   114

Phemister, Pauline   157, 166
Pierini, Tommaso   69
Pigliucci, Massimo   247, 271, 285
Pinkard, Terry   112, 114, 116 f., 131
Pinker, Steven   278, 284
Pippin, Robert   114, 116, 127
Piro, Francesco   25
Pittendrigh, Colin   34
Plato   14, 249, 275, 315, 317–320, 325 f.
Plessner, Helmuth   293, 297 f., 300, 307
Pope, Alexander   270
Popper, Karl   195
Post, David   285
Price, Huw   30
Prigogine, Ilya   48
Proudhon, Pierre-Joseph   148 f.
Przylebski, Andrzej   74

Quante, Michael   138 f., 141 f.
Quarfood, Marcel   66
Quine, William Van Orman   47, 266, 325

Rémond, Nicolas   158
Rey, Anne-Lise   158 f., 251, 253
Richards, Robert J.   170, 244
Ritchie, David George   5, 177 f., 186–195
Rochat, Philippe   194
Rockmore, Tom   8, 315
Rohde, Marieke   41
Rose, Steven   265, 272, 284, 286
Rosenblueth, Arturo   254–256
Rouse, Joseph   92
Rousseau, Jean-Jacques   315

Sachs, Carl   128 f., 132
Salmon, Wesley   34 f., 59
Sanguinetti, Federico   4, 7, 111
Sartre, Jean-Paul   9, 270
Schelling, Friedrich Wilhelm   5 f., 142, 153 f., 159–167, 169 f., 172–174, 220, 320
Schlick, Moritz   324 f.
Schlosser, Markus   40
Schmid, Hans Bernhard   310
Schmidt, Alfred   145, 150
Schrödinger, Erwin   48
Sell, Annette   74

Sellars, Wilfrid    11–14, 28–30, 245, 324 f.
Shani, Itay    41
Simpson, George    37
Skewes, Joshua    41
Smith, Adam    189
Smith, Justin E. H.    251
Smolin, Lee    297
Sober, Elliot    281
Sophie Electress of Hannover    156, 158
Soresi, Sergio    94, 102
Souche-Dagues, Denise    69
Spencer, Herbert    177, 180, 185 f., 189, 279
Spieker, Michael    74
Stahl, Georg Ernst    15, 248, 251–253, 259
Stalin, Iosif    270, 283
Stanguennec, André    69
Stederoth, Dirk    249
Steigerwald, Joan    66, 244
Sterelny, Kim    39
Stern, Robert    114
Stirling, James Hutchison    5, 177–184, 186, 195
Stroud, Barry    245
Sultan, Sonia    288
Svensson, Erik I.    285, 288

Takayama, Iwao    200
Tanabe, Hajime    6, 210 f., 213–240
Taylor, Charles    294, 304
Tennyson, Alfred    189
Thagard, Paul    191
Toepfer, Georg    252
Tomasello, Michael    302, 305
Tosaka, Jun    210
Trémaux, Pierre    268
Trudgill, Peter    179
Tversky, Amos    43

Ueda, Shizuteru    208
Unger, Roberto M.    297

Van den Berg, Hein    244

Van Fraassen, Bas    315, 323 f., 326 f.
Van Helmont, Jan Baptist    13
Van Valen, Leigh    279
Varela, Francisco    35, 51, 58, 243–245, 247, 258
Varzi, Achille    266
Venter, Craig    294
Verhulst, Jos    304
Verra, Valerio    69
Vieillard-Baron, Jean-Louis    74
von Haller, Albrecht    251
von Ranke, Leopold    236
Vygotsky, Lev Semyonovich    306

Wallace, Alfred Russell    181, 277
Wallace, William    185
Walsh, Denis    35, 40, 45, 49, 247
Watkins, Eric    244
Weber, Andreas    58, 243–245, 258
Wei, Yan    49
Whewell, William    177, 180 f., 184, 193, 195
Wiener, Norbert    254–256
Wilkins, Burleigh Taylor    69
Wilson, Catherine    29, 154
Wilson, Edward O.    273
Wimsatt, William    39
Wittgenstein, Ludwig    324 f.
Wohlfahrt, Günter    69
Wolfe, Charles    247–249, 251 f., 260
Wolff, Caspar Friedrich    251
Wolff, Christian    153, 244
Woodward, James    45
Wright, Larry    34, 43
Wright, Sewall    36
Wunsch, Matthias    249

Zammito, John H.    159, 244, 252
Zebina, Márcia    71, 81
Zuckert, Rachel    244
Zumbach, Clarck    244

www.ingramcontent.com/pod-product-compliance
Lightning Source LLC
Chambersburg PA
CBHW030521230426
43665CB00010B/706